THE
LOUDSPEAKER
DESIGN
COOKBOOK

The **L**OUDSPEAKER **D**ESIGN **C**OOKBOOK

7th *Edition*

BY

Vance Dickason

Published By

KCK Media Corp.

SEVENTH EDITION
2006
Copyright 2006 by Vance Dickason

Distribution Agents:

KCK Media Corp.

Library of Congress Card Catalog Number 87-060653

ISBN: 1-882580-47-8

First Printed January 2006

By Distribution Agents:
OLD COLONY SOUND LAB
&
AUDIO AMATEUR

This work is affectionately dedicated to my family,

to my lovely mother and departed father,

to my sister Jeanne and brother Steve,

to my children Jason and Jennifer,

and to my grandchildren Jackson and Belle

and their father Arch,

with the remembrance that

the only two things that really count in life

are love and the pursuit of knowledge.

LDC 7 INTRODUCTION

This 7th Edition of the *Loudspeaker Design Cookbook*, first published in 1977, marks its 28th anniversary. Each new edition has brought to light some different aspect of state-of-the-art loudspeaker technology. Several of the editions contained material that was directly the result of new software that allowed the computer simulation of ideas that would be difficult to communicate in any other way. The 4th Edition depended heavily on the release of LinearX's LEAP 4.0 software, and the 6th Edition featured important transducer simulations using Red Rock Acoustics' Spea*D* program. This 7th edition was made possible to a large extent by the use of the LEAP 5.0 and Dr. Wolfgang Klippel's Klippel distortion analyzer.

It's been five years since the previous edition was published. As has been the case with all the previous editions, the latest edition has added substantial new material. In terms of references and graphs, the 6th Edition added 83 new references and 214 new graphs. Following this tradition, the 7th edition includes some 42 new references and 341 new graphs, plus 318 graphs on the CD-ROM that is being made available. The CD-ROM has the 129 additional graphs from the Chapter 6 diffraction study (not available in the printed version), plus all the graphs published in print in Chapter 6 that will make the LEAP 5.0 polar plots much easier to read when enlarged on a computer screen and displayed in color.

While the *LDC* has been mostly a précis of available technology, a certain amount of the information has always been produced exclusively for the book, but never more so than in this 7th Edition. The explication on diffraction in Chapters 5 and 6 answers a number of questions concerning the significance of enclosure shape and where to locate drivers on a baffle. However, all this information was presented as computer simulations of single point microphone measurements, and while this gives you an excellent reference for the measurement consequences, it still leaves the nagging question of just how important diffraction and reflection issues are to the subjective sound quality of a speaker. Since I couldn't find any published information on this, I set up a subjective evaluation study in Chapter 6 for the various types of diffraction and reflection phenomena encountered in speaker design. While the results were not really unexpected, they were nevertheless very interesting and informative.

The second original piece of work being offered is a rather comprehensive study of woofer linearity. Working with Pat Turnmire, CEO of Red Rock Acoustics, we had eleven 10″ woofers manufactured for the project, starting with a very nonlinear design, then changing one aspect of the design for each driver sample until the final woofer had most of the bells and whistles of any high performance transducer. The entire group of drivers was characterized using the Klippel DA-2 distortion analyzer to reveal the consequences of each iterative change, and the study also investigated the thermal characteristics pole vent sizes.

The last new subject, voicing, is a frequently used term among loudspeaker manufacturers. This new section for Chapter 7 gives the reader a solid protocol for making final subjective adjustments to any new design.

As always, producing a new edition of the *Loudspeaker Design Cookbook* is a very rewarding and exciting experience for me, so I sincerely hope you enjoy reading and applying the information in this new edition as much as I have had writing and researching it.

Vance Dickason
August 2005

CONTENTS

Chapter 12 TWO SYSTEM DESIGNS: HOME THEATER AND A STUDIO MONITOR

CHAPTER 0

HOW
LOUDSPEAKERS
WORK

0.10 ELECTRODYNAMIC SPEAKERS.

This Seventh Edition of the *Loudspeaker Design Cookbook*, like its predecessors, aims to describe the operation, application, and measurement of electrodynamic loudspeakers and their associated enclosures and crossover networks. Electrodynamic drivers, the woofers, midranges, and tweeters found in the vast majority of loudspeakers are all based on the same basic concept: a diaphragm set in motion by the mechanical movement of a modulated electromagnetic field. As Mark Gander of JBL put it, "To make sound, you must move air."[1]

This mechanism is analogous to the electric motor, the rotating armature of a motor being replaced by the moving coil system of a speaker. *Figure 0.1* illustrates a cutaway view of a typical dynamic moving coil loudspeaker. As current is applied to the voice coil, an electromagnetic field is produced at right angles to the flow of current and to the permanent magnetic field. The resulting mechanical force causes the cone or dome diaphragm to move in a motion perpendicular to the gap field and consequently activate the air on either side of the diaphragm.

Three separate but interrelated systems operate in unison in a functional electrodynamic driver:

1. The Motor System: composed of the magnet, pole piece, frontplate/gap, and voice coil.

2. The Diaphragm: usually a cone and dust cap or a one–piece dome.

3. The Suspension System: made up of the spider and surround.

0.20 THE MOTOR SYSTEM.

The motor assembly is composed of five basic parts. These include the frontplate and pole piece, which together form the gap, the magnet, the voice coil, and the backplate. The backplate, frontplate, and pole piece are made from a highly permeable material, such as iron, which provides a path for the magnetic field of the magnet. The magnet is usually made of ceramic/ferrite material and shaped like a ring. The magnetic circuit is completed at the gap, causing an intense magnetic field to exist in the air space between the pole piece and the frontplate.

If an AC current is applied to the voice coil in the form of a sine wave at some given frequency, such as 60Hz, the flow of current in one direction on the positive half of the cycle will produce voice coil motion in one direction. When the current flow reverses on the negative half of the cycle, the polarity of the coil field reverses, and the motion of the voice coil changes direction as a consequence of the alternately attracting and repelling of the two fields.

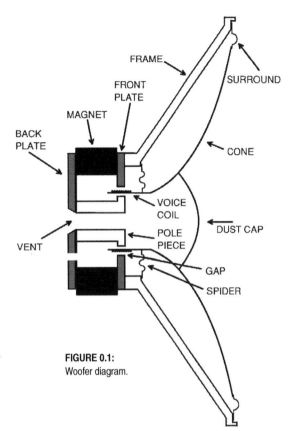

FIGURE 0.1:
Woofer diagram.

In order to accurately reproduce the motion induced by the sine wave, the voice coil has to move equally in both directions through the gap. For this to happen, it is important for the magnetic field to be as symmetrical as possible, so that motion in one direction will be applied with the same equal force as motion in the other direction. If this were not so, distortion of the signal would result.

If the flow of magnetic flux was confined only in the narrow space across the air gap, field symmetry would be assured and not be of concern. However, the magnetic lines of force "overflow" the gap area and produce stray fields on either side of the gap, known as fringe fields. Several methods are commonly used to ensure the symmetry of the fringe field, and are illustrated in *Fig. 0.2*. The straight pole piece in *Fig. 0.2a* illustrates an uneven fringe field caused by the non-symmetrical gap structure. Although adequate for many applications, this would be the least desirable method of construction. *Figure 0.2b* shows a symmetrical fringe field being created as the result of an undercut pole piece. *Figure 0.2c* depicts the effect of an angled pole piece on the fringe field,[2] which, like the undercut type, results in a more symmetrical fringe field.

The mechanical force developed by the current flowing through the voice coil is represented by the term "Bl." B × l is the force produced by a given number of turns (feet) of wire, l, being subjected to a given flux density per square centimeter, B.[3] Bl is a measurement of motor strength and is expressed in Tesla Meters/Newton. Directions on how to measure Bl are described in *Chapter 8, Loudspeaker Testing.*

0.21 GAP GEOMETRIES AND Bl.

Two different basic gap/coil geometries are used in loudspeakers, the underhung voice coil and the overhung voice coil. Of the two formats, illustrated in *Fig. 0.3*, the overhung coil is by far the most common. The distance labeled X_{MAX} in the diagram represents the distance over which the coil can travel in one direction and maintain a constant number of turns in the gap. X_{MAX} can be calculated by taking the voice coil length, subtracting the air gap height,

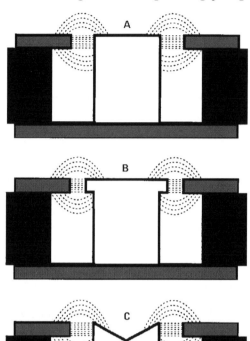

FIGURE 0.2: Fringe field effects for different pole geometries.

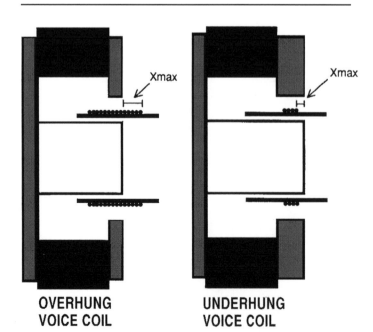

OVERHUNG VOICE COIL   **UNDERHUNG VOICE COIL**

FIGURE 0.3: Overhung and underhung coil geometries.

and dividing by 2.

Figure 0.4 shows the graphic comparison of Bl change with increasing excursion between the two gap geometries (this diagram represents movement of the voice coil in one direction through the gap). As increasing voltage is applied to the speaker, the coil moves further and further out of the gap, the number of turns of wire in the gap decreases, and the total Bl motor strength decreases. A speaker is said to be operating in a linear fashion if the number of turns in the gap is constant, and in a nonlinear fashion if the number of turns in the gap is decreasing and changing.

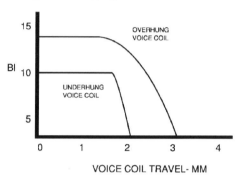

FIGURE 0.4: Comparison of Bl response for different coil geometries.

The underhung coil gives extreme linearity over a short distance, but generally has lower Bl than the overhung coil (due to the increased gap height and requirement for a greater magnetic field) and because of the short coil, a lower voice coil mass. The overhung arrangement has the advantage of reasonably good linearity and better efficiency (even with greater coil mass), which accounts for its popularity among manufacturers.

Different combinations of gap height and voice coil length will give the same X_{MAX} number, but behave quite differently in terms of nonlinear (beyond X_{MAX}) behavior. For example, a 12mm voice coil length and an 8mm gap have the same X_{MAX}, 2mm, as an 8mm voice coil length and a 4mm gap height. Although the X_{MAX} of these geometries is identical, the ratio of gap height to X_{MAX} is quite different, 4:1 in the case of the 12mm voice coil, and only 2:1 in the case of the 8mm voice coil. This ratio governs the rate at which Bl decreases as the coil rides out of the gap.

The curves in *Fig. 0.5* illustrate the variation in nonlinear behavior for geometries with the same X_{MAX} but with different ratios of gap height to X_{MAX}, as in the above example (from conversations with Chris Strahm, President of LinearX Systems and author of the Loudspeaker Enclosure Analysis Program, a/k/a LEAP). Looking at the diagram, Bl starts to gradually decrease beyond X_{MAX} to a point which is about two times X_{MAX} (double X_{MAX}) and then dramatically decreases. When the ratio of air gap height to X_{MAX} is large, the rate of Bl decrease is slower than when the ratio is low. At the furthermost limits of excursion, at the point where the coil is riding a large distance out of the gap, increased excursion does not change Bl significantly, and the

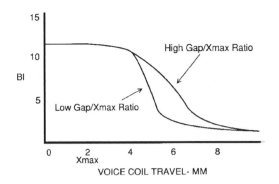

FIGURE 0.5: Comparison of Bl response with identical X_{MAX} and different coil/gap dimensions.

curve becomes more shallow and levels out as it approaches zero.

Although Bl tends to decrease slowly up to about twice the X_{MAX} distance, measurable distortion begins much sooner. In terms of distortion, the peak displacement limit of a moving coil can generally be taken as the X_{MAX} travel distance plus about 15%. Maximum excursion can be determined with a distortion analyzer setup to measure third-harmonic distortion as increasing voltage is applied to the voice coil. As excursion increases and the limits of X_{MAX} are exceeded, the third-harmonic distortion product increases. The $X_{MAX} + 15\%$ point will tend to coincide with an increase in third-harmonic distortion to a level of about 3%.[1]

0.22 SHORTED TURNS AND FARADAY LOOPS.

The current induced motion of the voice coil also causes an additional current flow, in the opposite direction of the drive current, which is known as back EMF (electromotive force). This EMF current is induced in the voice coil as it acts like the armature of a generator. This effect, along with the AC field generated by the program drive current in the voice coil, causes modulation of the magnetic gap field. This phenomenon, identified by W. J. Cunningham in 1949,[4] results in significant second-order harmonic distortion. Further investigation of this effect has shown that the modulation of the field is different depending on which direction the coil is traveling through the field. It is a nonsymmetrical effect.[5]

This nonsymmetrical phenomenon occurs in part because the pole piece, acting like a transformer core, is coincident with the coil throughout its rearward travel, and only partially coincident on the forward travel of the coil as it excurs beyond the limits of X_{MAX}. It has also been suggested that the voice coil flux interacts with and modifies the shape of the fringe field. This observation, at least in part, explains the benefit gained from push/pull configurations discussed in *Chapters 1* and *2*.

The most obvious solution is to use a sufficiently high level of permeability in the iron next to the voice coil so the metal is always saturated, which results in negligible modulation of the magnetic circuit. This technique is not often used because high permeability metals are relatively expensive. The most common technique devised to counteract this field modulation/eddy current problem is known as a shorted turn, or Faraday loop. Shown in *Fig. 0.6*, the application of the shorted turn has several variations, but all accomplish the same task by generating a field equal and opposite to fields induced by the voice coil. *Figure 0.6a* takes the form of a conductive coating, such as copper, over the pole tips; *Fig. 0.6b* shows a copper cap over the pole piece; *Fig. 0.6c* has a copper cylinder surrounding the pole piece;[2] and *Fig. 0.6d* illustrates the positioning of a shorted ring (flux stabilizer ring), sometimes made of aluminum, around the base of the pole piece.

The shielded pole piece method has the added benefit of causing a decrease in effective voice coil inductance, which results in a rise in high frequency response. The location and amount of shielding can be juggled to control midband and upper range driver response. The shorted ring at the base of the pole piece acts to reduce second-harmonic distortion, like the shielded pole methods, but does not affect the voice coil inductance and upper range response nearly as much. Although decreasing distortion is one of the benefits of the shorted turn method, controlling the mid and upper range frequency response behavior is more often the consideration.

Figure 0.7 illustrates the upper frequency response changes from using a T-pole and copper-shorting ring. The comparison is of the same driver, a Bravox 5.5″ poly cone woofer, with and without the T-pole/shorting ring combination[6]. Note that the response of the T-pole/shorting ring version begins rising

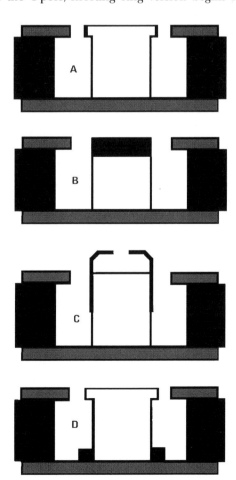

FIGURE 0.6:
Comparison of different shorted turn configurations.

above 500Hz and has as much as 3–4dB greater SPL than the version without the pole enhancements due to the reduction in losses from induced eddy currents. *Figure 0.8* depicts the same comparison at 30° off-axis, showing the effect to be not only across the bandwidth above 500Hz, but through the entire radiation angle of the driver, as would be expected from an effect that lowered the voice coil inductance.

0.23 VOICE COILS—FORMER MATERIALS AND WINDING CONFIGURATIONS.

Voice coils can be wound on a variety of materials, each having an effect not only on the T/S parameter set for a given driver, but also on the upper frequency response. There are two basic types of former materials used for loudspeakers, conductive and non-conductive. Conductive formers are by far the most common and are made from thin sheets

of aluminum or Duraluminum (Duraluminum has a higher strength to avoid such voice coil problems as neck deformation during long excursions). Since aluminum is an electrically conductive material, it develops eddy currents in the same fashion as the parts of the motor system (plates and pole). These parasitic "currents" cause losses that are reflected in terms of heat and distortion.

Aluminum formers are not a continuous cylinder, but have a small slit along the length of the former such that it does not act as a shorting element (incidentally, removing the slit and using a continuous aluminum loop does not have the same effect as a shorting ring, although it does lower Q_{MS} by about 10%). A non-shorting conductive former performance differs in two important aspects when compared to non-conductive former materials such as fiberglass or Kapton™ (a proprietary high temperature plastic material introduced by Dupont). The

FIGURE 0.7: On-axis comparison of two 5.5″ woofers, with and without copper shorting ring and shaped T-pole (solid = normal motor; dash = motor with shorting ring and T-pole).

FIGURE 0.8: Same as *Fig. 0.7*, but 30° off-axis.

FIGURE 0.9: On-axis comparison of two 5.5″ woofers, one with an aluminum voice coil former and one with a Kapton voice coil former (solid = woofer with aluminum voice coil former; dash = woofer with Kapton voice coil former).

FIGURE 0.10: Same as *Fig. 0.9*, but 30° off-axis.

FIGURE 0.11: On-axis comparison of two 5.5″ woofers, one with a two-layer voice coil, and one with a four-layer voice coil (solid = 2L voice coil; dot = 4L voice coil).

FIGURE 0.12: Same as *Fig. 0.11*, but 30° off-axis.

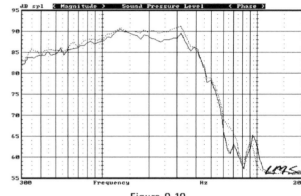

Figure 0.7

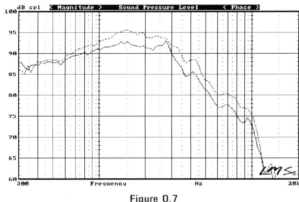

Figure 0.10

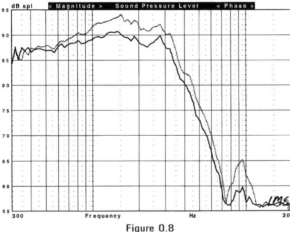

Figure 0.8

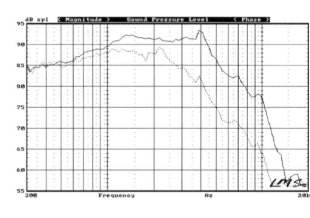

Figure 0.11

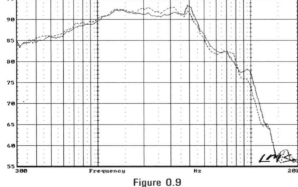

Figure 0.9

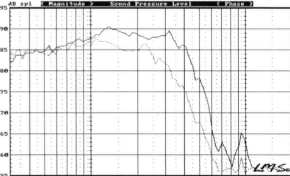

Figure 0.12

primary difference is that Q_{MS} (mechanical "Q") numbers using conductive formers are generally lower in the 2–4 range compared to non-conductive types. For non-conductive formers, higher Q_{MS} numbers between 4 and 12 are typical (the eddy current losses cause the Q_{MS} to be lower in conductive formers). Since non-conductive formers do not exhibit the induced eddy current problem, they also exhibit somewhat lower distortion.

The other performance difference between conductive and non-conductive former materials occurs in the upper frequency response. *Figure 0.9* depicts the comparison of two nearly identical Bravox 5.5"[6] woofers with identical cones, suspensions and motor structures, except that one driver has a Duraluminum voice coil former and the other has a Kapton former. As can be seen, the woofer with the Kapton former has 1–2dB greater output above 1.5kHz. *Figure 0.10* shows the same comparison but at 30° off-axis showing the effect to be somewhat more prominent. Again, this is due mostly to the difference in eddy current losses of the two materials. It should also be noted that part of this effect is due to mass differences (i.e., Kapton is a lighter material than Duraluminum).

The other notable response variation caused by voice coils is the manner in which they are wound. Obviously, larger voice coils with longer winding lengths have more turns of wire and hence greater inductance that will rolloff a driver's upper frequency response in the same fashion as a series inductor in a crossover. While there are different voice coil inductances for every possible combination of turns of wire and diameters of the former and pole piece, the biggest generalized difference in driver inductance is dependent on the number of layers of wire wound on to the former. The most common layer formats in woofers are two-layer and four-layer. Four-layer formers are frequently used on subwoofers to achieve the required Bl for the target response.

However, this can also be looked at in terms of upper frequency response control. *Figure 0.11* shows two nearly identical drivers, both Bravox 5.5" woofers[6], with the same motor, cone, and suspension, but one woofer with a two-layer voice coil, and the other using a four-layer coil (see *Fig. 0.12* for the same comparison at 30° off-axis). As can be seen, the woofer with the four-layer coil has a much lower frequency low-pass rolloff (–3dB at 2.5kHz) than the two-layer version (–3dB at 4.5kHz). Some manufacturers have developed two-way loudspeakers by taking advantage of this effect and controlling the inductance of a four-layer voice coil to produce a natural low-pass rolloff that will work with a particular tweeter. By doing this, the woofer does not require a separate low-pass crossover filter section and can be operated "wide open" with a crossover comprised of just the tweeter high-pass filter. I did a series of two-way prototypes for MB Quart several years ago using a 5.5" Bravox woofer and a 13mm MB Titanium tweeter. One prototype used a woofer with four-layer coil and a 3kHz mechanical rolloff and the other a standard two-layer coil with a higher rolloff. Both crossovers were computer optimized,

but the four-layer woofer prototype had no low-pass filter on the woofer and a third-order high-pass topography on the tweeter while the two-layer woofer prototype had a second-order low-pass on the woofer and a third-order high-pass filter on the tweeter. In a subjective comparison between these two prototypes, once the levels were adjusted, both models sounded quite good and were very comparable in overall sound quality with the advantage that the four-layer model was cheaper to manufacture and had fewer parts in the crossover. The other difference was that the four-layer model was lower in overall efficiency by 2–3dB due to the extra weight in the four-layer coil, and therein lies the tradeoff.

0.30 THE DIAPHRAGM.
Explaining the physics of speaker cones generally begins with the theoretical discussion of the radiation of an infinitely rigid piston pushing against the air. The transference of motion from the piston to the air would be bounded, in terms of frequency, at the low end of the spectrum by its resonance frequency (below which its ability to transfer energy is limited by mechanical constraints), and the upper frequency limit by the nature of the radiation impedance of the air. Air has resistance to motion, radiation impedance, which decreases with frequency to a point where any additional increase in frequency will be met with the same amount of resistance.

This upper frequency point below which energy transfer will exhibit a steady decrease is a function of both the nature of the radiation impedance of air and the radius of the radiating surface. Smaller radiating surfaces can reproduce higher frequencies than larger radiating surfaces, a fact of nature which accounts for the advent of specialized speakers which cover different frequency ranges.

Real-world cones are not infinitely rigid and will flex to some degree depending upon the characteristics of the material from which they are constructed. Cone flexing has a critical effect upon the high frequency efficiency, SPL response, and polar response of a driver. While different materials have different degrees of stiffness and transmit vibration at different speeds internally, they all tend to produce the same types of flexing, usually referred to as "modes."

0.31 CONE RESONANCE MODES.
Two mode classifications, radial and concentric, are used in analyzing speaker cone vibration, depicted in *Fig. 0.13* (after Beranek with changes). Radial modes extend from the cone center to the edge, occurring mostly at low frequencies and considered secondary in nature. Concentric modes form a collection of waves or ripples that spread outward from the center of the cone. These concentric modes, made visible using holographic techniques, look similar to what

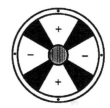

RADIAL CONE MODES

CONCENTRIC CONE MODES

COMBINATION OF RADIAL
AND CONCENTRIC MODES

FIGURE 0.13:
Cone vibration
modes.

you see when a pebble is dropped into the center of a bowl of water.

The number of waves varies with frequency, and as frequency changes, some of the ripples are reflected back to the center, forming interference patterns. These waves and ripples push against the air in a complex fashion, and some are in phase with the voice coil signal, while some are out of phase. The "+" and "-" areas in *Fig. 0.6* represent areas of the cone with opposite phase. This complex relationship of addition and cancellation, referred to as cone breakup, creates the many peaks and valleys in the typical loudspeaker SPL curve.

As frequency increases, the effective radiating area of the cone decreases so that very high frequencies tend to radiate only from the center area of the cone. At some frequency the effective radiating mass of the cone becomes small and a steep decrease in output begins which is described as the high frequency rolloff. To achieve a high cutoff frequency, the ratio of the voice coil mass and the cone mass must be as small as possible.[7] Upper frequency rolloff is also controlled by voice coil inductance.

0.32 CONE DIRECTIVITY.
As frequency increases all speakers become more directive and the high frequencies begin to "beam" like the light from an automobile headlight. At frequencies where the wavelength of sound (wavelength being equal to the speed of sound divided by the frequency = c/f, i.e., 1kHz has a wavelength of 1.13 feet) is large compared to the circumference of the cone (about 3 times the diameter), the radiation is spherical. As the frequency increases to the point where wavelength is equal to the circumference of the driver or smaller, the radiation pattern becomes progressively narrower. The chart in *Fig. 0.14* gives the -6dB off-axis points for different diameter speaker diaphragms (after Daniels with changes, JBL Pro Soundwaves, Fall 1988).

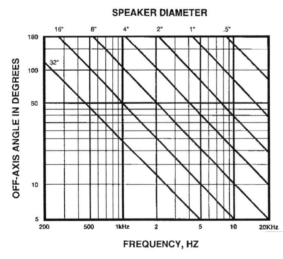

FIGURE 0.14: Diaphragm dispersion chart.

0.33 CONE SHAPE.
Different shaped cones have different response characteristics. There are two basic shapes used in cone design: conical or flat, and convex. Conical shaped cones tend to have a high peak at the extreme high end of the response range, the location of the peak being in part determined by the angle of the cone. Compared to the convex shape the bandwidth is somewhat wider. Convex cones tend to have a smoother frequency response and only a moderate peak in the upper response (less high frequency efficiency), but with a somewhat reduced bandwidth compared to flat cones.[8]

The frequency response of the convex cones can be altered and controlled by changing cone curvature.

0.34 DUST CAPS.
Gap widths in loudspeakers can vary from several tenths of an inch for large diameter speakers to the thickness of a piece of heavy gauge paper for small cone tweeters. The width is as narrow as practical to maximize flux density while allowing for variations in voice coil alignment and swelling due to heating. When the voice coil is attached to the cone, the area between the pole piece and the voice coil is usually shimmed to accurately align the assembly. This procedure leaves the gap between the coil and pole piece exposed to foreign particles. This being so, it would be possible for small particles to become lodged between the two areas and create obvious problems. The traditional solution is to affix a seal, known as a dust cap, over this area.

Putting a dust cap over the junction of the cone and voice coil solves one problem and draws attention to several others. Two basic types of dust caps are used with speaker cones; solid and porous. A solid dust cap does not allow air to pass through its surface, and creates a small acoustic chamber that will generate air pressure changes as the cone moves back and forth over the pole piece. This compression and rarefaction can have detrimental effects on speaker operation.

Since the area between the voice coil and pole piece is too small to effectively relieve this pressure caused by the motion of the dust cap, manufacturers use two practical solutions to the problem. One is to vent the pole piece, which requires a small hole to be drilled through the pole piece so that air can pass out an opening in the backplate. The other is to punch vent holes into the voice coil former where it attaches to the cone. This will allow air to flow out of the small chambered area and relieve the pressure between the pole piece and the dust cap.

Porous dust caps also readily relieve the air pressure created above the pole piece, but create other problems. First they provide a leakage path from the inside of the enclosure. This is not terribly significant since the volume of air leakage through the gap is small, especially compared to that of a lossy surround. The other problem area happens as the cone moves inward on the pole piece, and air is forced through the dust cap toward the cone's radiating surface. This sudden squirt of air will be out of phase with cone radiation and can cause a frequency response problem.[9] It is probably not a good idea to seal porous dust caps that cause offensive response anomalies, since the original design may have specified the porous dust cap for cooling purposes. The air flow through the gap area can provide significant cooling for the heat generated by the voice coil. Sealing the dust cap can also cause compliance and Q changes which may or may not be desirable.

Dust caps also modify a driver's upper range frequency response. Since the cone tends to radiate near the center at high frequencies, the dust cap can play a critical part in shaping the upper end response of a driver depending on its material composition and shape. Solid caps tend to cause greater changes in frequency response than porous ones. Occasionally you see solid caps which have small round vents with screens to relieve air pressure, giving them the benefits (or detriments) of both methods.

Figure 0.15 depicts the frequency response comparison for the exact same Bravox 5.25″ woofer (same motor, suspension, cone, voice coil, and so on) but with five different types of dust caps: porous cloth; doped cloth; soft PVC (Poly Vinyl Chloride); hard polypropylene; and an inverted hard poly type dust cap (all other dust caps in this study were standard convex types). Since this graph is somewhat difficult to read with this much information, *Figs. 0.16–0.19* give a more meaningful comparison, with each of the graphs comparing the standard cloth porous dust cap to the four other types of dust caps.

Figure 0.16 compares the porous cloth dust cap with a doped cloth dust cap (the same cloth cap with a soft damping material painted onto the surface). Not quite intuitive, the doped cloth actually increased the output in the upper frequencies with somewhat more attenuation above 4kHz with an overall smoother response. The overall response

of the undoped dust cap is also fairly smooth and even and has the added advantage of providing increased voice coil cooling by providing a passage for air moving past the voice coil.

Figure 0.17 compares the cloth dust cap to a soft PVC dust cap, which is a favorite among many manufacturers. As can be seen, the response of the PVC dust cap is smooth and even with no significant response anomalies, but a little less extension above 4kHz, probably due to mass and density of the material. Not only do manufacturers frequently choose this type of dust cap for its benign response characteristics, but also for its cosmetic appearance which gives a more coherent "look" in this era where so much emphasis is placed on industrial design esthetics.

Figure 0.18 compares the cloth cap to a hard poly dust cap. The hard plastic material in this case has a prominent resonance that is producing over 6dB more output centered on 5kHz, not a real convenient location for such an anomaly if you were trying to cross this woofer over at 3kHz to a tweeter. This is not necessarily typical of all hard poly dust caps as the nature of this anomaly is dependent on the diameter, shape, and density of the dust cap, but this is somewhat typical of what I have seen over the years and generally speaking, I never order a woofer sample deliberately with a hard plastic dust cap for use in a two-way application. For subwoofers or woofers intended to cross over to a smaller diameter driver at frequencies at least one or two octaves below the dust cap induced response anomaly, it really is not relevant and in these applications hard plastic dust caps are fine.

Figure 0.19 gives the last comparison between a

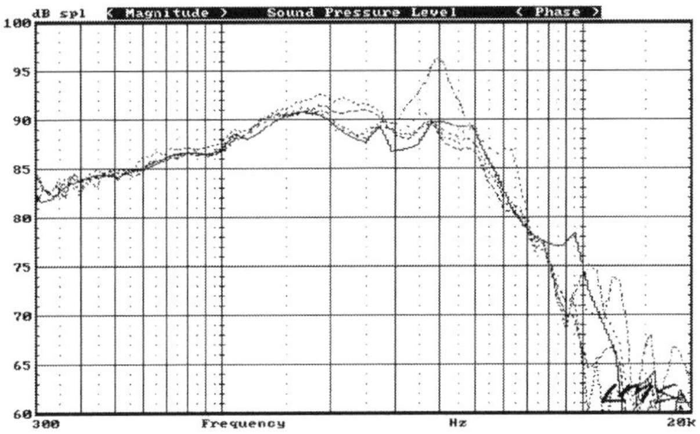

FIGURE 0.15: On-axis comparison of five identical 5.25″ woofers, each with a different style dust cap (solid = cloth dust cap; dot = doped cloth dust cap; dash = soft PVC dust cap; dash/dot = hard poly dust cap; wide dash = inverted hard poly dust cap).

FIGURE 0.16: On-axis comparison of two 5.25″ woofers, one with a cloth dust cap, and one with a doped cloth dust cap (solid = cloth dust cap; dot = doped cloth dust cap).

FIGURE 0.17: On-axis comparison of two 5.25″ woofers, one with a cloth dust cap, and one with a soft PVC dust cap (solid = cloth dust cap; dot = soft PVC dust cap).

FIGURE 0.18: On-axis comparison of two 5.25″ woofers, one with a cloth dust cap, and one with a hard convex poly dust cap (solid = cloth dustcap; dot = hard poly dust cap).

Figure 0.15

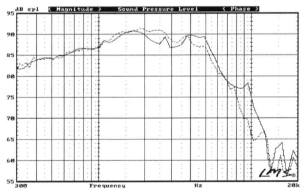

Figure 0.17

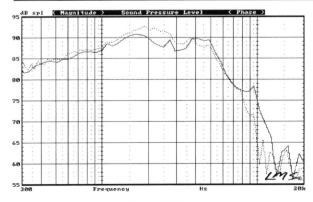

Figure 0.16

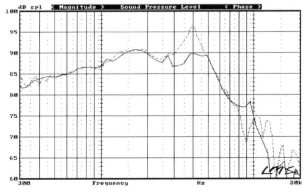

Figure 0.18

cloth dust cap and a hard plastic inverted dust cap. The inverted hard plastic dust cap has a number of advantages and has gained in popularity over the years. As can be seen in the response comparison, the inverted cap has a similar and smooth response as does the cloth dust cap. The other attribute that an inverted cap has is that, if they are made reasonably small so that they fit into the area near the neck joint of the cone (the junction of the voice coil former and cone), the cap can help strengthen the joint area and decrease the tendency of the cone neck to deform on hard excursions.

0.35 DOME SHAPES.

Dome tweeters and midranges have problems characteristic of those in cones. The two basic shapes are convex and concave. Concave dome radiators usually have much greater efficiency in the high frequency range, but a narrower directivity pattern. The higher efficiency is due in part to a wide peak caused by cavity resonances (although this can be damped to some extent) and the fact that convex

domes are usually made of hard materials. Convex domes have a wider directivity pattern in the upper frequency range, and lack the efficiency of concave domes in that range.

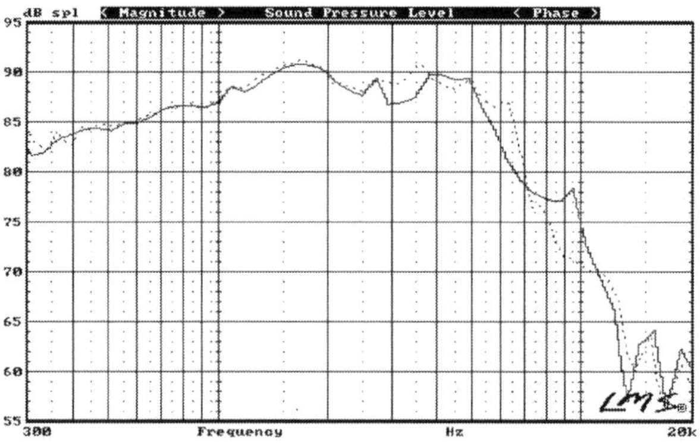

0.40 THE SUSPENSION SYSTEM.

The suspension system in any speaker is comprised of two elements, the surround and the spider. The surround, usually made of rubber, foam, or treated linen, performs several tasks. The surround helps keep the cone centered and provides a portion of the restoring force that keeps the coil in the gap. The surround also provides a damped termination for the cone edge. The spider, usually made of corrugated linen, likewise keeps the voice coil centered on the pole piece and also provides the restoring force that keeps the coil in the gap.

0.41 THE SURROUND.

The stiffness provided by the surround and spider is usually presented in terms of ease of motion, or compliance (compliance is the reciprocal of stiffness). In terms of the total compliance of the speaker, the spider provides about 80% and the surround perhaps 20% of the total compliance. The surround has two important functions. Its primary job is to keep the voice coil centered over the pole piece; however, damping the vibration modes at the outer edge of the cone is also critically important. The choice of thickness and type of material used in a surround can dramatically alter the response of the speaker. The ability of the surround to damp cone modes and prevent reflections back down the cone can alter both the amplitude and phase of modes combinations, making it an integral element of cone design and a viable response shaping tool.

Figure 0.20 illustrates the response comparison of different surrounds materials attached to the three samples of the same 5.25″ Bravox woofer[6] (same cone, spider, voice coil, motor, and so on). The three different materials used were rubber (in this case, not pure butyl, but a commonly used rubber compound for surrounds called NBR), foam, and injected Santoprene (this is a sophisticated process that Bravox uses to over-mold the Santoprene surrounds by injecting them onto the cone edge). Rubber gives the smoothest response with the least anomalies and overall one of the best edge damping materials to use for a surround where smooth upper frequency response is a consideration. The only downside to using rubber as a surround material is that it has to be made using a vulcanizing process that is slow and more costly than the heat forming process used with foam-type surrounds. The foam surround response is not quite as smooth by comparison and at around 10kHz it has obvious problems damping the cone's upper frequency modes, which is of course not too relevant to a woofer that is being crossed over between 2–3kHz. Foam surrounds are relatively easy to fabricate and inexpensive, but suffer from degradation over time due to exposure to light and various air impurities found in large metropolitan areas (hey, even in small towns sometimes). Overall, they do not perform as well as rubber when it comes to edge damping, but foam is

FIGURE 0.19: On-axis comparison of two 5.25″ woofers, one with a cloth dust cap, and one with an inverted concave hard poly dust cap (solid = cloth dust cap; dot = inverted hard poly dust cap).

FIGURE 0.20: On-axis comparison of three 5.25″ woofers, one with a rubber surround, one with a foam surround, and one with a Santoprene surround (solid = rubber surround; dot = foam surround; dash = Santoprene surround).

FIGURE 0.21: Same as *Fig. 0.20*, but 30° off-axis.

Figure 0.19

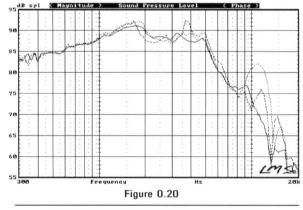
Figure 0.20

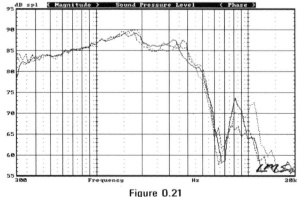
Figure 0.21

still one of the most frequently employed materials for surrounds. Less common, but becoming more prevalent, is Santoprene. Santoprene looks like rubber, and can be either heat formed (which is done a lot for subwoofer surrounds) like foam surrounds, or injection molded. This material is as inexpensive as foam, but does generally not have really good edge damping characteristics at upper frequencies, as can be seen in the graph. *Figure 0.21* shows the same comparison of all three surround materials, but at 30° off-axis. From this it is obvious that a lot of the response problems are relatively on-axis and that many of the response anomalies are less pronounced viewed off-axis.

0.42 THE SPIDER.

The spider has several functions. Its secondary tasks are to keep the voice coil centered over the pole piece and provide a barrier that keeps foreign particles away from the gap area. The primary purpose, however, is to provide the main restoring force (compliance) for the speaker. It is the stiffness of the spider which determines the speakers' resonance. Speaker resonance is a function of compliance and mass and can be related by:

$$f_s = [6.28(C_s \times M_D)^{1/2}]^{-1}$$

Where f_s = the driver free air resonance frequency, C_s = the driver compliance, and M_D = the total mass of the driver (the weight of the entire cone assembly, voice coil, cone, spider and surround, plus free-air mass load).

0.43 LINEAR AND PROGRESSIVE SUSPENSION SYSTEMS.

It seems intuitively evident that the best type of suspension would be one which would provide uniform restoring force throughout its range of travel. While this can be true for closed box type speakers where the compliance of the air within the box acts as a restoring force on the cone, the exact opposite is true of drivers in vented cabinets. This anomaly, dubbed the "oil-can effect" by Don Keele,[11] results in dynamic offset of the voice coil. The offset problem, a nonlinear phenomenon, occurs as the driver is being driven towards its X_{MAX} limitation. As the coil moves to a position where more turns are out of the gap, Bl decreases, back EMF decreases, and the coil draws more current, pushing the coil even farther out of the gap and thus creating distortion.

A progressive suspension system can counteract this nonlinear offset problem. This type of spider and surround combination provides increasing stiffness at the same time the Bl is decreasing (*Fig. 0.5*). If the breakpoint for the increase in stiffness coincides closely with the breakpoint for Bl decrease, then the voice coil is prevented from accelerating out of the gap. Suspension systems of this sort are found frequently in professional sound woofers intended for high SPL applications. Unfortunately, many amateur audio designers seem to be unaware of this fact, since it is not uncommon to find a woofer with an extremely linear suspension system being used in a vented application.

0.50 MODELING LOUDSPEAKER IMPEDANCE.

All of the systems I have described can be mathematically modeled by an electrical circuit diagram which is analogous to their operation. This technique is at the heart of all the box calculation methods described in *Chapters 1–4*. The electrical analogy of a driver is represented by a circuit with an impedance which duplicates that of the actual driver. The details of the actual measured impedance of a typical loudspeaker are shown in *Fig. 0.22*. *Figure 0.23* gives the analogous electrical model of a speaker. The circuit's elements are as follows:

R_E = DC resistance of the speaker
R_{EVC} = frequency dependent resistive component of the voice coil reactive rise (real part of voice coil inductance)
L_{EVC} = frequency dependent inductive reactance component of the voice coil reactive rise (imaginary part of voice coil inductance)
M_D = mechanical parameters due to mass
C_S = mechanical parameters due to compliance
R_{ES} = mechanical parameters associated with damping
Z_B = rear radiation impedance of the driver
Z_F = front radiation impedance of the driver

This model is similar to the one described by Beranek,[12] with the exception that the voice coil reactance was taken to be a fixed value as opposed to being frequency dependent[13] as shown in this diagram.

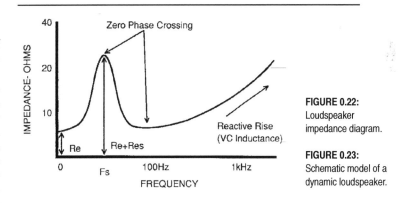

FIGURE 0.22:
Loudspeaker impedance diagram.

FIGURE 0.23:
Schematic model of a dynamic loudspeaker.

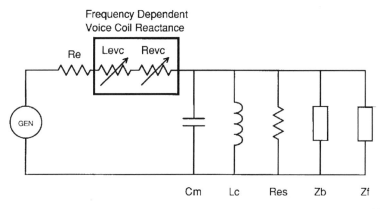

11

0.60 POWER, EFFICIENCY, AND ROOM SIZE.

The loudness produced by a given amount of amplifier power is a direct function of the efficiency of the loudspeaker and the volume of air it is trying to excite. Deciding how much loudspeaker capability you need to achieve a target volume level in a given room is an important question you should consider prior to building your own speakers. Since most loudspeakers are rather inefficient devices, usually on the order of 0.5 to 2%, coming up with the appropriate amount of acoustic power may not be calculated simply. If we consider a typical infinite baffle driver with efficiency of 0.5% (calculation of loudspeaker efficiency is discussed in *Chapters 1* and *2*), and an amplifier capable of delivering 50W RMS, then the acoustic power available from this system would be 0.25 acoustic watts (0.005 × 50W = 0.25W).

The graph in *Fig. 0.24* can be used to establish the approximate program material SPL produced for a specified amount of acoustic output in a given room volume. If we take our 0.25 acoustic watts and put it into a typical 20′ × 22′ × 8′ livingroom, which is approximately 100 cubic meters of space, it would achieve an SPL of about 97dB. To produce an additional 3dB for 100dB SPL, we would have to double the amplifier power to 100W.

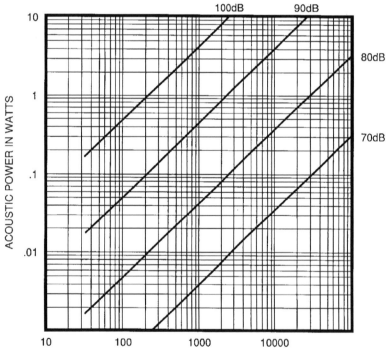

FIGURE 0.24:
Acoustic power vs.
room size chart.

Depending on the RMS rating of the speaker, and providing the rating is in some way adequately related to the thermal capabilities of the driver (many are not), our single 0.5% efficiency speaker may not be sufficient. Rather than double the amplifier power, another solution would be to use an additional driver. The addition of a parallel connected second driver doubles the cone area and increases acoustic power by a factor of four. The one acoustic watt of the combined drivers could produce nearly 103dB in the same room driven by the original 50W of output power. To reach the same SPL level with a single driver would require 200W.

0.70 ADVANCED TRANSDUCER DESIGN TOPICS.

Most all of the discussion about designing speaker enclosures in the *LDC* (see *Chapters 1–4*) and elsewhere in the loudspeaker industry literature is from the perspective of predicting box performance from a specific set of driver parameters. This, however, is the perspective only of a system designer. The other perspective not frequently considered in print is that of the transducer engineer, whose job is to come up with the combination of parts, cones, voice coils, magnets, top and bottom plates, dust caps, surrounds, and spiders to produce a woofer that will perform in a specific enclosure type.

For the most part, transducer engineers are a rare breed, and good ones are hard to find. There is no specific curriculum at the university level that teaches this skill, so becoming a professional in this field is either self-taught or passed from one practitioner to another, usually within the corporate confines of either the larger speaker companies or OEM driver manufacturers.

System design was at one time considered a "black art" until the advent of the professional CAD software packages (such as the LinearX LEAP software, which I rely upon for most of the simulations in this book) for doing this type of work. Until recently, transducer engineering still fell into the "black art" category, because the only way to develop a new woofer was a lot of cut-and-try and years of experience. This process has been radically changed by the introduction of new transducer modeling software released about the time I was putting together the 6th edition of the *LDC*. A design sequence that once would require a lot of experience, often at least three to four sample iterations, and perhaps months of R&D time to produce now can be done much faster in a computer simulation.

The first of these new CAD programs was authored by Red Rock Acoustics and is titled Spea*D* (Speaker Designer) (see *Chapter 9* for details about Spea*D* and another example of this type of software, WinMotor). Spea*D* consists of a set of software tools that makes transducer design much faster and easier. Essentially, it lets you play "what if" with your designs and build a woofer in your computer.

The Spea*D* tools include two separate programs:

• Reverse Spea*D*—software that can synthesize the required T/S parameters required to achieve a set of performance targets such as f_3, box size, enclosure type, and tuning.
• Spea*D*—generates speaker parts specifications that will produce a driver whose performance matches the desired target T/S parameters.

The exciting part of this type of transducer simulation software is that you can make changes in seconds and easily explore possibilities that were once simply too time consuming or difficult. With the availability of software that can predict parts specifications for building woofers, transducer engineering will become less of a cut-and-try game and more of a computer game, but that is the era in which we live.

Armed with Spea*D* and its counterpart, Reverse Spea*D*, I present the following short tutorial that will show how the various motor and cone assembly parts in a woofer change as the box design requirements change. The enclosure designs that I consider all use a 12″ woofer, with the first computer-designed woofer optimized for a moderate 2.0ft³ sealed box.

The next design shows the motor and cone part changes required to optimize the same woofer to work into a 2.0ft³ vented box. This is followed by the motor/cone changes for the woofer to function in a 2.0ft³ sealed rear-chamber bandpass enclosure. The last simulation shows the motor and cone part requirements for the same 12″ woofer to work in a compact 0.75ft³ sealed box.

All the data and graphs (and some of the text) used for this tutorial were provided by Red Rock Acoustics and specifically by Red Rock's CEO, Pat Turnmire, who happens to be one of those "hard to find" transducer engineers.

0.71 WOOFER DESIGN FOR SEALED ENCLOSURES.

This first example generates the woofer parts specifications for a sealed box. Specifications for this project begin with the box performance criteria. For the purposes of this tutorial, imagine that you are the speaker engineer in a small car audio company and the marketing department asks you to develop a new sealed box 12″ subwoofer.

The criteria you are given are pretty limited. Marketing department research reveals that a 12″ car audio subwoofer producing an f_3 of 40Hz in a 2.0ft³ box, handling 200W RMS, incorporating a rubber surround and polypropylene cone, and with a box QTC of 0.9 would do well in the marketplace. In other words, a moderate-sized sealed box 12″ car sub with a fairly warm sound to the bottom end and enough excursion to become loud and low (given the lift in a car compartment, 40Hz is low cutoff frequency, but see *Chapter 11* for details on that). *Table 0.1* summarizes all of this information, including assumptions about X_{MAX} for a 12″ subwoofer that would need to come from experience or observation and some help estimating Q_{MS} using one of

the Spea*D* utilities.

The design procedure begins with generating the T/S parameter set that will produce the results required by the marketing department's product definition. Spea*D*'s reverse box synthesis section, Reverse Spea*D*, is designed to do just that. Looking at the data in *Table 0.1*, the only target parameter that Reverse Spea*D* needed that wasn't immediately obvious was the M_{MD}, which represents the total moving mass (less air load—with air load this is called M_{MS}) of the speaker.

M_{MD} for any given driver is a combination of four items: the cone weight; half of the surround edge weight; voice-coil weight (former and wire); and the various miscellaneous weights that include the dust cap, spider, and adhesives used in putting together the cone assembly. The cone and surround weight are relatively easy to find. The poly cone and attached rubber surround used in this example weigh in at 72 grams after cutting off half of the edge.

The voice-coil assembly weight is the next part of the M_{MD} equation. From experience (note that experience still plays a big part in transducer design, even with sophisticated software and a fast computer), a designer knows that a 2″ diameter voice coil would likely yield the thermal power handling to achieve the target of 200W established by the marketing department. Experience also suggests that a four-layer coil is required to produce the extra Bl needed to push around a heavy 12″ woofer cone.

Given the 8mm X_{MAX} of *Table 0.1* and the educated guess that the frontplate thickness would be in the range of 10mm, the total voice-coil winding height would be 26mm (voice-coil winding height = frontplate thickness + 2[X_{MAX}]). You can then calculate the target voice-coil weight by using a utility provided in Reverse Spea*D*. In this case the 2″ diameter four-layer voice coil with a 26mm winding height and a 3.5Ω DCR should weigh in the vicinity of 51 grams. Adding the cone weight of 72 grams to the estimated voice-coil weight of 51 grams plus another 17 grams for the dust cap and other miscellaneous weight, you get a target M_{MD} of 140 grams.

After you enter all of this information into Reverse Spea*D*, the program automatically calculates the parameter set shown in *Fig. 0.25*. As you can see, the final speaker would be fairly efficient with an SPL of 89.66dB at 2.83V.

After you have determined the woofer parameter targets, the next step is to use the main body of the Spea*D* software to come up with a set of woofer com-

TABLE 0.1: SEALED BOX CRITERIA.

R_{EVC}	3.5Ω	Standard for a 4Ω car audio woofer
S_D	0.049m³	Standard 12″ cone area
Q_{MS}	10	An estimate based on the voice coil and cone type and the data chart in the Spea*D* help system
Box volume	2.0ft³	Given by marketing
Q_{TC}	0.9	A compromise between efficiency and a slightly "warm" bottom end that car audio enthusiasts like
M_{MD}	????	
X_{MAX}	8mm	The upper range excursion for most 12″ subwoofers
f_3	40Hz	Given by marketing

ponent parts that will produce these parameters. Step 1 is to use the Spea*D* Voice Coil optimizer to finalize the voice-coil specification, entering voice-coil data already produced with the Reverse Spea*D* Coil Tool into the Coil Optimizer (four layer coil, 2″ ID, Kapton former, 26mm winding height, and a DCR of 3.5Ω). The Coil Optimizer calculates that the wire size would be 25 AWG and the actual winding height with a 3.5 DCR would be 24.767mm—close enough to the target of 26mm (*Fig. 0.26*).

The next step is the motor design. Again, experience would lead a transducer engineer to start with a 30 oz magnet that had likely been successfully used on other similar 12″ woofer projects. The dimensions for a 30 oz magnet are 120mm outside diameter × 20mm thick × 60mm inside diameter. As a rule of thumb, the ideal frontplate and backplate diameter is generally calculated to be the mag-net OD less the frontplate thickness—in this case, 110mm would be perfect.

Entering 110mm for the frontplate OD, 10mm for the thickness, and then using Spea*D*'s recommended inside diameter of 54.25mm, you've finished the frontplate design. To do this, the program takes into consideration the diameter of the coil, former thickness, and winding depth plus a typical gap spacing that would produce maximum Bl and still have room for mechanical travel.

The backplate/polepiece is just as simple. Beginning any design, a designer can always assume the backplate to be the same OD and thickness as the frontplate. For the polepiece, a standard pole diameter of 48.95mm would give an appropriate amount of clearance between the pole and the voice-coil former inside diameter.

Also incorporated into this speaker motor design is a 5mm extended polepiece (see the later section on extended polepiece dynamics) that would, according to the Spea*D* Help menu, be ideal for the 10mm frontplate thickness. The extended pole not only provides for more linear flux in both movement directions for the voice coil, but also provides a guide for the coil on extreme excursions that take the coil completely out of the gap. The extended pole prevents the possibility of the voice coil jamming into the front plate and causing catastrophic failure. The last detail is a 25mm vent hole in the polepiece that was added for cooling and pressure release behind the solid dust cap.

After you enter all of this data into the software, the program immediately reveals a substantial amount of information about the motor design. First, each of the metal parts—frontplate, backplate, and

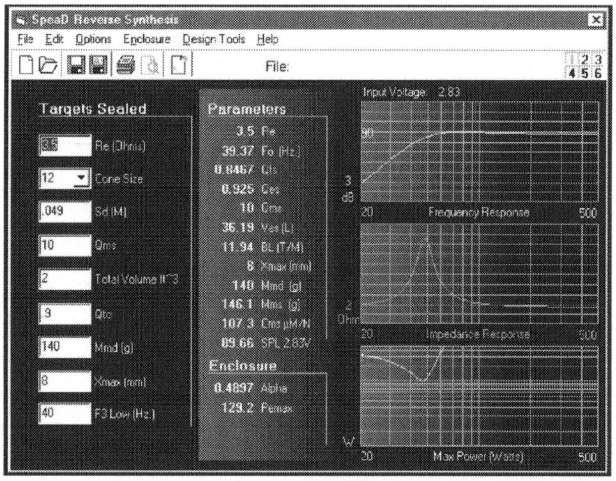

FIGURE 0.25:
Calculating parameters with Reverse Spea*D*.

FIGURE 0.26: Spea*D* Voice Coil Optimizer (left).

FIGURE 0.27: Motor design using Spea*D* (right).

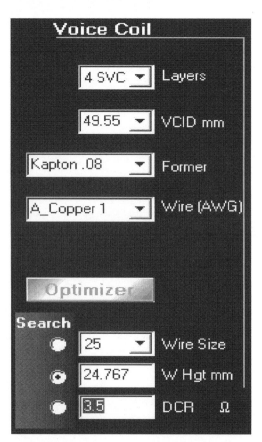

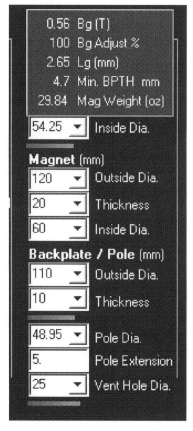

polepiece—has a "saturation bar," an indicator that shows to what degree the part is magnetically saturated. Normally, the most efficient (and least expensive) way to design woofers is with the components at or just below their saturation levels. In this case, none of the metal parts in the motor simulation was close to saturation, which shows that the motor was not losing any energy because the parts were too thin (oversaturated).

Next, from the drop-down box in *Fig. 0.27*, you can see that the gap flux has a reading of 0.56 Tesla (Bg). Spea*D* also indicates that the minimum backplate thickness is less than 6mm for this design. This means that if the motor design meets the required goals, the company could save money in parts costs by reducing the backplate thickness from 10mm to 6mm.

At this point, most of the information for the Soft Parts section of Spea*D* is available and can be entered into the program in order to get the final result. This includes the cone area (S_D) of 0.0491m³, the cone f_O of 41Hz supplied by the manufacturer (cone f_O and the weight of the cone can be used to define the cone compliance), the weight of half the surround and cone measured at 72 grams, plus the estimated miscellaneous 17 grams of dust cap/adhesive mass. The spider deflection number comes from the spider supplier for a standard 6″ diameter spider with a 2″ coil opening. Deflection for this spider was 0.5mm using a 50 grams test weight. The Q_{MS} value of ten that came from the Spea*D* help system is the last input value required.

The motor parts data and the soft parts data entered into the program produced the T/S parameter set in *Table 0.2*. From this it is obvious that at least one problem still exists, as indicated by the starting data, which shows that the f_O prediction is about 34Hz, 6Hz lower than the target f_O of 40Hz. Since the easiest way to raise f_O is to decrease the spider compliance (use a stiffer spider), the first change is to experiment with the spider deflection number.

After a few trials, a deflection of 0.15 caused the parameter set to fall in line with the original targets as depicted in *Fig. 0.28*. After this analysis, a designer would be ready to order parts and build a prototype woofer with a high degree of certainty that this design will produce the required result.

0.72 WOOFER DESIGN FOR VENTED ENCLOSURES.

Keeping with the theme of being a transducer engineer in a small car audio company, imagine you just finished prototyping the sealed box product when the marketing department decides that maybe a 2ft³ vented box with an f_3 about half an octave lower in frequency would be better. Marketing also wishes to consider doing a bandpass speaker in the same size box as well and has asked for two new prototypes to test.

The first thing is to check the viability of using the current woofer prototype in a 2ft³ vented box. Even though the Q_{TS} is much too high (see *Chapter 1, Section 1.40 Woofer Selection and Enclosure Construction*), running a LEAP simulation using the sealed box woofer parameters and a 2ft³ box tuned to 40Hz confirms this (*Fig. 0.29*). The 7dB peak in the response indicates very poor damping for the

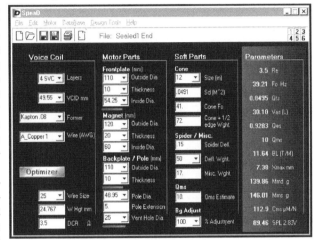

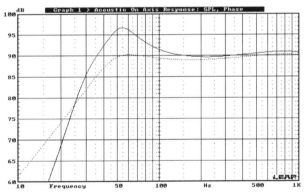

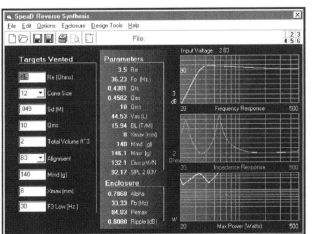

TABLE 0.2 SEALED BOX MOTOR AND SOFT PARTS.

Target Parameters		Starting Data	1st Iteration
R_{EVC}	3.5	3.5	3.5
f_O	39.7	33.77	39.21
Q_{TS}	0.8467	0.7404	0.8495
Q_{ES}	0.9250	0.7996	0.9283
Q_{MS}	10	10	10
V_{AS}	36.19	51.47	38.18
Bl	11.94	11.64	11.64
X_{MAX}	140.00	139.86	139.86
SPL 2.83V	90.69	89.46	89.46

Key Parts

Spider deflection	0.5mm @ 50g	0.15mm @ 50g

FIGURE 0.28: Final woofer parameters in Spea*D*.

FIGURE 0.29: LEAP simulation for sample woofers.

FIGURE 0.30: Sample vented woofer parameters using Reverse Spea*D*.

woofer/box combination and is unacceptable.

Using Reverse Spea*D* and a little computer "cut and try," I achieved a moderately damped B3 type response by lowering the f_3 target to 30Hz (*Fig. 0.30*). Marketing liked the idea of the sub going lower in frequency. I entered the Reverse Spea*D* information into the main body of the program and after three iterations of changes produced the data presented in *Table 0.3*.

The first step, as with the sealed box design, is to adjust the compliance to match its target value. This is again easily accomplished by adjusting the spider deflection and making it stiffer. Decreasing the spider compliance with a new deflection of 0.25mm

raised the f_O sufficiently to match the new vented box target.

The next step is to increase Bl to a level that lowers the Q_{TS} to its target value. Increasing the magnet OD is the most obvious approach to try initially. The next size available is 140mm. Along with the increase in the magnet OD, it is necessary to also increase the frontplate and backplate OD to match the magnet.

This 140mm magnet change brought the Q_{TS} very close to the target, but Q_{TS} is still too high. Increasing the magnet OD beyond 140mm is one possible technique that could be used to increase Bl enough to achieve the target Q_{TS}, but an alternate approach is to try increasing the magnet thickness. This would give the voice coil additional travel in the rear direction of coil travel and help prevent "bottoming" (the back of the voice-coil former slamming into the backplate on a rearward stroke).

Given the excursion potential of all vented loudspeakers below the box tuning frequency, this would definitely be an appropriate choice, especially given the somewhat well-known fact that car audio customers are notorious for playing their systems at high volume levels with the bass full up. A change from 20mm to 25mm in the magnet thickness increased Bl sufficiently to provide an exact match with the target Q_{TS}.

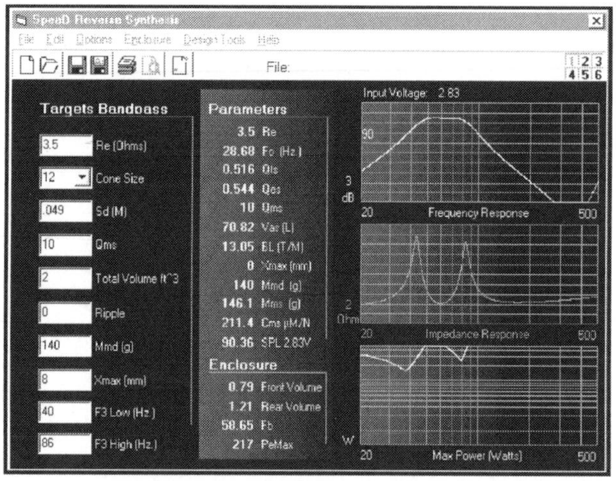

0.73 WOOFER DESIGN FOR BANDPASS ENCLOSURES.

Bandpass designs are more complex and require more information to define the targets in Reverse Spea*D*. Typical additional specs (besides a total volume of 2ft³ and an f_3 in the vicinity of 40Hz) include a flat response with 0dB ripple and a -3dB high-frequency cutoff of between 80 and 90Hz. After entering the appropriate data, Reverse Spea*D* calculated the parameter targets for a box with a 0.77ft³ rear volume and a 1.2ft³ front volume tuned to 60Hz (*Fig. 0.31*).

The woofer described has a driver efficiency of 90.6dB at 2.83V and a system gain for the bandpass enclosure of 4.25dB. Entering this data into the

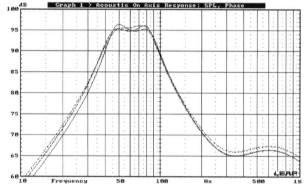

FIGURE 0.31:
Sample bandpass design using Reverse Spea*D*.

FIGURE 0.32:
LEAP bandpass simulations.

TABLE 0.3: VENTED BOX DESIGN

Target Parameters		Starting Data	1st Iteration	2nd Iteration	3rd Iteration
R_{EVC}	3.5	3.5	3.5	3.5	3.5
f_O	36.23	33.77	36.20	36.20	36.20
Q_{TS}	0.4381	0.7404	0.7859	0.4740	0.4210
Q_{ES}	0.4582	0.7996	0.8572	0.4976	0.4395
Q_{MS}	10	10	10	10	10
V_{AS}	44.53	51.47	44.79	44.79	44.79
Bl	15.94	11.64	11.64	15.28	16.26
X_{MAX}	8.00	7.38	7.38	7.39	7.39
M_{MD}	140.00	139.86	139.86	139.9	139.90
SPL 2.83V	92.17	89.46	89.46	91.82	92.36
Key Parts					
Spider deflection		0.5mm @ 50g	0.25mm @ 50g	.25mm @ 50g	.25mm @ 50g
Magnet OD		120	120	140	140
FP & BP OD		110	110	130	130
Magnet TH		20	20	20	25

main program menu results in the data displayed in *Table 0.4*, which includes the four iterations of driver part variations to achieve the final result.

The first iteration required a new surround (lower cone f_O). Instead of stiffening the suspension as was done with the spider change for the vented design, what was needed was a surround that was less stiff than the original design. I used a medium stiffness (also called durometer) material to achieve a cone f_O of 35Hz. This brought the woofer f_O to 29.62Hz, just a bit higher than the target. A minor adjustment to the spider deflection, as shown in the second iteration, was all that was additionally needed to obtain the target speaker f_O.

The next step is to manipulate Q_{TS} so that it matches the target value. Since Q_{TS} is greatly affected by the Q_{ES}, which in turn is strongly related to Bl, I made changes in the magnet thickness.

As you can see, the target Q_{TS} is achieved with a new magnet thickness of 40mm, which is twice the starting size, although the motor strength is somewhat high. This also seems like an expensive option, so by experimenting with the magnet OD, it was easy to determine that changing the magnet from 120mm to 130mm produced good results.

After looking at a magnet supplier's catalog, I inserted 130mm-OD and 65mm-ID magnet dimensions to reduce the total magnetic area and bring the parameters close to the desired target. You can see how each of these iterations affected the box performance in *Fig. 0.32*, which is a LEAP box simulation series that uses the starting T/S parameter set given in this exercise plus each of the four iteration T/S parameters sets.

0.74 WOOFER DESIGN FOR A COMPACT SEALED ENCLOSURE.

After all of this work, the marketing department for the small car audio company finally woke up and smelled the coffee to realize that most of its competitors were selling car audio subwoofers that performed reasonably in small enclosures of 1.25–0.75ft³. Two new performance parameters were added to the mix—a 0.75ft³ box with an f_3 of 50Hz. Given the low-frequency response "lift" supplied by the average car compartment, 50Hz is more than adequate (see *Chapter 11, Section 11.4 Computer Simulation of Closed-Field Performance*).

These box criteria are typical of current car subwoofers on the market and describe many of the 12″ car subs I have reviewed for *Car Audio and Electronics* magazine over the past two years. The Reverse Synthesis program determined a target parameter set (*Fig. 0.33*) which I then entered into the main menu system of Spea*D*. You can follow the various changes made through four separate iterations done for the driver in *Table 0.5*.

Just as with all of the other designs, the process begins by getting the f_O to match the target param-

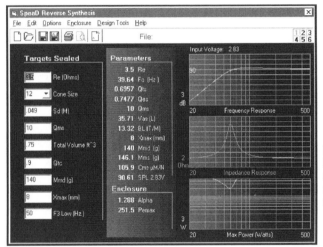

FIGURE 0.33:
Parameters for compact design using Reverse Spea*D*.

TABLE 0.4: BANDPASS ENCLOSURE DESIGN

Target Parameters		Starting Data	1st Iteration	2nd Iteration	3rd Iteration	4th Iteration
R_{EVC}	3.5	3.5	3.5	3.5	3.5	3.5
f_O	28.68	33.77	29.62	28.79	28.79	28.79
Q_{TS}	0.5160	0.7404	0.6553	0.6381	0.5161	0.5366
Q_{ES}	0.5440	0.7996	0.7013	0.6816	0.5442	0.567
Q_{MS}	10	10	10	10	10	10
V_{AS}	70.82	51.47	66.91	70.83	70.83	70.83
Bl	13.05	11.64	11.64	11.64	13.03	12.77
X_{MAX}	8.00	7.38	7.38	7.38	7.38	7.38
M_{MD}	140.00	139.86	139.86	139.86	139.86	139.86
SPL 2.83V	90.36	89.46	89.46	89.46	90.43	90.26
Vb1 S	0.79					
Vb2 V	1.20					
Key Parts						
Cone f_O		41	35	35	35	35
Spider deflection		0.5mm @ 50g	0.5mm @ 50g	0.7mm @ 50g	0.7mm @ 50g	0.7mm @ 50g
Magnet OD		120	120	120	120	130
Magnet TH		20	20	20	40	20
Magnet ID		60	60	60	60	65
FP & BP OD		110	110	110	110	120
					60 oz magnet	35 oz magnet

eter. f_O on this speaker was similar to the first design and required only a change to the spider deflection to hit the free-air resonance target.

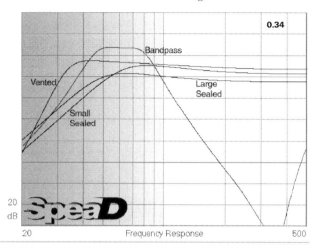

20 dB

20   Frequency Response   500

FIGURE 0.34:
Comparing design responses.

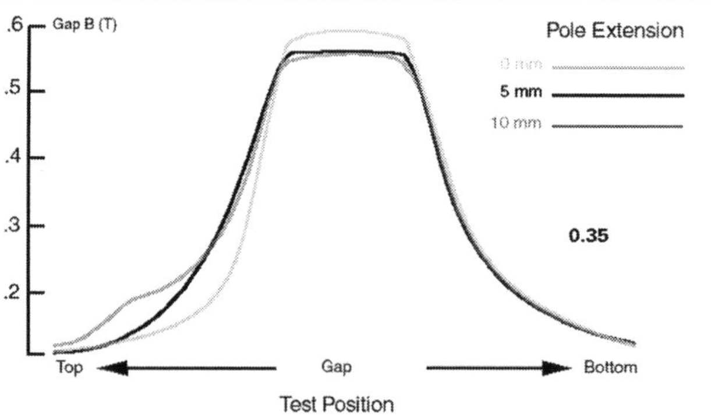

FIGURE 0.35:
Pole extension comparison.

The second change was again to increase the magnet and plate ODs. This time changing the OD produced a little more magnet energy than needed, so in the last iteration the frontplate ID was modified to fine-tune the Bl and thereby adjust the Q_{TS}.

Using the Design Overview feature in Reverse SpeaD allows you to compare the four different response curves generated for this exercise (*Fig. 0.34*). It is interesting to note that the first three designs all have the same basic targets for box size but very different responses. The first sealed system has the lowest efficiency, while the vented system is slightly more efficient and includes more low-frequency extension. The bandpass system is very efficient, but has a somewhat narrow bandwidth.

For use as a car audio subwoofer, the small sealed box is probably the best choice out of the four. It is efficient, with good power handling, and a reasonably small magnet structure to keep its cost down.

0.75 EXTENDED POLE DYNAMICS.

Section 0.20, The Motor System, discussed some of the basics of different pole shapes and their effects upon the magnetic linearity of a driver motor. For the specific case of extended pole configurations, SpeaD offers some unique opportunities to play "what if" with different degrees of pole extension.

One of the most important aspects of this software is the way that the magnet modeling system tracks the strength of the magnetic flux in the gap (Bg), and, more important, the shape of the magnetic field that determines the amount of energy applied to the coil. An interesting way to apply this program feature is to compare what happens to the driver's

TABLE 0.5: COMPACT SEALED BOX DESIGN.

Target Parameters		Starting Data	1st Iteration	2nd Iteration	3rd Iteration
R_{EVC}	3.5	3.5	3.5	3.5	3.5
f_O	39.64	33.77	39.21	39.21	39.21
Q_{TS}	0.6957	0.7404	0.8495	0.6461	0.6820
Q_{ES}	0.7477	0.7996	0.9283	0.6907	0.7368
Q_{MS}	10	10	10	10	10
V_{AS}	35.71	51.47	38.18	38.18	38.18
Bl	13.32	11.64	11.64	13.50	13.07
X_{MAX}	8.00	7.38	7.38	7.38	7.38
M_{MD}	140	139.86	139.86	139.86	139.86
SPL 2.83V	90.61	89.46	89.46	90.74	90.46

Key Parts					
Spider deflection		0.5mm @ 50g	0.15mm @ 50g	0.15mm @ 50g	0.15mm @ 50g
Magnet OD		120	120	130	130
FP & BP OD		110	110	120	120
Frontplate ID		54.25	54.25	54.25	54.75

TABLE 0.6: POLEPIECE EXTENSION DYNAMICS.

	Pole Extension mm								
	0	**1**	**2**	**3**	**4**	**5**	**6**	**8**	**10**
B_g (T)	0.61	0.60	0.59	0.58	0.57	0.56	0.56	0.56	0.55
Bl (TM)	10.90	11.02	11.16	11.30	11.47	11.64	11.60	11.51	11.42
Q_{TS}	0.958	0.939	0.919	0.897	0.874	0.850	0.856	0.868	0.881
Q_{ES}	1.060	1.036	1.012	0.985	0.958	0.928	0.936	0.951	0.966
$X_{MAX}f$ (mm)	9.58	9.77	9.98	10.23	10.50	10.82	10.82	10.82	10.82
SPL @ 2.83V	88.88	88.98	89.08	89.2	89.32	89.46	89.42	89.35	89.29

parameters when the pole piece is extended from a point that is flush with the top of the frontplate to an extension that equals the gap height.

To help illustrate how this works, *Fig. 0.35* shows a plot of the magnetic flux measurements (Bg curves) that reflect the measurement of motor strength in the middle of the gap. The three curves show the frontplate gap Bg (flux density) in Tesla for 0mm (flush pole), 5mm, and 10mm pole extensions.

A flush 0mm pole is by far the most common design used today and actually has the highest peak reading in the center of the gap. However, the shape of the flux field is not symmetrical on both sides of the gap, and a large amount of stray flux is lost on the topside of the magnet structure.

A 5mm extension, which is half of the frontplate thickness, offers a very symmetrical field. The peak Bg in the gap is slightly less than the 0mm extension because the stray flux is better contained; however, the actual usable energy has increased. As the extension increases past the ideal of half the frontplate thickness, the improved symmetry continues, although the usable energy decreases because of the losses caused by the magnetic having to "drive" the extra metal of the longer polepiece. Loss caused by the extra extension can be minimized by making a cup in the top of the pole to reduce the amount of metal volume (*Fig. 0.2c*).

Other advantages provided by extra pole extension include improved alignment on extreme excursions and most notably an increase in thermal power handling as a result of the extra heatsink area next to the voice coil. The illustrations depicted in *Figs. 0.36* and *0.37* show FEA (Finite Element Analysis) models of a woofer motor cross-section. This shows the intensity of the magnetic field in and around the gap and makes it very easy to see the shape of the flux in the gap. *Figure 0.36* describes a flush pole, while *Fig. 0.37* depicts a pole with the ideal extension. Note the symmetry of the magnetic field above and below the gap on the extended pole as compared to the flush pole.

The data in *Table 0.6* gives the predictions for incrementally increased pole extensions (only the parameters affected by the pole extension changes are shown). Notice how the efficiency and Bl increase as Q_{TS} decreases between 0mm and 5mm. Once you go past this "ideal" half of the frontplate length pole extension (5mm), the relationships reverse and efficiency and Bl decrease as Q_{TS} increases as the length increases to 10mm. Spea*D* is able to make these predictions because of its ability to model the shape of the flux in and around the gap.

This motor model also allows the program to determine the real "working" X_{MAX} of the speaker. Spea*D* displays both the commonly used X_{MAX} (based on the measured voice coil X_{MAX} = gap height - [voice coil length/2]) and the working X_{MAX} that is based on the software's magnetic geometry modeling, referred to as $X_{MAX}f$. $X_{MAX}f$ is a more realistic representation of how far the coil can travel before the total number of voice-coil "turns" (the "l" in Bl) starts decreasing and the speaker becomes uncontrolled and distorted.

The criteria used for determining the linear operating envelope for a speaker system used elsewhere in this book use a rough approximation of field extension by applying X_{MAX} + 15%, as explained in *Section 1.62, Displacement Limited Acoustic Power Output*. This approximation does not, however, distinguish between more linear configurations such as pole extensions, as Red Rock Acoustic's software is capable of. The $X_{MAX}f$ values for the various pole extensions are included in *Table 0.6*.

0.80 ADVANCED TRANSDUCER ANALYSIS STUDIES.

There is no doubt that the Klippel Distortion Analyzer is the most significant new tool for loudspeaker characterization to come along in the past few years. Coupling a sophisticated distortion analyzer (*Fig. 0.38*) with a positioning laser (*Fig. 0.39*), Dr. Wolfgang Klippel and his group have given the industry a powerful weapon for understanding the dynamic functioning of woofers, midranges, and tweeters (for more, visit the Klippel website at www.klippel.de, and pay specific attention to the "Know-How" section). The reason this device is so significant is that a small signal linear model such as the familiar Thiele/Small model cannot describe the behavior of electrodynamic loudspeakers at high amplitudes.

Because the standard small signal model doesn't deal with the dynamics of increasing input levels (the LEAP model is an exception and has always dealt with dynamic and nonlinear changes in some fashion), it is oblivious to the thermal variations and the various other nonlinear mechanisms that cause distortion and ultimately limit acoustic output of any device. With its ability to measure dynamic Bl, compliance, inductance, temperature, and a wide variety of other speaker characteristics, the Klip-

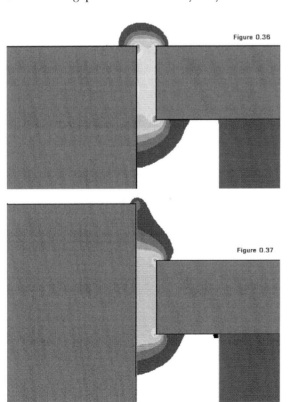

Figure 0.36

Figure 0.37

FIGURE 0.36:
Shape of the flux in the gap with a flush pole.

FIGURE 0.37:
Extended pole.

pel Analyzer was the perfect choice for a series of short studies that were designed for the 7th Edition of the *Loudspeaker Design Cookbook*. Topics that are illuminated using the Klippel DA include Bl linearity, pole extensions and linearity, shorting rings and

FIGURE 0.38:
Klippel DA-2 distortion analyzer hardware.

FIGURE 0.39:
Klippel test stand with positioning laser and microphone attached.

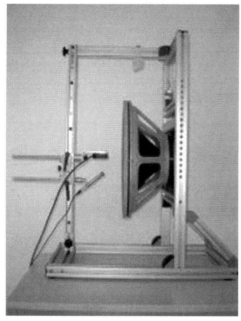

distortion, voice coil temperature and motor mass, and voice coil temperature and pole vents.

0.81 Bl AND COMPLIANCE LINEARITY.
If you have read through and absorbed the first part of this chapter (*Sections 0.21–023*), you know that Bl describes the electromagnetic "horsepower" of a loudspeaker motor, and is essentially the number of turns of wire exposed to the concentrated magnetic field in the gap area. Every loudspeaker designer's goal is to have a woofer that is perfectly linear and works equally well in both directions of travel and faithfully tracks the input signal with zero distortion. For the most part, this speaker doesn't exist, but it is the goal.

To one extent or another, all woofers behave, or misbehave as it were, in a manner that leads to some level of distortion of the input

waveform. Getting them to behave as linear as possible is the task at hand. To that end, Pat Turnmire, CEO of Red Rock Acoustics and author of the SpeaD and RevSpeaD software (*Section 0.70*), and I put together a short study to illuminate just how it is that transducer engineers such as Mr. Turnmire make speakers operate in a more linear fashion with lower levels of distortion.

The concept was to build up a series of 10″ woofers, starting with a purposely nonlinear design and gradually change one "item" per iteration until we have gone from the initially low performance woofer to one that has all the bells and whistles available to improve performance. In total, Pat generated ten woofers for the Bl, pole extension, shorting ring, and voice coil temperature explications that follow. For this section, however, only five of the ten woofers were used. All five used the same frame, cone, surround, four-layer type voice coil wound on the same diameter and thickness aluminum former, the same carbon content steel in the plates, all using Y33 magnet material, and each with a 10mm gap height. The physical specifications for each driver are summarized in *Table 0.7*.

Before using the Klippel analyzer, all of the Thiele/Small parameters were measured using a MLSSA analyzer and the results are summarized in *Table 0.8*.

Looking at *Table 0.7*, you can see that the initial change in woofer samples 1–3 was in voice coil length. From the data given in *Table 0.8*, you can also see some of the criteria used in assembling this group. Obviously, an attempt was made to keep Re and Fs reasonably constant so that all the woofers would have about the same tuning characteristics in a common sealed box volume. However, since all five samples used the same cone, the cone assembly weight (Mmd) would then vary with voice coil size. Because Mmd varied significantly, you will also notice that changes were made in the magnet size to keep Qts values within a fairly narrow range, with the exception of woofer sample #1, which has a somewhat higher Qts than the rest of the group.

Following the initial increase in Xmax in the first

TABLE 0.7

Sample #	Magnet Dim.	VC Length	Xmax	Spider Type	Pole Extension
1	110mm × 15mm	14mm	2mm	Cupped	No
2	140mm × 20mm	24mm	7mm	Cupped	No
3	140mm × 40mm	40mm	15mm	Cupped	No
4	140mm × 40mm	40mm	15mm	Flat	No
5	140mm × 40mm	40mm	15mm	Flat	Yes (5mm)

TABLE 0.8

	1	2	3	4	5
R_E	3.10Ω	3.00Ω	2.96Ω	3.06Ω	3.07Ω
F_S	30.49Hz	30.76Hz	28.13Hz	28.19Hz	27.82Hz
Q_{MS}	6.86	6.52	6.06	7.34	7.92
Q_{ES}	0.78	0.65	0.58	0.58	0.56
Q_{TS}	0.70	0.59	0.53	0.54	0.52
V_{AS}	34.23 liters	39.51 liters	36.92 liters	37.21 liters	38.20 liters
M_{MD}	88.9 grams	122.1 grams	157.3 grams	157.3 grams	157.3 grams
Bl	9.47 Tm	10.56 Tm	12.07 Tm	12.22 Tm	12.44 Tm

three woofers, the next change was in the suspension system, which means that the first three drivers had cupped or elevated type spiders and samples #4 and #5 had a more linear flat type spider that lacked the asymmetrical problems generally associated with the "hinge" on a cupped spider. The last iterative change to make the motor system more linear was with the addition of a pole extension that was equal to one-half of the gap height for woofer sample #5. All the data from *Table 0.8* was entered into the LEAP 5 Enclosure Shop software and the five woofers simulated in a 2ft³ enclosure with 50% fiberglass fill material.

The SPL and impedance curves are shown in *Figs. 0.40* and *0.41*, respectively. Again, with the exception of woofer sample #1, the remaining four samples had at least similar performance in the same enclosure volume. However, using computer box simulations, all you can see is the similar box performance, some SPL difference, and some obviously higher voice coil inductance in some of the woofers, but you really don't have much of an idea how these drivers are performing dynamically, and even if you rerun the simulations at higher input voltages, you will only see the relative increase in SPL and changes in overall damping in the woofer/box combination.

The best way to reveal the changes in linearity as we increased the voice coil length, improved the suspension linearity, and increased the forward fringe field linearity by adding a pole extension is using the Klippel Analyzer. Because this is a very expensive test instrument that ranges from $20,000 to $30,000, it is unfortunately out of the reach of most amateur designers, and even many smaller loudspeaker manufacturers. However, Dr. Klippel's group has graciously provided a complete Klippel setup that Patrick Turnmire and I use for driver testing in *Voice Coil* magazine, product reviews in *Car Audio and Electronics* magazine, and for this section of the 7th Edition of the *LDC*.

I chose three graphs from the selection of data produced by the Klippel Analyzer, Bl (X), Kms (X), and Le (X). This is the same basic curve set that I use in the Test Bench section of *Voice Coil* each month. Bl (X) is a dynamic depiction of Bl change as the woofer reaches the extremes of excursion in both the forward and the rear directions of travel. Kms (X) is the stiffness of the suspension curve and is the mathematical inverse of a compliance curve. The Le (X) curve shows the changes in inductance that occur as the voice coil travels in both directions.

Testing woofers using the Klippel Distortion Analyzer involves running a noise stimulus with a preprogrammed series of level changes. This test takes approximately 5–10 minutes per driver and yielded the Klippel data curves for the five woofer samples as follows:

	Bl (X)	Kms (X)	Le (X)
Sample #1	0.42	0.43	0.44
Sample #2	0.45	0.46	0.47
Sample #3	0.48	0.49	0.50
Sample #4	0.51	0.52	0.53
Sample #5	0.54	0.55	0.56

Looking at the set of curves for woofer sample #1, you can see that the Bl curve has a rather pointed shape and falls off quickly in both directions of cone travel. Actually, this is very similar to what you often see in high efficiency pro sound woofers where excursion is not as important of an issue as efficiency. Both the Bl (X) and Kms (X) curves show a forward offset of the voice coil, which while not extreme in this instance, the goal is generally to have these curves centered at the rest position[14] and symmetrical in both directions. However, offsets in the voice coil at rest position can be both good and bad. In most circumstances, an offset voice coil only produces additional signal distortion, while a deliberate offset that compensates for an asymmetrical magnetic field can actually improve a woofer's performance[15].

The Bl curve for woofer sample #1 is actually quite symmetrical, but the cupped spider Kms (X) curve shows an obvious lack of symmetry with different curve slopes for the rear direction of travel compared to the forward direction. While this is typical of cupped (elevated) spiders, it should be said that it is also possible to configure this type of spider such that its compliance is more symmetrical.

Another byproduct of Klippel analysis is for the software to generate excursion limitations for the individual Bl, compliance, and inductance components that define how these components contribute to a particular level of distortion[16]. Because the woofers produced for this test are all basically 4Ω subwoofers, the criterion for a 20% distortion level is Bl decreasing to a minimum of 70%, compliance decreasing to a minimum of 50%, and inductance decreasing to a minimum of 10%. A reasonable criterion for subwoofers is 20% distortion, which is subjectively not readily discernible at frequencies below 100Hz. For full-range devices, the criteria

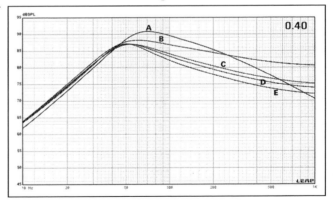

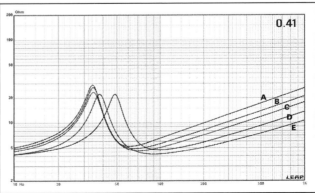

FIGURE 0.40:
Computer box simulation SPL curves for woofer samples #1–5 (A = woofer 1; B = woofer 2; C = woofer 3; D = woofer 4; E = woofer 5).

FIGURE 0.41:
Computer box simulation impedance curves for woofer samples #1–5 (A = woofer 2; B = woofer 4; C = woofer 3; D = woofer 1; E = woofer 5).

are adjusted from 20% to a lower 10% level of distortion.

For woofers and subwoofers, Le (X) isn't really relevant and primarily refers more to hearable intermodulation distortion in midrange-type speakers. *Table 0.9* gives displacement limiting Bl and compliance numbers for the five samples. For woofer sample #1, even though Xmax is only 2mm physically, this woofer can give satisfactory performance out to 5mm of travel before Bl derived distortion became slowly audible.

TABLE 0.9

Sample#	Bl (X)	Cms (X)
1	5mm	7.3mm
2	10mm	8.4mm
3	15.2mm	11.3mm
4	15.2mm	10.8mm
5	16.5mm	10.4mm

Woofer sample #2 has as its primary change an increase in voice coil length from 14mm to 24mm giving the driver an Xmax of 7mm, up from 2mm for woofer sample #1. Looking at the Bl (X) curve in *Fig. 0.45*, it's immediately obvious what the effect of additional voice coil length has on the excursion capability of the woofer. The Bl curve for this woofer has more of a broad plateau shape and only gradually decreases Bl with increasing travel in either direction. As with woofer sample #1, you still see a moderate forward offset in the Bl (X) curve, with not much in the way of change in the Kms (X) or Le (X) curves, as would be expected. Displacement limiting numbers likewise increased, but now are somewhat less than the physical Xmax for this driver.

Woofer sample #3 also has a substantial increase in the voice coil length, again accompanied by an increase in motor size. Xmax for sample #3 is now 15mm, more than double the Xmax of woofer sample #2. The Bl (X) curve in *Fig. 0.48* is now more

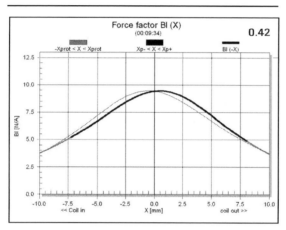

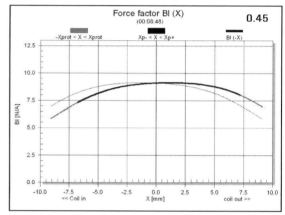

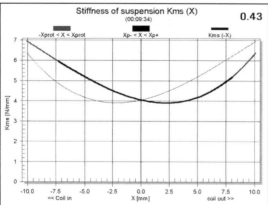

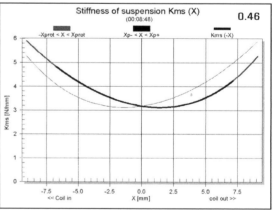

FIGURE 0.42: Klippel Bl (X) curve for woofer sample #1.

FIGURE 0.43: Klippel Kms (X) curve for woofer sample #1.

FIGURE 0.44: Klippel Le (X) curve for woofer sample #1.

FIGURE 0.45: Klippel Bl (X) curve for woofer sample #2.

FIGURE 0.46: Klippel Kms (X) curve for woofer sample #2.

FIGURE 0.47: Klippel Le (X) curve for woofer sample #2.

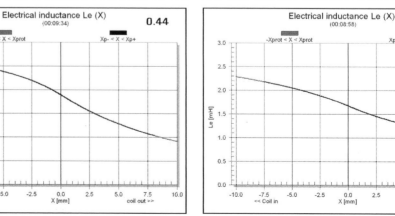

defined and has a wider, less shallow, plateau, but also has a definite asymmetrical tilt and a substantial rearward offset of 7mm at rest due mostly to the gap field asymmetry and the longer voice coil. Again, with no changes in the suspension, the Kms (X) curve is similar to woofer sample #2.

The only change made in woofer sample #4 was to trade the asymmetrical cupped spider for a more linear flat spider. The Bl (X) curve for sample #4 shows little change, but the Kms (X) curve is now more symmetrical and has less forward offset than was present with the first three woofer samples that featured the cupped-type spider.

The easiest way to make the fringe field in the forward direction more like the rear fringe field is to extend the pole (see *Section 0.75* and *Figs. 0.36–0.37*). This was the only change made to sample #5, and the effect is very pronounced. Looking at the Bl (X) curve in *Fig. 0.54*, the Bl plateau is now very symmetrical with practically no offset whatsoever.

Because the Klippel analyzer will also output dis-

tortion curves for the individual distortion components caused by Bl, compliance, or inductance, Bl distortion curves were generated to compare performance with and without the pole extension. *Figures 0.57* and *0.58* depict the distortion component of the Bl function for woofer sample #4 and woofer sample #5, respectively, where the only difference is

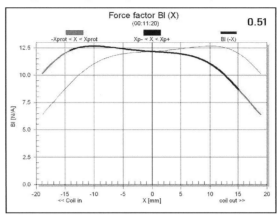

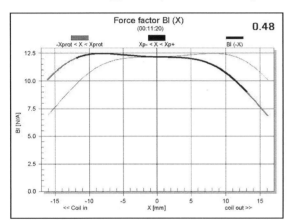

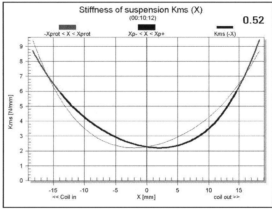

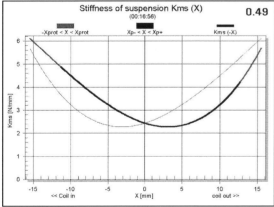

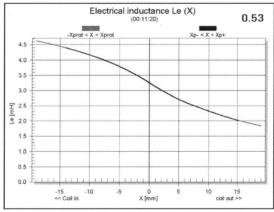

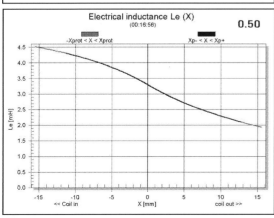

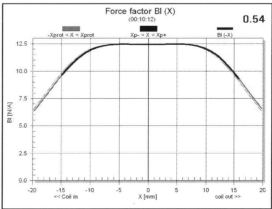

FIGURE 0.48: Klippel Bl (X) curve for woofer sample #3.

FIGURE 0.49: Klippel Kms (X) curve for woofer sample #3.

FIGURE 0.50: Klippel Le (X) curve for woofer sample #3.

FIGURE 0.51: Klippel Bl (X) curve for woofer sample #4.

FIGURE 0.52: Klippel Kms (X) curve for woofer sample #4.

FIGURE 0.53: Klippel Le (X) curve for woofer sample #4.

FIGURE 0.54: Klippel Bl (X) curve for woofer sample #5.

the 5mm pole extension. Combine this with the reasonably symmetrical compliance in woofer sample #5, and you now have a substantially more linear driver with significantly greater excursion and lower distortion at high output levels compared with the first woofer sample #1.

0.82 SHORTING RINGS AND DISTORTION.

Section 0.22 discusses the benefits of adding a shorting ring to a woofer or midrange to reduce the AC eddy currents generated by voice coil motion, a parasitic current flow that is responsible for flux modulation of the magnetic field and sometimes a substantial amount of inductive heating in large motor woofers and subwoofers. The change in inductance using a shorting ring is immediately obvious when you examine the Klippel Le (X) curves for woofer sample #6 and woofer sample #10, essentially the same driver but with sample #10 having a large aluminum shorting ring similar to Type D in *Fig. 0.6*.

If you compare the MLSSA derived T/S parameters in *Table 0.10*, you see virtually no difference due to the shorting ring. However, the difference in voice coil inductance can easily be observed in Le (X) curves shown in *Fig. 0.59* for woofer sample #6 and *Fig. 0.60* for woofer sample #10. Not only has inductance decreased in the rearward direction with the introduction of a large aluminum shorting ring, but it has also resulted in an overall decrease in total motor inductance.

TABLE 0.10

	Sample #6	Sample #10
R_E	3.08	3.06
F_S	24.86	24.67
Q_{MS}	6.81	6.23
Q_{ES}	0.50	0.50
Q_{TS}	0.46	0.47
V_{AS}	40.82 liters	40.43 liters
M_{MD}	158.85	158.85
Bl	12.52	12.36

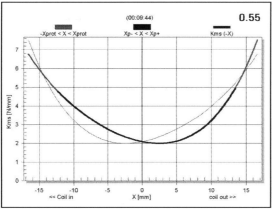

FIGURE 0.55: Klippel Kms (X) curve for woofer sample #5.

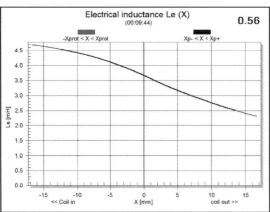

FIGURE 0.56: Klippel Le (X) curve for woofer sample #5.

FIGURE 0.57: Klippel Bl product distortion curve for woofer sample #4 (no pole extension).

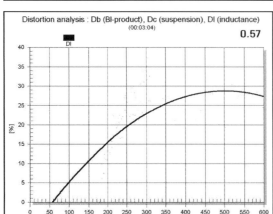

FIGURE 0.58: Klippel Bl product distortion curve for woofer sample #5 (with 5mm pole extension).

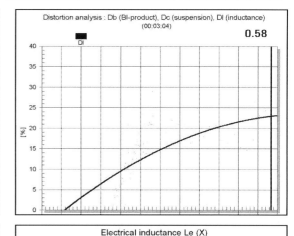

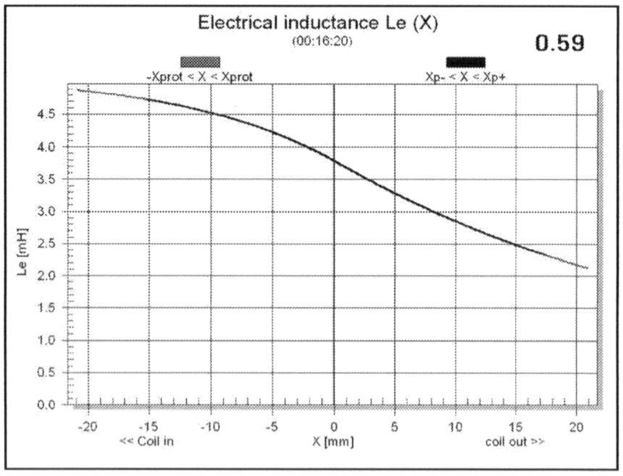

FIGURE 0.59: Klippel Le (X) curve for woofer sample #6 (no shorting ring).

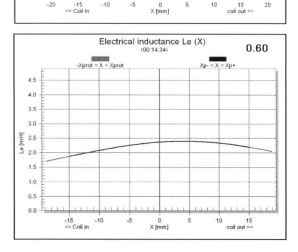

FIGURE 0.60: Klippel Le (X) curve for woofer sample #10 (with shorting ring).

In terms of distortion due to both Bl and inductance, *Fig. 0.61* gives the distortion curves for woofer sample #6 and *Fig. 0.62* shows the same data for woofer sample #10. The decrease in distortion level and decrease in inductance is substantial for just a few cents worth of aluminum.

0.83 VOICE COIL TEMPERATURE AND MOTOR MASS.

The secret to making woofers as indestructible as possible is to somehow deal with the voice coil current induced heating caused by the voice coil resistance (OK, superconductor voice coil wire would be one answer). Maintaining motor temperatures below a level where your parts start to melt and the adhesives break down increases the power handling and lowers distortion in woofers when operating at high SPL. The most obvious factor to consider is the amount of mass in the motor. Larger motors can hold and radiate more heat than smaller woofer motors.

Another test procedure available with the Klippel Distortion Analyzer software modules is the Power Test Module that produces the Klippel PWT temperature test[17]. This is a power test that measures voice coil temperature change (delta T) over time. All ten of the test woofers were put through a high voltage 60-minute PWT procedure. For woofer samples 1–3, the results are shown in *Figs. 0.63–0.65*. Magnet dimensions are given in *Table 0.7*, but beyond that woofer sample #1 has a 17.61 oz. magnet with 102mm diameter front and rear plates, woofer sample #2 has a 38.61 oz magnet with 120mm diameter plates, and woofer sample #3 has a 77.23 oz magnet with 120mm diameter plates.

While there are other factors to consider, such as the size of the voice coil, the number of turns, and gauge of wire (woofer sample #1 has 21.33m of 0.37 wire; woofer sample #2 has 30.85m of 0.45 wire; and woofer sample #3 has 41.74m of 0.55 wire), woofer sample #1 rapidly increased voice

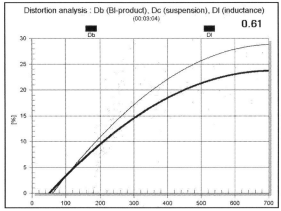

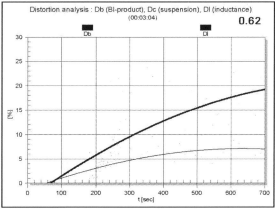

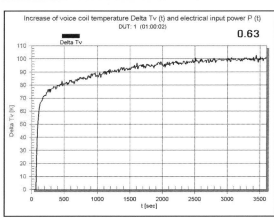

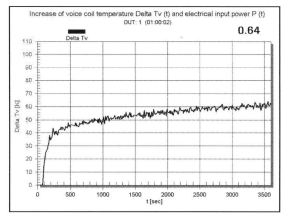

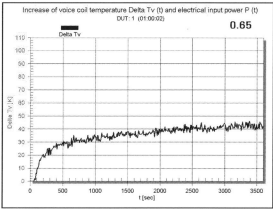

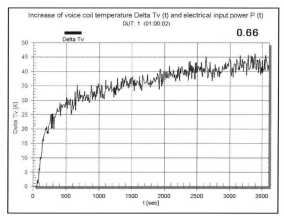

FIGURE 0.61: Klippel Bl product and inductance distortion curves for woofer sample #6 (no shorting ring).

FIGURE 0.62: Klippel Bl product and inductance distortion curves for woofer sample #10 (with shorting ring).

FIGURE 0.63: Klippel PWT temperature change graph for woofer sample #1 (17.6 oz magnet).

FIGURE 0.64: Klippel PWT temperature change graph for woofer sample #2 (38.6 oz magnet).

FIGURE 0.65: Klippel PWT temperature change graph for woofer sample #3 (77.2 oz magnet).

FIGURE 0.66: Klippel PWT temperature change graph for woofer sample #6 (30mm pole vent).

coil temperature from an ambient temperature of 75° Fahrenheit by 100 Kelvin. This change is a 180° F temperature increase, meaning that after 60 minutes of pink noise at about 30W, the voice coil was operating at 255°F. With the larger magnet and plates, woofer sample #2 finished its power test at 184°F, and the largest magnet, woofer sample #3, ended its test at 151°F.

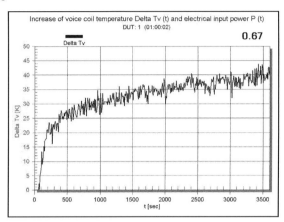

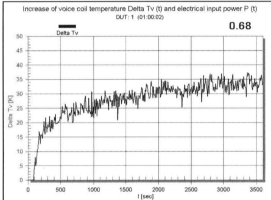

FIGURE 0.67: Klippel PWT temperature change graph for woofer sample #7 (22.5mm pole vent).

FIGURE 0.68: Klippel PWT temperature change graph for woofer sample #8 (16.9mm pole vent).

FIGURE 0.69: Klippel PWT temperature change graph for woofer sample #9 (12.7mm pole vent).

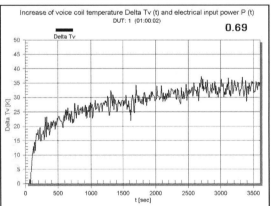

0.84 VOICE COIL TEMPERATURE AND POLE VENTS.

A very traditional way to provide voice coil cooling for high power woofers has been to put the largest diameter hole possible in the motor pole piece. The concept is that the vent hole encourages air to travel away from the voice coil area carrying heat out the rear of the motor and thereby providing an efficient convention path. However, this conventional wisdom was challenged in a paper presented at the 114th AES Convention titled "Nonlinear Modeling of the Heat Transfer in Loudspeakers," by Dr. Wolfgang Klippel[18].

While the paper discussed the limits of traditional thermal modeling and offered a new extended thermal model, it also gave an example of optimal thermal design. The optimal design was a direct comparison of a woofer with a typical vented pole and the same woofer with the pole blocked so that no air could travel through the pole vent. The result was a decrease in the driver Qms accompanied by a significant decrease in the change in temperature over time. Essentially, blocking the vent forces more air past the voice coil and out the top of the gap and into the area beneath the spider and provides more heat transfer than the large pole vent. Because I have done subwoofer reviews in *Car Audio and Electronics* magazine for a number of years, I have been able to observe some of the trends in thermal management in high power drivers, and one of the currently popular techniques is to provide substantial venting below the spider mounting shelf and the front plate.

When I suggested doing a Klippel woofer study to Mr. Turnmire, he thought it would be interesting to include a series of woofers starting with a typically large pole vent and then decreasing the vent size over four woofers to see what the temperature differential actually was and whether there was an optimal vent size short of blocking the vent. Woofer samples #6–9 were produced especially for this purpose. All four woofers had the same mechanical description as woofer sample #5 in *Table 0.7*, while *Table 0.11* gives the pole vent sizes plus the MLSSA derived T/S parameters.

You can observe from *Table 0.11* that these woofers are all identical, with only minor variations in their parameters, but with a pole vent size that decreases from 1.2″ in diameter to about 0.5″ in diameter and the resulting different amounts of temperature change over time. Temperature change over time

TABLE 0.11.

	Sample #6	Sample #7	Sample #8	Sample #9
R_E	3.08	3.06	3.08	3.10
F_S	24.86Hz	24.90Hz	23.72Hz	24.55Hz
Q_{MS}	6.81	5.67	6.24	6.83
Q_{ES}	0.50	0.43	0.42	0.43
Q_{TS}	0.46	0.40	0.39	0.43
V_{AS}	40.82 liters	40.67 liters	44.82 liters	41.72 liters
M_{MD}	158.85 grams	158.85 grams	158.85 grams	158.85 grams
Bl	12.52	13.53	13.37	12.98
Pole Vent Diameter	30mm	22.5mm	16.9mm	12.7mm
Delta T	47 K	44 K	38 K	38 K

graphs are depicted in *Figs. 0.66–0.69* for woofer samples #6–9, respectively. The Delta T numbers are taken from a statistical best line fit for the data in these graphs, but it is clear that decreasing the pole vent area seems to stop decreasing the voice coil temperature past a certain point, and although not shown, when the vent was blocked completely, the temperature change was the same as woofer samples #8 and #9.

Looking at all of this, I would say that if you have a woofer with a very large pole vent, the engineer in charge of the design very likely was not aware of the possibility of optimizing the pole vent hole size for maximum voice coil cooling. Also, you should be aware that this represents only one of many methods available for improving convection-based temperature reduction in a woofer design.

REFERENCES

1. Gander, Mark, "Moving Coil Loudspeaker Topology as an Indicator of Linear Excursion Capability," *JAES*, Jan./Feb. 1981.

2. Lian, R., "Distortion Mechanisms in the Electrodynamic Motor System," 84th AES Convention, March 1988, preprint #2572.

3. Jordan, E.J., Loudspeakers, Focal Press Ltd., 1963. (from the *Encyclopedia of High Fidelity*, John Borwick, editor).

4. Cunningham, W. J., "Nonlinear Distortion in Dynamic Loudspeakers Due to Magnetic Effects," *Journal Acoustical Society of America*, Vol. 21, May 1949.

5. Birt, David, "Nonlinearities in Moving Coil Loudspeakers with Overhung Voice Coils," 88th AES Convention, March 1990, preprint #2904.

6. BRAVOX S/A, a large OEM driver manufacturer in Brasil (Brazil if you live in North America) that I have been representing for the last five years, provided all driver samples for this section. Bravox Chief Engineer, and good friend, Sergio Pires, created all the samples for this section of the new edition of *LDC*.

7. Frankfort, F. J. M., "Vibration Patterns and Radiation Behavior of Loudspeaker Cones," *JAES*, September 1978.

8. Pierce, Richard, "The Dust Cap Solution," *Speaker Builder* 1/91.

9. Shindo, Yashima and Suzuki, "Effect of Voice-Coil and Surround on Vibration and Sound Pressure Response of Loudspeaker Cones," *JAES*, July/August 1980.

10. Suzuki and Tichy, "Radiation and Diffraction Effects by Convex and Concave Domes," *JAES*, December 1981.

11. Keele, Jr., D.B., "Equipment Profile: The Paradigm SE Loudspeaker," *Audio*, September 1990, p. 90, sidebar.

12. Beranek, Leo, *Acoustics*, McGraw-Hill, 1954, Acoustical Society of America, 1986.

13. Wright, J. R., "An Empirical Model for Loudspeaker Motor Impedance," *JAES*, October 1990.

14. Klippel Application Note, "AN-01 Voice Coil Rest Position," April 25, 2002.

15. Klippel Application Note, "AN-21 Reduce Distortion by Shifting Voice Coil," June 26, 2003.

16. Klippel Application Note, "AN-05 Displacement Limits Due to Driver Nonlinearities," June 25, 2002.

17. Klippel Application Note, "AN-18 Thermal Parameter Measurement," March 14, 2003.

18. Klippel, W., "Nonlinear Modeling of the Heat Transfer in Loudspeakers," *JAES* January/February 2004.

CHAPTER ONE

CLOSED-BOX LOW-FREQUENCY SYSTEMS

1.10 DEFINITION.

The closed-box is the simplest of all loudspeaker designs, consisting of an enclosed volume of air and the loudspeaker or driver. Its electrical and pneumatic circuits are analogous to a second-order high-pass filter with the response controlled by the resonance and the associated damping. There are two basic types of closed-box systems: the infinite baffle (IB) and the air suspension (AS).

The IB enclosure is made large so the compliance of the air within the enclosure (its pneumatic "spring" quality) is greater than the compliance of the driver suspension. A closed-box loudspeaker becomes an acoustic suspension speaker when the compliance of the air volume inside the box is less than the compliance of the woofer by a factor of three or more.[1,2] This design combination of a loose woofer surround and a small box was popularized in the 1950s by Acoustic Research, and is still frequently used by loudspeaker manufacturers.

Because of its highly controllable response shape and transient characteristics, and because of the relative ease of achieving correct box parameters, the closed-box design is probably the best for home construction, especially if you are a beginner.

1.15 DEFINITION OF TERMS.

f_3 minus three decibel half-power frequency (designates the beginning of low end rolloff).

f_s resonance frequency of driver.

f_c resonance frequency of the closed-box system.

Q ratio of reactance to resistance (series circuit) or resistance to reactance (parallel circuit).

Q_{ts} total Q of driver (woofer) at f_s, considering all driver resistances.

Q_{tc} total Q of speaker system at f_c, including all system resistances.

V_{as} volume of air having the same acoustic compliance as the driver suspension.

V_{ab} volume of air having the same acoustic compliance as the enclosure.

X_{max} peak linear displacement of driver cone.

S_d effective surface area of a driver cone.

V_d peak displacement volume of driver cone.

V_b net internal volume of enclosure.

α compliance ratio.

η_0 reference efficiency.

C_{as} acoustic compliance of the driver suspension.

C_{ab} acoustic compliance of the air in the enclosure.

1.20 HISTORY.

Infinite baffle closed-box designs were popular from the very beginning until the early 1950s. However, after the air suspension design was patented in 1949 by Harry Olson and his associate J. Preston, things began to change. This change was largely brought about by the work of an early proponent of AS design, Edgar Villchur. In 1954,[F1] he began a series of articles in *Audio* magazine that established the AS as the ultimate speaker design. It was during this same period that Acoustic Research introduced the classic AS designed speaker, the AR-3. Henry Kloss, the co-founder of AR (with Villchur), also went on to found two other successful companies which continued to popularize the AS design, namely KLH and Advent.

In 1972, Richard Small published the most definitive work on closed-box design to date.[3,4] The following presentation on closed-box design relies upon this monumental work of clarity and simplicity.

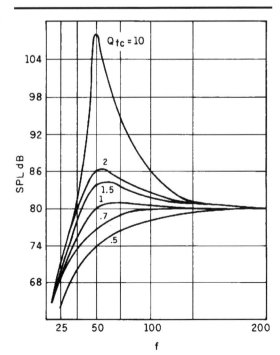

FIGURE 1.1: Frequency response of closed-box systems with different Qs.

1.30 DRIVER "Q" and ENCLOSURE RESPONSE.

The whole point of fitting a woofer to a particular box volume is to control the response characteristics of the combination. The objective method is to measure and align the Q factor. Q is a composite term, used, in this case, to describe resonant magnification in speaker boxes. It represents the degree to which the electrical, mechanical, and pneumatic circuits of the woofer/box combination interact to control resonance. *Figure 1.1* [5,6] illustrates the relationship between different values of Q and frequency response.

Several observations can be made from this family of response curves depicted in *Fig. 1.1*. First, closed-boxes exhibit a fairly shallow rolloff of about 12dB/octave. Compared to vented and passive-radiator designs with 24dB/octave rolloff slopes, a closed-box with the same f_3 will yield more low bass and have better transient stability.

Second, certain values of Q_{tc} have specific response characteristics, which can be categorized as follows:

$$Q_{tc} = 0.5$$

Critically Damped–transient perfect

$$Q_{tc} = \frac{1}{\sqrt{3}} = 0.577$$

Bessel response (D_2)–max flat delay

$$Q_{tc} = \frac{1}{\sqrt{2}} = 0.707$$

Butterworth response (B_2)–max flat amplitude response with minimum cutoff

$$Q_{tc} \geq \frac{1}{\sqrt{2}}$$

Chebyshev (elliptical–C_2) Equal Ripple response–max power handling and max efficiency, somewhat degraded transients.

Although these particular Q_{tc}s are just points on a continuum, we can generalize about their subjective sound quality. High Qs in the vicinity of 1 tend to have a warm, if somewhat robust, quality that audio marketing people describe as "saleable." Lower Q values around 0.8 sound more detailed (due in part to improved transients) and, by comparison, somewhat shallow. $Q_{tc} = 0.5$ is usually regarded as excessively taut and overdamped. Some authorities, however, still consider this value (0.5 –0.6) to be optimal.[7] Occasionally, a quality loudspeaker, such as the Rogers LS3/5A, is designed with a very high value of Q (1.2). In this case, it was to give the minimonitor more apparent bass in a mobile sound-van situation. Q_{tc} greater than 1.2 or so, however, should be regarded as undesirable.

Table 1.0 illustrates the relationship between Q_{tc} and a peak in frequency response above a flat magnitude. The frequency of this peak, f_{gmax}, is given as a ratio with the box resonance, f_c. Also included is the ratio f_{xmax}, which gives the frequency of maximum cone displacement.

TABLE 1.0

Q_{tc}	Peak dB	f_{gmax}/f_c	f_{xmax}/f_c
0.5	0	—	0
0.577	0	—	0
0.707	0	—	0
0.8	0.213	2.138	0.468
0.9	0.687	1.616	0.619
1.0	1.249	1.414	0.707
1.1	1.833	1.305	0.766
1.2	2.412	1.238	0.808
1.3	2.974	1.192	0.839
1.4	3.515	1.159	0.863
1.5	4.033	1.134	0.882

where:

$$\text{Peak dB} = 20\log_{10}\sqrt{\frac{Q_{tc}^4}{Q_{tc}^2 - 0.25}}$$

$$f_{gmax} = \frac{1}{\sqrt{1 - \frac{1}{2Q_{tc}^2}}}$$

$$f_{xmax} = \sqrt{1 - \frac{1}{2Q_{tc}^2}}$$

The results of a survey of US, British, and European closed-box systems, done in 1969 by Professor Richard Small[4], revealed that most AS speakers fall into one of two categories:

1. cutoff frequency below 50Hz; Q_{tc} up to 1.1; size greater than 1.4ft³.
2. cutoff frequency above 50Hz; Q_{tc} from 1.2–2.0; size less than 2ft³.

Category 1 boxes tend to produce good low bass for orchestral and organ music, while category 2 boxes had "demonstrably stronger bass" with pop electric music.

1.35 ALTERNATE METHODS OF ANALYZING BOX "Q."

Along with the changes in the peak height, f_{xmax} and f_{gmax}, several other factors describe box Q. As the mechanical damping on the driver changes, and the impulse response degrades (as the bass goes from "tight" to "tubby"), the whole spectrum of changes occurs. These are

TABLE 1.1

Q_{tc}	dB Peak	Phase Angle	Slope dB/oct.	f_3	V_b ft²
0.7	0	90°	10.60	35Hz	2.6
0.9	0.69	97°	11.95	39Hz	1.6
1.0	1.25	100°	12.08	43Hz	1.18
1.1	1.83	103°	12.82	46Hz	0.92
1.2	2.41	106°	13.19	50Hz	0.74
1.5	4.0	110°	13.96	64Hz	0.420

the phase angle at the −3dB frequency, cone excursion curves, the shape and magnitude of the group delay curve, the shape of the imped-ance curve, and the cone velocity and volume current. The math needed to calculate this type of information makes for great mental exercise, but using one of the CAE (Computer Aided Engineering) programs available for loudspeaker design is not only a great deal eas-ier and faster, but provides information that is much too laborious, if not entirely impossible for hand-calculation methods.

LEAP 4.0 (Loudspeaker Enclosure Analysis Program by LinearX Systems) was selected for computer simulations and is used throughout this book. At the time of publication, LEAP 4.0 is the most flexible and powerful tool available for professional design and the only program which can readily import and export data from comput-er-based analyzers like the DRA Labs MLSSA FFT and the Audio Precision System 1.

The program was used to model a series of closed boxes with a Q_{tc} range of 0.7 through 1.5, and the results used to illustrate these concepts. A 10″ driver (the Audio Concepts AC-10) with parameters suitable for infinite baffle type enclo-sures was used for the simulation. No series resis-tance was included, and the box was modeled to include 50% fiberglass fill (50% fill generally cor-responds to having the back and four sides of the box lined with 3″ thick 1 lb/ft³ fiberglass). Enclosure volumes were calculated to provide the appropriate amount of peaking for each Q value and the other data read from the graphs generat-ed. The results are shown in *Table 1.1*.

From the information given in *Table 1.1*, and illustrated in the simulated SPL and acoustic phase curves in *Fig. 1.2*, it is obvious that as Q increases, the −3dB phase angle, rolloff slope and −3dB frequency also increase. "Phase," as it applies to loudspeakers, is a function of the slope of the magnitude response. If there is no more phase shift than what is dictated by the magnitude response, the device is called mini-mum phase, which is what loudspeakers are generally considered to be. Phase is measured by determining the time difference between the input signal from the signal source and the output signal occurring at the cone surface. The more time delay, the greater the measured phase angle at that point. The family of phase curves in *Fig. 1.2* shows a primary change in

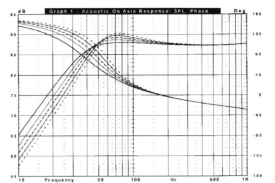

FIGURE 1.2

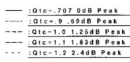

— :Qtc=.707 0dB Peak
······ :Qtc=.9 .69dB Peak
— — :Qtc=1.0 1.25dB Peak
—·— :Qtc=1.1 1.83dB Peak
- - - :Qtc=1.2 2.4dB Peak

FIGURE 1.3

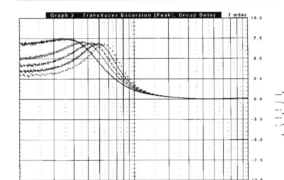

— :Qtc=.707 0dB Peak
······ :Qtc=.9 .69dB Peak
— — :Qtc=1.0 1.25dB Peak
—·— :Qtc=1.1 1.83dB Peak
- - - :Qtc=1.2 2.4dB Peak

FIGURE 1.4

—— :Qtc=.707 0dB Peak
······ :Qtc=.9 .69dB Peak
—·—— :Qtc=1.0 1.25dB Peak
—·—— :Qtc=1.1 1.83dB Peak
- - - :Qtc=1.2 2.4dB Peak

FIGURE 1.5

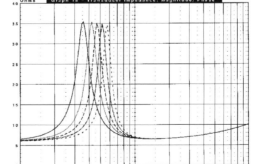

—— :Qtc=.707 0dB Peak
······ :Qtc=.9 .69dB Peak
— — :Qtc=1.0 1.25dB Peak
—·— :Qtc=1.1 1.83dB Peak
- - - :Qtc=1.2 2.4dB Peak

FIGURE 1.6

—— :Qtc=.707 0dB Peak
······ :Qtc=.9 .69dB Peak
— — :Qtc=1.0 1.25dB Peak
—·—— :Qtc=1.1 1.83dB Peak
- - - :Qtc=1.2 2.4dB Peak

TABLE 1.2

Closed-Box Q_{tc} = 1.0

	Driver 1	Driver 2
f_s	31.5Hz	38Hz
Q_{ts}	0.45	0.54
V_{as}	2.97ft³	1.92ft³
V_b	0.75ft³	0.79ft³
f_c	69.9Hz	70.4Hz

V_b should be within 10%, which in this case, it is.

TABLE 1.3

$Q_{tc} = 0.5$
Second-Order, Critically Damped

Q_{ts}	α	f_c/f_s
0.1000	24.0000	5.0000
0.1200	16.3611	4.1667
0.1300	13.7929	3.8462
0.1400	11.7551	3.5714
0.1500	10.1111	3.3333
0.1600	8.7656	3.1250
0.1700	7.6505	2.9412
0.1800	6.7161	2.7778
0.1900	5.9252	2.6316
0.2000	5.2500	2.5000
0.2100	4.6689	2.3810
0.2200	4.1653	2.2727
0.2300	3.7259	2.1739
0.2400	3.3403	2.0833
0.2500	3.0000	2.0000
0.2600	2.6982	1.9231
0.2700	2.4294	1.8519
0.2800	2.1888	1.7857
0.2900	1.9727	1.7241
0.3000	1.7778	1.6667
0.3100	1.6015	1.6129
0.3200	1.4414	1.5625
0.3300	1.2957	1.5152
0.3400	1.1626	1.4706
0.3500	1.0408	1.4286
0.3600	0.9290	1.3889
0.3700	0.8262	1.3514
0.3800	0.7313	1.3158
0.3900	0.6437	1.2821
0.4000	0.5625	1.2500
0.4100	0.4872	1.2195
0.4200	0.4172	1.1905
0.4300	0.3521	1.1628
0.4400	0.2913	1.1364
0.4500	0.2346	1.1111
0.4600	0.1815	1.0870
0.4700	0.1317	1.0638
0.4800	0.0851	1.0417
0.4900	0.0412	1.0204

(A.S. marked between 0.2300 and 0.2500; INF. BAFFLE marked from 0.3700 downward)

TABLE 1.4

$Q_{tc} = 0.577$
Second-Order Bessel (D_2)
Max Flat Delay Response

Q_{ts}	α	f_c/f_s
0.1000	32.2929	5.7700
0.1100	26.5148	5.2455
0.1200	22.1201	4.8083
0.1300	18.6700	4.4385
0.1400	15.9862	4.1214
0.1500	13.3797	3.8467
0.1600	12.0050	3.6063
0.1700	10.5200	3.3941
0.1800	9.2756	3.2056
0.1900	8.2224	3.0368
0.2000	7.3232	2.8850
0.2100	6.5494	2.7476
0.2200	5.8787	2.6227
0.2300	5.2936	2.5087
0.2400	4.7800	2.4042
0.2500	4.3269	2.3080
0.2600	3.9250	2.2192
0.2700	3.5669	2.1370
0.2800	3.2465	2.0607
0.2900	2.9587	1.9897
0.3000	2.6992	1.9233
0.3100	2.4644	1.8613
0.3200	2.2513	1.8031
0.3300	2.0572	1.7485
0.3400	1.8800	1.6971
0.3500	1.7178	1.6486
0.3600	1.5689	1.6028
0.3700	1.4319	1.5595
0.3800	1.3056	1.5184
0.3900	1.1889	1.4795
0.4000	1.0808	1.4425
0.4100	0.9805	1.4073
0.4200	0.8874	1.3738
0.4300	0.8006	1.3419
0.4400	0.7197	1.3114
0.4500	0.6441	1.2822
0.4600	0.5734	1.2543
0.4700	0.5071	1.2277
0.4800	0.4450	1.2021
0.4900	0.3866	1.1776
0.5000	0.3317	1.1540
0.5100	0.2800	1.1314
0.5200	0.2312	1.1096
0.5300	0.1852	1.0887
0.5400	0.1417	1.0685
0.5500	0.1006	1.0491
0.5600	0.0616	1.0304
0.5700	0.0247	1.0123

(A.S. marked near 0.2200; INF. BAFFLE marked from 0.4300 downward)

the region of the response where the slope of the rolloff is changing.

Figure 1.3 depicts the change in cone excursion with changes in box Q. As the box size for a given woofer increases, the amount of excursion required also increases, and the maximum level at which the driver produces acceptable levels of distortion decreases. Looking at these curves, the tradeoff between better damping and power handling is readily apparent.

Another indication of damping changes (Q changes) comes from the shape and absolute level of the enclosure group delay curve. Group delay is calculated from the phase response of the driver/enclosure combination. By definition[8], group delay is the slope of the phase response (specifically, the negative derivative of the phase slope). It describes the relative delay of the spectral components of a waveform. Mathematically:

$$\text{Group Delay} = -(\text{phase at } f_2 - \text{phase at } f_1) / (f_2 - f_1)$$

A flat group delay curve indicates that all frequencies are arriving simultaneously, while a peak in group delay shows some frequencies arriving later. Better damping quality is associated with simultaneous arrival. *Figure 1.4* depicts the group delay curves for the various Q_{tc}s. Notice that as Q increases, the shape of the curve goes from almost flat at $Q_{tc} = 0.7$ (a

TABLE 1.5

$Q_{tc} = 0.707$
Second-Order Butterworth (B$_2$)
Max Flat Amplitude Response

	Q_{ts}	α	f_c/f_s
	0.1500	21.2155	4.7133
	0.1600	18.5254	4.4188
	0.1700	16.2958	4.1588
	0.1800	14.4274	3.9278
	0.1900	12.8462	3.7210
	0.2000	11.4962	3.5350
	0.2100	10.3344	3.3667
A.S.	0.2300	8.4489	3.0739
	0.2400	7.6779	2.9458
	0.2500	6.9976	2.8280
	0.2600	6.3942	2.1792
	0.2700	5.8566	2.6185
	0.2800	5.3756	2.5250
	0.2900	4.9435	2.4379
	0.3000	4.5539	2.3567
	0.3100	4.2013	2.2806
	0.3200	3.8813	2.2094
	0.3300	3.5900	2.1424
	0.3400	3.3240	2.0794
	0.3500	3.0804	2.0200
	0.3600	2.8569	1.9639
	0.3700	2.6512	1.9108
	0.3800	2.4616	1.8605
	0.3900	2.2863	1.8128
	0.4000	2.1241	1.7675
	0.4100	1.9735	1.7244
INF. BAFFLE	0.4200	1.8336	1.6833
	0.4300	1.7033	1.6442
	0.4400	1.5819	1.6068
	0.4500	1.4684	1.5711
	0.4600	1.3622	1.5370
	0.4700	1.2628	1.5043
	0.4800	1.1695	1.4729
	0.4900	1.0818	1.4429
	0.5000	0.9994	1.4140
	0.5100	0.9218	1.3863
	0.5200	0.8486	1.3596
	0.5300	0.7795	1.3340
	0.5400	0.7142	1.3093
	0.5500	0.6524	1.2855
	0.5600	0.5939	1.2625
	0.5700	0.5385	1.2404
	0.5800	0.4859	.1.2190
	0.5900	0.4359	1.1983
	0.6000	0.3885	1.1783

TABLE 1.6

$Q_{tc} = 0.8$
Second-Order Chebychev (C$_2$)
Equal Ripple Response

	Q_{ts}	α	f_c/f_s
	0.2000	15.0000	4.0000
	0.2100	13.5125	3.8095
A.S.	0.2200	12.2231	3.6364
	0.2300	11.0983	3.4783
	0.2400	10.1111	3.3333
	0.2500	9.2400	3.2000
	0.2600	8.4675	3.0769
	0.2700	7.7791	2.9630
	0.2800	7.1633	2.8571
	0.2900	6.6100	2.7586
	0.3000	6.1111	2.6667
	0.3100	5.6597	2.5806
	0.3200	5.2500	2.5000
	0.3300	4.8770	2.4242
	0.3400	4.5363	2.3629
	0.3500	4.2245	2.2857
	0.3600	3.9383	2.2222
	0.3700	3.6749	2.1622
	0.3800	3.4321	2.1053
	0.3900	3.2078	2.0513
	0.4000	3.0000	2.0000
	0.4100	2.8073	1.9512
	0.4200	2.6281	1.9048
	0.4300	2.4613	1.8605
	0.4400	2.3058	1.8182
	0.4500	2.1605	1.7778
	0.4600	2.0246	1.7391
	0.4700	1.8972	1.7021
	0.4800	1.7778	1.6667
	0.4900	1.6656	1.6327
	0.5000	1.5600	1.6000
INF. BAFFLE	0.5100	1.4606	1.5686
	0.5200	1.3669	1.5385
	0.5300	1.2784	1.5094
	0.5400	1.1948	1.4815
	0.5500	1.1157	1.4545
	0.5600	1.0408	1.4286
	0.5700	0.9698	1.4035
	0.5800	0.9025	1.3793
	0.5900	0.8386	1.3559
	0.6000	0.7778	1.3333

$Q_{tc} = 0.5$ would have a flat group delay) to developing a sharp knee at $Q_{tc} = 1.2$.

The shape of the impedance "hump" at resonance also changes with changes in Q. As Q_{tc} increases, the shape gets more narrow and sharp. *Figure 1.5* shows the comparison between the impedance curves of the same Q set. Notice also that the height of the impedance peak is decreasing with increasing Q_{tc}.

Figure 1.6 illustrates the cone velocity and volume current curves. Cone velocity curves give the relationship between frequency and driver acceleration, and are expressed in meters/seconds. Volume current curves represent a similar concept, except that the relationship is between frequency and the amount of air (current) displaced as the driver accelerates and is given in cubic meters/seconds. Notice that the shape of these curves, like the impedance curve, gets more narrow and sharp as Q increases. The frequency of maximum acceleration occurs approximately at the box resonance frequency f_c.

1.40 WOOFER SELECTION and ENCLOSURE CONSTRUCTION.

Woofers intended for use in closed-box systems are characterized by a low free-air resonance (f_s), relatively high cone mass, and long voice coils. This description fits many of the raw drivers currently available to the experimenter.

TABLE 1.7

$Q_{tc} = 0.9$
Second-Order Chebychev (C2)
Equal Ripple Response

	Q_{ts}	α	f_c/f_s
A.S.	0.2000	19.2500	4.5000
	0.2100	17.3673	4.2857
	0.2200	15.7355	4.0909
	0.2300	14.3119	3.9130
	0.2400	13.0625	3.7500
	0.2500	11.9600	3.6000
	0.2600	10.9822	3.4615
	0.2700	10.1111	3.3333
	0.2800	9.3316	3.2143
	0.2900	8.6314	3.1034
	0.3000	8.0000	3.0000
	0.3100	7.4287	2.9032
	0.3200	6.9102	2.8125
	0.3300	6.4380	2.7273
	0.3400	6.0069	2.6471
	0.3500	5.6122	2.5714
	0.3600	5.2500	2.5000
	0.3700	4.9167	2.4324
	0.3800	4.6094	2.3684
	0.3900	4.3254	2.3077
	0.4000	4.0625	2.2500
	0.4100	3.8186	2.1951
	0.4200	3.5918	2.1429
	0.4300	3.3807	2.0930
	0.4400	3.1839	2.0455
	0.4500	3.0000	2.0000
	0.4600	2.8280	1.9565
	0.4700	2.6668	1.9149
	0.4800	2.5156	1.8750
	0.4900	2.3736	1.8367
INF. BAFFLE	0.5000	2.2400	1.8000
	0.5100	2.1142	1.7647
	0.5200	1.9956	1.7308
	0.5300	1.8839	1.6981
	0.5400	1.7778	1.6667
	0.5500	1.6777	1.6364
	0.5600	1.5829	1.6071
	0.5700	1.4931	1.5789
	0.5800	1.4078	1.5517
	0.5900	1.3269	1.5254
	0.6000	1.2500	1.5000

TABLE 1.8

$Q_{tc} = 1.0$
Second-Order Chebychev (C2)
Equal Ripple Response

	Q_{ts}	α	f_c/f_s
A.S.	0.2500	15.0000	4.0000
	0.2600	13.7929	3.8462
	0.2700	12.7174	3.7037
	0.2800	11.7551	3.5714
	0.2900	10.8906	3.4483
	0.3000	10.1111	3.3333
	0.3100	9.4058	3.2258
	0.3200	8.7656	3.1250
	0.3300	8.1827	3.0303
	0.3400	7.6505	2.9412
	0.3500	7.1633	2.8571
	0.3600	6.7160	2.7778
	0.3700	6.3046	2.7027
	0.3800	5.9252	2.6316
	0.3900	5.5746	2.5641
	0.4000	5.2500	2.5000
	0.4100	4.9488	2.4390
	0.4200	4.6689	2.3810
	0.4300	4.4083	2.3256
	0.4400	4.1653	2.2727
	0.4500	3.9383	2.2222
	0.4600	3.7259	2.1739
	0.4700	3.5269	2.1277
	0.4800	3.3403	2.0833
	0.4900	3.1649	2.0408
	0.5000	3.0000	2.0000
	0.5100	2.8447	1.9608
	0.5200	2.6982	1.9231
INF. BAFFLE	0.5300	2.5600	1.8868
	0.5400	2.4294	1.8519
	0.5500	2.3058	1.8182
	0.5600	2.1888	1.7857
	0.5700	2.0779	1.7544
	0.5800	1.9727	1.7241
	0.5900	1.8727	1.6949
	0.6000	1.7778	1.6667

In terms of driver Q_{ts}, closed-box loudspeakers generally require woofers with a fairly high Q_{ts} of greater than 0.3. This implies using drivers with moderate-sized magnet structures, although excessively small magnet (underdamped) drivers should be avoided.[9,10] R. Small suggested a good rule-of-thumb which he called Efficiency Bandwidth Product (EBP).[11]

EBP is quantified by:

$$EBP = \frac{\text{Resonance Frequency}}{\text{Driver Electrical Q}} = \frac{f_s}{Q_{es}}$$

EBP in the vicinity of 50 or less indicates a sealed enclosure would be more suitable, while EBPs of about 100 suggest a vented enclosure.[F2]

The other important criterion is the amount of voice coil overhang. Because sealed box woofers must excurse farther than their vented counterparts, they generally require longer voice coils. In absolute terms, this means at least 2–4mm for small diameter woofers (6–8″) and 5–8mm for larger diameter woofers (10–12″). If the manufacturer does not specify the amount of voice coil overhang (X_{max}), it is possible to check it yourself. Hold the driver near a strong light and look through the spider. Most spiders are transparent enough to allow a view of the coil (*Fig. 1.7*).

Enclosures for closed-box speakers should be air tight. A good way to ensure an air-tight box, regardless of box construction method (butt joint, mitered, etc.), is to apply silicone adhesive to the inside of all enclosure joints. Care should also be taken to keep air from leaking past connection terminals. If you use a commercially available connector "cup," make certain it has a gasket seal. If the cup does not have a gasket, use silicone to seal any

TABLE 1.9

$Q_{tc} = 1.1$
Second-Order Chebychev (C₂)
Equal Ripple Response
Max PWR Handling/Max Efficiency

Q_{ts}	α	f_c/f_s
0.2500	18.3600	4.4000
0.2600	16.8994	4.2308
0.2700	15.5981	4.0741
0.2800	14.4337	3.9286
0.2900	13.3876	3.7931
0.3000	12.4444	3.6667
0.3100	11.5911	3.5484
0.3200	10.8164	3.4375
0.3300	10.1111	3.3333
0.3400	9.4671	3.2353
0.3500	8.8776	3.1429
0.3600	8.3364	3.0556
0.3700	7.8386	2.9730
0.3800	7.3795	2.8947
0.3900	6.9556	2.8205
0.4000	6.5625	2.7500
0.4100	6.1981	2.6829
0.4200	5.8594	2.6190
0.4300	5.5441	2.5581
0.4400	5.2500	2.5000
0.4500	4.9753	2.4444
0.4600	4.7183	2.3913
0.4700	4.4776	2.3404
0.4800	4.2517	2.2917
0.4900	4.0396	2.2449
0.5000	3.8400	2.2000
0.5100	3.6521	2.1569
0.5200	3.4749	2.1154
0.5300	3.3076	2.0755
0.5400	3.1495	2.0370
0.5500	3.0000	2.0000
0.5600	2.8584	1.9643
0.5700	2.7242	1.9298
0.5800	2.5969	1.8966
0.5900	2.4760	1.8644
0.6000	2.3611	1.8333

TABLE 1.10

$Q_{tc} = 1.2$
Second-Order Chebychev (C₂)
Equal Ripple Response

Q_{ts}	α	f_c/f_s
0.2500	22.0400	4.8000
0.2600	20.3018	4.6154
0.2700	18.7531	4.4444
0.2800	17.3673	4.2857
0.2900	16.1225	4.1379
0.3000	15.0000	4.0000
0.3100	13.9844	3.8410
0.3200	13.0625	3.7500
0.3300	12.2231	3.6364
0.3400	11.4567	3.5294
0.3500	10.7551	3.4286
0.3600	10.1111	3.3333
0.3700	9.5186	3.2432
0.3800	8.9723	3.1579
0.3900	8.4675	3.0769
0.4000	8.0000	3.0000
0.4100	7.5663	2.9268
0.4200	7.1693	2.8571
0.4300	6.7880	2.7907
0.4400	6.4380	2.7273
0.4500	6.1111	2.6667
0.4600	5.8053	2.6087
0.4700	5.5188	2.5532
0.4800	5.2500	2.5000
0.4900	4.9975	2.4490
0.5000	4.7600	2.4000
0.5100	4.5363	2.3529
0.5200	4.3254	2.3077
0.5300	4.1264	2.2642
0.5400	3.9383	2.2222
0.5500	3.7603	2.1818
0.5600	3.5918	2.1429
0.5700	3.4321	2.1053
0.5800	3.2806	2.0690
0.5900	3.1367	2.0339
0.6000	3.0000	2.0000
0.6100	2.8699	1.9672
0.6200	2.7461	1.9355
0.6300	2.6281	1.9048
0.6400	2.5156	1.8750
0.6500	2.4083	1.8462

potential air leaks. Air leakage caused by a speaker's lossy surround or porous dust cap should probably be ignored, since attempts at correction can create as many problems as they solve. Lossy pleated edge surrounds are generally not appropriate for closed boxes and should not be used at all. Porous dust caps can be responsible for air leakage, but usually have been employed to enhance voice coil cooling. Sealing such dust caps could cause premature driver failure and nonlinear changes in driver Q.

1.50 BOX SIZE DETERMINATION and RELEVANT PARAMETERS.

Box size determination for a closed-box speaker is a fairly straightforward process. The following driver parameters are required:

1. f_s driver free-air resonance
2. Q_{ts} total Q of the driver
3. V_{as} volume of air equal to driver compliance
4. X_{max} amount of voice coil overhang in millimeters
5. S_d effective driver radiating area in square meters
6. V_d displacement volume = $S_d (X_{max})$ in cubic meters

Parameters 4, 5, and 6 can be supplied by the manufacturer. You should recalculate parameters 1, 2, and 3 using the procedures described in *Chapter 8, Loudspeaker Testing.* Not only is the manufacturer's data subject to production line changes, which you may or may not know about, but you must account for the various series resistances that can have a critical effect on your final results. These include amplifier source resistance, the connecting cable (between the amp and speaker), and series crossover resistances.

The best way to choose a box size and associat-

35

TABLE 1.11

$Q_{tc} = 1.5$
Second-Order Chebychev (C₂)
Equal Ripple Response

Q_{ts}	α	f_c/f_s
0.3000	24.0000	5.0000
0.3100	22.4131	4.8387
0.3200	20.9727	4.6875
0.3300	19.6612	4.5455
0.3400	18.4637	4.4118
0.3500	17.3673	4.2857
0.3600	16.3611	4.1667
0.3700	15.4354	4.0541
0.3800	14.5817	3.9474
0.3900	13.7929	3.8462
0.4000	13.0625	3.7500
0.4100	12.3849	3.6849
0.4200	11.7551	3.5714
0.4300	11.1687	3.4884
0.4400	10.6219	3.4091
0.4500	10.1111	3.3333
0.4600	9.6333	3.2609
0.4700	9.1856	3.1915
0.4800	8.7656	3.1250
0.4900	8.3711	3.0612
0.5000	8.0000	3.0000
0.5100	7.6505	2.9412
0.5200	7.3210	2.8846
0.5300	7.0100	2.8302
0.5400	6.7160	2.7778
0.5500	6.4380	2.7273
0.5600	6.1747	2.6786
0.5700	5.9252	2.6316
0.5800	5.6885	2.5862
0.5900	5.4637	2.5424
0.6000	5.2500	2.5000
0.6100	5.0468	2.4590
0.6200	4.8533	2.4194
0.6300	4.6689	2.3810
0.6400	4.4932	2.3438
0.6500	4.3254	2.3077
0.6600	4.1653	2.2727
0.6700	4.0123	2.2388
0.6800	3.8659	2.2059
0.6900	3.7259	2.1739
0.7000	3.5918	2.1429

A.S. (marked beside rows 0.4700–0.4800)

ed response is to generate a design table which specifies data for Q_{ts} from 0.5 through 1.1, and then consider the various possibilities. Use *Tables 1.3* through *1.12*, or the design equations which follow, to find the values of α (alpha) and f_c, the resonance of the driver in the box. The f_3, –3dB cutoff point can be taken from *Table 1.12*.

CLOSED-BOX DESIGN EQUATIONS

$$\alpha = \left(\frac{Q_{tc}}{Q_{ts}}\right)^2 - 1$$

$$f_c = \frac{Q_{tc} \times f_s}{Q_{ts}}$$

TABLE 1.12

–3dB Rolloff Point f_3

Q_{tc}	f_3/f_c
0.500	1.5538
0.577	1.2725
0.707	1.0000
0.800	0.8972
0.900	0.8295
1.000	0.7862
1.100	0.7567
1.200	0.7358
1.500	0.6993

$$f_3 = \left[\frac{\left(\frac{1}{Q_{tc}^2} - 2\right) + \sqrt{\left(\frac{1}{Q_{tc}^2} - 2\right)^2 + 4}}{2}\right]^{1/2} \times f_c$$

Then, using the design tables:

Box Volume $\qquad V_b = \dfrac{V_{as}}{\alpha}$

–3dB Point $\qquad f_3 = \left(\dfrac{f_3}{f_c}\right) \times f_c$

Box Resonance $\qquad f_c = \left(\dfrac{f_c}{f_s}\right) \times f_s$

Remember that air suspension operating systems exist between values of α from 3–10. Values of α less than that venture into the realm of the infinite baffle.[4] Also, for bookshelf sized speakers, $f_c = 50$Hz or less is a reasonable figure for a closed-box system.[12]

1.51 Q_{ts}, F_s, and V_{as} VARIANCES.
For a given production run of identical drivers (same cone, voice coil, surround, dust cap, magnet, gap and so on), values of Q_{ts}, f_s, and V_{as} can exhibit considerable variation. Consequently, if the measured values for a small sample (usually two for most of us) are quite different, don't be overly alarmed. Although individual driver parameters can have a wide range of values, the ratios of f_s/Q_{ts} and the products of $V_{as}f_s^2$ are fairly constant. Thus, the final and essential results will also tend to be constant. Consider the data for two identical drivers (*Table 1.2*).

1.60 ADDITIONAL PARAMETERS.
Three other factors are useful in evaluating the prospective performance of a closed-box system: the reference efficiency, the displacement limited acoustic power output, and the electrical input power required to produce the displacement limited output.

1.61 REFERENCE EFFICIENCY(η_0).
Reference efficiency is, primarily, dependent upon the driver parameters and *not* upon the

enclosure. It is often expressed as a percent, or more commonly, as a sound pressure level (SPL). Basically, η_0 is most useful for comparing the efficiency of drivers in a multi-driver speaker system. As such, it can be used to determine the amount of attenuation needed for mid and high frequency drivers (for reasons of practicality, woofers are seldom, if ever, attenuated to match other drivers in a multiway speaker).

Free-air reference efficiency can be determined by:

$$\eta_0 = \frac{K(f_s^3 V_{as})}{Q_{es}}$$

where:

K = 9.64×10^{-10} for V_{as} in liters
9.64×10^{-7} for V_{as} in cubic meters
2.70×10^{-8} for V_{as} in cubic feet

For this calculation, V_{as} should be measured on a baffle or enclosure the same size as the speaker cabinet. This equation gives η_0 as a decimal equivalent. To convert η_0 to:

Percentage $\% = \eta_0 \times 100$

SPL, 1W/1M   dB = $112 + 10\log_{10}\eta_0$

η_0 can vary from low values around 0.35% to higher values up to 1.5%. *Table 1.13* lists efficiency data for well-known drivers, which will give a perspective to the situation.

For comparison purposes, η_0 in a sealed enclosure (unfilled) can be determined by:

$$\eta_{0fc} = \frac{Kf_c^3 V_{as} V_b}{Q_{ec}(V_{as} + V_b)}$$

1.62 DISPLACEMENT LIMITED ACOUSTIC POWER OUTPUT (P_{ar}).

P_{ar} is the maximum output which can be produced for a driver working within its linear operating range, without appreciable levels of distortion. You can determine the linear operating range of voice coil outside (overhangs) the magnetic circuit (gap) (*Fig. 1.7*). To calculate this

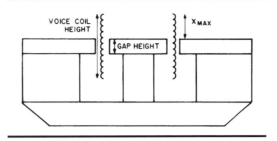

FIGURE 1.7: Voice coil "overhang."

length, designated X_{max}, subtract the magnet gap height from the total voice coil length, and then divide by two. Setting maximum linear excursion to X_{max} + 15% is rather conservative, and a somewhat greater excursion, and hence higher output levels, can be allowed if you consider such factors as the relative insensitivity of the human ear to low frequency distortion.[13]

P_{ar} can be calculated in terms of RMS sine wave power by:

$$P_{ar(cw)} = K_p f_3^4 V_d^2$$

K_p is a power rating constant which varies with box Q_{tc}.

Table 1.14 gives values of K_p for different values of Q_{tc}.

V_d is the volume of air displaced by the cone traveling through its maximum excursion limits, and is expressed in cubic meters. Thus, $V_d = S_d \times X_{max}$. S_d, the effective radiation area of the driver cone, is listed for different diameter cones in *Table 1.15*. S_d is calculated by:

$$S_d = \frac{3.1416 \times D^2}{4}$$

Where D is the diameter of the cone plus one-third of the surround at both ends of the measurement.

$P_{ar(cw)}$ is included in this discussion because it gives a relative comparison of output for various values of Q_{tc}. This coincides with the fact that maximum power handling occurs at Q_{tc} = 1.1 and then decreases with decreasing values of Q_{tc}. The frequency at which X_{max} occurs, however, is below f_3 for all Q_{tc}s lower than 1.1. Since most of the program material is above f_3, it is practical to

TABLE 1.13

Brand Name	Free Air %	η_0 SPL
Altec 411-8A	1.44	94dB
KEF B-139	0.62	90dB
Polydax HD-20 B 25	0.42	88dB
Focal 8N401	0.48	89dB
Focal 10C02	0.62	90dB

TABLE 1.14

Q_{tc}	K_p
0.500	0.06
0.577	0.15
0.707	0.39
0.800	0.57
0.900	0.75
1.000	0.84
1.100	0.85
1.200	0.84
1.500	0.71

TABLE 1.15

Advertised Dia. (")	$S_d \, M^2$
5	0.0089
7	0.0158
8	0.0215
10	0.0330
12	0.0450
15	0.0855

take P_{ar} at the maximum value. Therefore, $P_{ar(p)}$, in watts, for program material equals:

$$P_{ar(p)} = 0.85 \, f_3^4 V_d^2$$

and

$$SPL \, 1W \, 1M = 112 + 10 Log_{10} \, P_{ar(p)}$$

1.63 ELECTRICAL INPUT MAXIMUM (P_{er}).
P_{er} is the input power required to produce P_{ar}. P_{er} is determined by:

$$P_{er} = \frac{P_{ar(cw)}}{\eta_0}$$

where:

P_{er} in watts
P_{ar} in watts
η_0 as a decimal

You can compare P_{er} with the thermal limited power-handling capacity given by the manufacturer. Caution is advised if the thermal rating of the driver is less than P_{er}.

1.70 EXAMPLE DESIGN CHARTS.
Tables 1.16 and *1.17* give examples of two drivers suitable for closed-box construction. Remember that, when calculating total box size, it is important to over-volume the enclosure to compensate for anything which will detract from its target volume. This includes:

- midrange cavities
- woofer basket and magnet structures
- fibrous damping material (about 10% of measure volume)
- crossover
- solid damping material—braces, felt and so on.

Last, always be sure to include all series resistances in the calculation of Q_{ts} (*Chapter 8*). Examples of closed-box construction articles can be found in *Speaker Builder*: "A Ceramic Speaker Enclosure," by David Weems, 6/88, p. 37; "A Small Two-Way System," by Fred Thompson, 2/90, p. 24; and "A Modular Three-Way Active Loudspeaker," by Fernando Ricart, 4/90, p. 36.

1.71 MINIMUM CUT-OFF FREQUENCY.
The popular notion persists that cut-off frequency for sealed enclosures decreases with increasing box volume. This is true, however, only for Q_{tc} greater than, or equal to 0.707. For values of Q_{tc} less than 0.707, increasing volume causes an increase in cut-off frequency.[14] It is obvious that the tradeoffs for the use of one of the so-called over-damped alignments will be a less than optimal f_3. *Table 1.18* illustrates the ratio of f_3/f_s for different box Q_{tc} with various driver Q_{ts}.

In all cases, f_3/f_s increases for values of Q_{tc} lower than 0.707. Looking at the design in *Tables*

TABLE 1.16
8″ WOOFER EXAMPLE

Q_{ts}	Q_{es}	Q_{ms}	f_s	X_{max}	$S_d(m^2)$	$V_d(m^3)$	V_{as}
0.45	0.53	3.0	31.5	3.5mm	2.15×10^{-2}	7.525×10^{-5}	$2.97 ft^3$

Q_{tc}	V_b ft³	f_3 Hz	f_c Hz	$P_{ar(cw)}$ watts	SPL (dB)	$P_{ar(p)}$ watts	SPL (dB)	η_0 %	SPL (dB)	P_{er} watts
0.5	12.70	54	35	0.0033	87	0.041	98	0.47	89	0.70
0.7	2.02	50	50	0.014	94	0.028	97	0.47	89	2.98
0.8	1.38	50	56	0.020	95	0.03	97	0.47	89	4.26
0.9	0.99	52	63	0.032	97	0.035	98	0.47	89	6.81
1.0	0.75	55	70	0.044	98	0.044	98	0.47	89	9.76

TABLE 1.17
10″ WOOFER EXAMPLE

Q_{ts}	Q_{es}	Q_{ms}	f_s	X_{max}	$S_d(m^2)$	$V_d(m^3)$	V_{as}
0.446	0.51	3.63	27.2	5.25mm	3.40×10^{-2}	1.79×10^{-4}	$5.79 ft^3$

Q_{tc}	V_b ft³	f_3 Hz	f_c Hz	$P_{ar(cw)}$ watts	SPL (dB)	$P_{ar(p)}$ watts	SPL (dB)	η_0 %	SPL (dB)	P_{er} watts
0.5	24.80	54	35	0.016	94	0.23	106	0.62	90	2.58
0.7	4.10	43	43	0.038	98	0.09	102	0.62	90	6.13
0.8	2.60	44	49	0.067	100	0.10	102	0.62	90	10.81
0.9	1.88	46	55	0.103	102	0.12	103	0.62	90	16.61
1.0	1.47	48	60	0.139	103	0.13	104	0.62	90	22.42

1.16 and 1.17, as would be expected, both drivers show an increase in cut-off frequency, f_3, for increasing box volume past $Q_{tc} = 0.707$.

1.72 DYNAMIC CHANGES IN FREQUENCY RESPONSE.

Although calculated large signal parameters such as P_{ar} and P_{er} give you some idea of a driver's excursion potential, these numbers do not adequately describe dynamic changes which occur in a loudspeaker as input power and operating temperature increase. When the Thiele/Small calculation methods are applied to design a loudspeaker enclosure, sealed or vented, the speaker will exhibit the predicted response only at a small signal level. The loudspeaker can be expected to perform in accordance with the design tables and formula calculations at 1W power input, but beyond that, as power increases and voice coil temperature increases the characteristics of the driver/box combination will undergo constant dynamic change.

Loudspeakers generally operate over a thermal range that extends from room temperature, about 25°C, to somewhere in the vicinity of 250°C, where adhesives begin to break down and failure occurs. Driver power ratings should somehow correlate the rating with the level where the driver will be closely approaching thermal failure. If a speaker is rated at 200W, it should be able to maintain a temperature somewhat below 250°C over a period of time. As a function of input power, dividing the maximum temperature, 250°C, by the speaker power rating gives a rough estimate of the temperature increase per watt of power increase (providing the manufacturer has somehow related the power rating to thermal failure, which is not always the case). A driver rated 150W, for instance, would increase its temperature about 1.667°C for each additional watt of power used to drive the speaker.

As voice coil temperatures increase, voice coil resistance increases, and the overall damping of the driver decreases. The effect is illustrated in *Fig. 1.8*. This computer simulation shows the different SPL curves for a box/driver combination with a calculated Q_{tc} = 0.7 at five increasing power levels starting at 1W (1, 5, 10, 20, and 40W). The results are summarized in *Table 1.19*.

The increasing phase angle (at −3dB) and slope indicates a decrease in overall damping and is expressed by the equivalent Q_{tc} number.

Graph 1 - Acoustic On Axis Response: SPL, Phase

———	:1W	29.2 C
·········	:5W	46 C
— — —	:10W	66.7 C
———	:20W	108.7 C
- - -	:40W	191.7 C

Changes in damping are also reflected in changes in cone excursion, *Fig. 1.9*, group delay, *Fig. 1.10*, impedance, *Fig. 1.11*, and transducer cone velocity, *Fig. 1.12*. The cone excursion and group delay curves look reminiscent of those in *Figs. 1.3* and *1.4* for different Qs and box volumes. Notice, however, that the impedance peak at f_c remains at about the same frequency and width, although the overall impedance level is constantly increasing with temperature. The peak in cone velocity also remains at the same frequency, and does not change with increasing input power.

When using hand-calculator methods to design, it is not possible to anticipate these dynamic deviations. The best advice is to err on the side of lower Q when other factors allow. As a perspective, several successful commercial designs have small signal Q_{tc}s as low as 0.5.

1.80 ENCLOSURE FILLING.

All the above box volume calculations were based on an unfilled enclosure or, at most, an enclosure with no more than a 1″ lining of

TABLE 1.18
f_3/f_s (Q_{tc} vs. Q_{ts})

		Q_{ts}			
		0.20	0.30	0.40	0.50
	1.5	5.2	3.5	2.6	2.1
	1.2	4.4	2.9	2.2	1.8
	1.1	4.2	2.8	2.1	1.7
	1.0	3.9	2.6	2.0	1.6
Q_{tc}	0.9	3.7	2.5	1.9	1.5
	0.8	3.6	2.4	1.8	1.4
	0.71	3.5	2.4	1.8	1.4
	0.58	3.7	2.5	1.9	2.0
	0.50	3.9	2.6	2.0	—

TABLE 1.19
DYNAMIC CHANGES IN DRIVER RESPONSE AT MULTIPLE POWER LEVELS

V	Power	Temp.	−3dB Phase Angle	Slope dB/oct.	Equiv. Q_{tc}	f_3 Hz
2.83	1	29.2°	90°	10.60	0.7	35
6.35	5	46°	93°	10.87	0.8	34
8.95	10	66.7°	95°	11.18	0.85	34
12.68	20	108.7°	100°	11.74	1.0	32
17.89	40	191.7°	106°	12.71	1.2	31

FIGURE 1.9

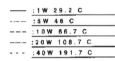

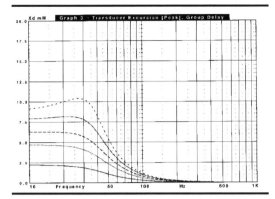

FIGURE 1.10

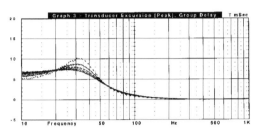

FIGURE 1.11

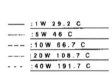

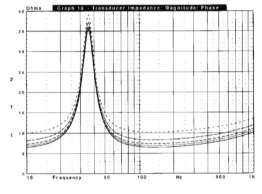

FIGURE 1.12

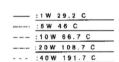

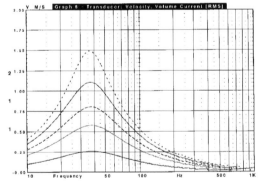

A. *Compliance Increase*. Use of low density, high specific heat material (fiberglass, Dacron and long fiber wool) will cause an increase in the acoustic compliance of the enclosure (C_{ab}). This is equivalent to increasing the box size and can amount to as much as a theoretical 40%. Practical equivalent volume increases of 15%–25% are quite attainable.

B. *Efficiency Increase*. Proper selection of the amount, type of material, and the location of the material within the enclosure can cause an increase in efficiency by as much as 15%.

C. *Mass Changes*. Filling material has the potential of changing the total moving mass of the system. This phenomenon is generally thought to be related to the restriction of air flow directly behind the driver. The increase in effective mass will cause a decrease in efficiency, but not as great as the increase in efficiency caused by compliance changes. Since a decrease in efficiency is undesirable, there are two techniques which will limit these effects. One, used in the Advent loudspeaker, is to use a brace directly behind the woofer to keep damping material (in this case, foam) away from the rear of the driver. The second is to use a non-compressed, low density material in the area immediately behind the driver basket as a sort of buffer between the higher density material and the driver.[16]

D. *Damping Losses*. If you pack filling material relatively densely, and close to the rear of the driver basket, frictional losses can be substantial.

The quantity of fill material required to change the pneumatic action of the enclosure from adiabatic to isothermic (which is what the above criteria describe) has been suggested by several authors,[17,18,19] but not really well-defined for the specific use of the most commonly used fill material, fiberglass. However, the effects of fiberglass damping material can be easily observed by computer simulation.

This simulation was set up using the parameters of the same 10″ speaker used in previous simulations. A somewhat small box volume, 1.75ft³, was selected since it has a sufficiently high Q_{tc} and renders the influence of the fill material readily observable. The simulation was then programmed to calculate the response of the 1.75ft³ box with various amounts of fiberglass fill material. The amounts, shown in *Table 1.20*, were of standard 1 lb/ft³ R19 household type fiberglass. The density of this type of material is somewhat hard to calculate since it expands and compresses depending on how it is handled. The 1 lb/ft³ measurement was made with the material unrolled and removed from its paper backing with a thickness of roughly 3″. The table shows the changes in f_3 and the phase angle for different fill percentages. *Figure 1.13* shows the changes in magnitude response and *Fig. 1.14* depicts the changes in the box impedance. As can be seen, a decrease in the f_3 rolloff frequency is accompanied by a decrease in Q_{tc}. The amount of

fiberglass to damp standing waves. This is as it should be, since a properly designed enclosure will be able to attain its design goal without any additional manipulation. In the real world, however, the "art" of box stuffing can be an invaluable tool to alter box response in order to achieve certain parameters of size and Q that can't be had any other way.

Beside the obvious advantage of increased suppression of the internal reflections that can produce strong colorations in the sound quality,[15] stuffing a box has the following effects on box parameters:

TABLE 1.20

%Fill	f_3	Phase	Q_{tc}
-0-	39.94	98.92°	1.19
50	38.31	96.42°	0.89
100	37.37	93.38°	0.73

change will not be exactly the same for every speaker and will depend somewhat on the compliance ratio of the box and driver.

1.81 DESIGN ROUTINE FOR ENCLOSURE FILLING.

The best way to determine the effects of stuffing on a particular project are to add material to the enclosure and then measure the changes in Q_{tc} at the terminals. Use the same procedure outlined for determining free-air Q as described in *Chapter 8, Loudspeaker Testing*. Small and Margolis[20] offered a hand-calculation method of approximating the effects of 100% fill of low-to-medium-density fiberglass material prior to building the enclosure. Routine 1 gives the approximation for the change in Q_{tc} keeping the volume constant, while Routine 2 gives the change in volume keeping Q_{tc} constant.

Routine 1—New Q_{tc} if box size remains the same:

Equations	Example (10″ driver from *Table 1.17*)
$V_{ab} = 1.2V_b$	$V_{ab} = 4.92$
$\alpha = \dfrac{V_{as}}{V_{ab}}$	$\alpha = 1.177$
$L = \sqrt{\alpha + 1}$	$L = 1.475$
$Q_{tc'} = LQ_{ts}$	$Q_{tc'} = 0.658$
$Q_{tc} = \left(\dfrac{1}{Q_{tc'}} + .2\right)^{-1}$	$Q_{tc} = 0.58$

Routine 2—New box volume, V_b, if Q_{tc} remains the same:

Equations	Example (10″ driver from *Table 1.17*)
$Q_{tc} = \left(\dfrac{1}{Q_{tc'}} - .2\right)^{-1}$	$Q_{tc'} = 1.098$
$L = \dfrac{Q_{tc'}}{Q_{ts}}$	$L = 2.461$
$\alpha = L^2 - 1$	$\alpha = 5.056$
$V_{ab} = \dfrac{V_{as}}{\alpha}$	$V_{ab} = 1.145$
$V_b = \dfrac{V_{ab}}{1.2}$	$V_b = 0.95\text{ft}^3$

1.82 EMPIRICAL COMPARISON OF BOX STUFFING MATERIALS.

The calculator method described above can give you a rough idea of Q or box size tradeoffs, but the end result can vary considerably. Not only can the measured f_3 and Q_{tc} change depending upon density and distribution within the enclosure, but a number of materials other than fiberglass, such as Dacron and acoustic foam, as well as combinations of these materials, can be used to effect changes in box responses.

Over the years, amateurs and professionals have used a wide variety of fill materials to alter box response. These are accompanied by an equally wide variety of claims as to the best choice in terms of subjective performance. Since computer simulation and hand-calculator routines are limited to standard types of fiberglass, the following information can be used as a guideline when attempting to determine the effects of these different combinations of materials.

The test methods were fairly straightforward, although the results are subject to interpretation. Six materials were used, Dacron, Acousta-Stuf® (a Dacron-like material with crimped fibers purportedly giving the product the same sonic qualities as long fiber wool—available from Mahogany Sound), 1 lb/ft³ density fiberglass (standard household type R19), 2 lb/ft³ fiberglass, 4 lb/ft³ fiberglass, and acoustic foam (the acoustic foam was the type supplied by Audio Concepts, but is otherwise similar to a number of other "egg crate" style foams on the market). These materials were tested with 50% fill (lining the box) and 100% fill in the test enclosure. Six other 50/50 combinations were tested (the first material listed was used to line the box, the second to fill it the rest of the way

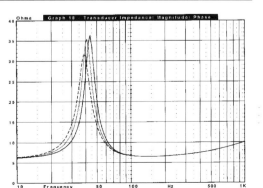

FIGURE 1.13

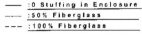

FIGURE 1.14

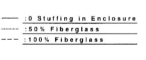

TABLE 1.21
COMPARISON OF BOX STUFFING MATERIALS

	Z_0	f_0	Q_m	Q_e	Q_{tc}	f_3
Empty Box	38.96	84.54	7.74	1.45	1.22	61.91
50%						
Fill Dacron	34.46	83.21	6.51	1.41	1.16	61.86
Acousta-Stuf	33.53	82.08	6.23	1.40	1.14	61.36
1lb/ft³ Fiberglass	27.77	79.51	4.93	1.40	1.09	60.37
2lb/ft³ Fiberglass	30.27	78.75	5.42	1.39	1.10	59.59
4lb/ft³ Fiberglass	31.52	82.06	5.78	1.40	1.13	61.52
Acoustic Foam	33.44	80.23	6.19	1.40	1.14	59.98
100%						
Fill Dacron	29.77	81.77	5.39	1.40	1.11	61.68
Acousta-Stuf	23.82	80.09	3.97	1.38	1.02	62.42
1lb/ft³ Fiberglass	19.87	76.79	3.20	1.44	0.99	60.64
2lb/ft³ Fiberglass	15.80	79.37	2.27	1.44	0.88	66.71
4lb/ft³ Fiberglass	16.71	78.66	2.37	1.42	0.89	65.67
Acoustic Foam	17.35	75.91	2.44	1.35	0.87	64.26
50/50 Combinations						
2lb/1lb Fiberglass	14.78	78.12	1.80	1.28	0.75	73.91
2lb FG/Acousta-Stuf	20.97	77.17	3.22	1.35	0.95	62.17
	14.67	75.86	1.79	1.29	0.75	71.76
4lb FG/Acousta-Stuf	21.96	79.11	3.49	1.33	0.96	63.40
Foam/1lb Fiberglass	18.11	73.55	2.58	1.33	0.88	61.83
Foam/Acousta-Stuf	24.28	76.91	4.04	1.37	1.02	59.94

to 100%): 2 lb/1 lb fiberglass, 2 lb/Acousta-Stuf, 4 lb/1 lb fiberglass, 4 lb/Acousta-Stuf, AC foam/1 lb fiberglass, and AC foam/Acousta-Stuf.

The material was placed in a 0.95ft³ standard rectangular test enclosure, constructed from 0.75″ particle board. The box was air tight and employed a closed cell foam gasket for the 8″ driver. Each material, quantity and combination were sequentially placed in the enclosure and two measurements performed on each category.

The first measurement was a swept sine wave impedance curve made using an Audio Precision System 1 analyzer. The System 1 was set up for the voltage divider impedance of the analyzer's generator set at 600Ω to provide the series resistance. The computer ASCII data files were then imported into LEAP 4.0 and converted to true impedance readings (impedance calculated with a large series resistance is similar, but not exactly the same as the driver impedance function). The data was then loaded into the LEAP automated Speaker Parameter Measurement module and the box Q_{tc} parameters calculated.

The second measurement performed was a near-field frequency response using the DRA

FIGURE 1.15A

FIGURE 1.15B

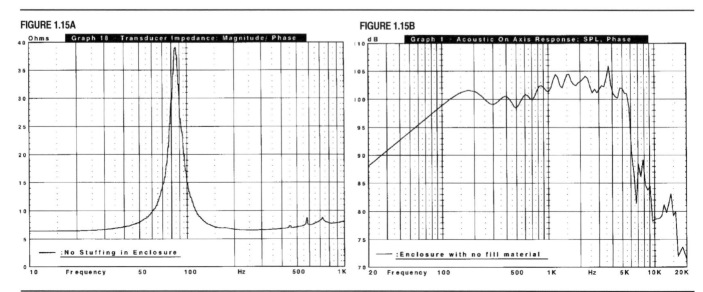

Labs MLSSA FFT analyzer and an ACO Pacific 7012 precision measurement microphone. The MLSSA was set up for a 2048 point acquisition length and a 20kHz bandwidth. A 2048 point FFT was performed on the impulse response and the results printed. The data was exported from the MLSSA software and imported back into LEAP 4.0 to obtain PostScript printouts. The response of the driver with each category of fill material was displayed along with the response of the driver in the enclosure with no fill material for comparison.

The results of this series of tests, shown in *Table 1.21*, summarize the measurements. The

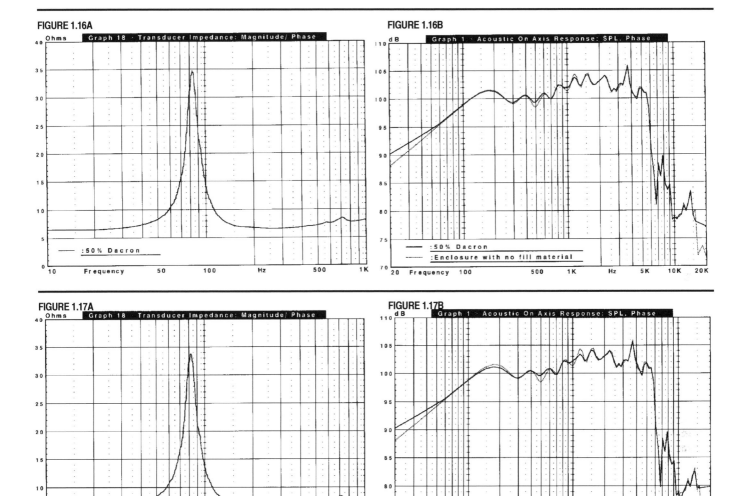

FIGURE 1.16A

FIGURE 1.16B

FIGURE 1.17A

FIGURE 1.17B

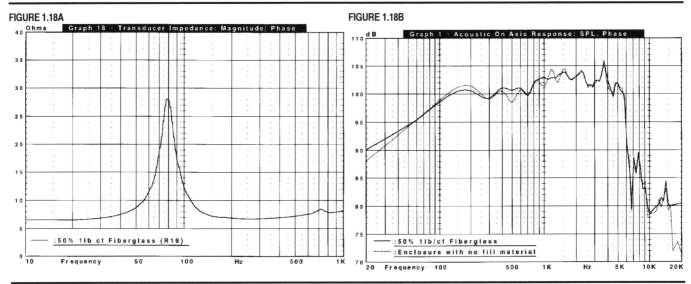

FIGURE 1.18A

FIGURE 1.18B

table is grouped into three sections, 50% fill, 100% fill, and 100% fill of 50/50 combinations. The height of the impedance peak, Z_0, changed from 38.96Ω with an empty enclosure to a low of 14.67Ω for the 4 lb/1 lb fiberglass combination. The box resonance frequency, f_0, ranged from 84.54Hz in the empty enclosure to 75.16Hz, with the 4 lb/1 lb combination. Mechanical Q, Q_{mc}, exhibited wide variation, from 7.74 for the empty box, to 1.79, again with the 4 lb/1 lb combination. Electrical Q, Q_{ec}, for the most part remained unchanged for the homogeneous 50% and 100% fill tests, but did show some variation with the different combinations of materials. Q_{tc} changed from 1.22 with no stuffing to a low of 0.75 for both the fiberglass 50/50 combinations. The 3dB down point, f_3, is dependent on both Q_{tc} and the box resonance, f_0, and seems independent of the other trends.

It would appear that the old standby of 100% fill of standard R19 fiberglass is still a reasonable choice for moderate damping and a low f_3. However several of the other materials deserve serious consideration. Both of the fiberglass 50/50 combinations yielded low Q values, with

only slightly higher f_3 frequencies. The foam/fiberglass combination also looks attractive with reasonable damping and a lower –3dB point.

The graph pairs in *Figs. 1.15–1.33* illustrate both the impedance curve and the associated near field (about 6″ from the dust cap) frequency response for the various combinations listed in *Table 1.21*. The impedance graphs in *Figs. 1.24–1.28, 1.30,* and *1.32* have been scaled to 20Ω to make them easier to read. All other impedance plots are scaled to 40Ω. The solid line in the MLSSA frequency response result is the curve of the material, while the dotted line represents the curve of the box with no fill material.

The response changes due to different fill percentages are interesting. For the 50% fill group, Dacron didn't seem to do very much in the way of box mode suppression. In order to have more of an effect, this material would have to be packed with a greater density. Comparing the 50% fill group to the 100% fill group, clearly the increased amount of material improved the suppression of box modes. The 50/50 group performed in a similar fashion to the

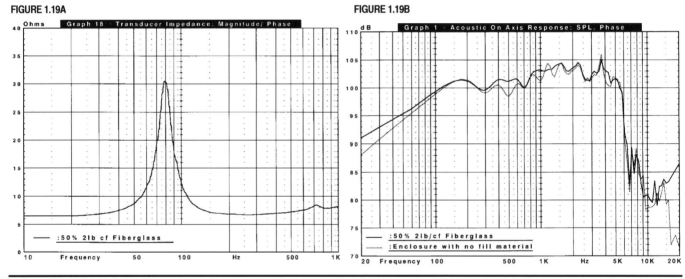

FIGURE 1.19A

FIGURE 1.19B

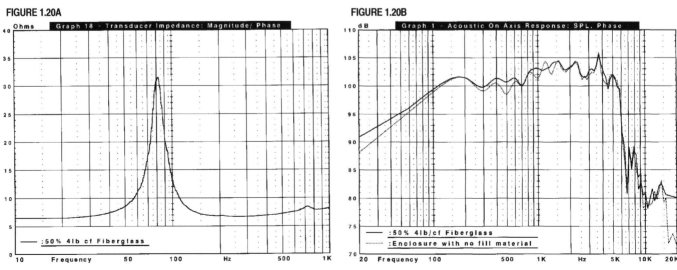

FIGURE 1.20A

FIGURE 1.20B

other 100% fill group. Using 100% fill to suppress internal box modes is just as important as changing f_3 and Q_{tc}.

1.90 MULTIPLE WOOFER FORMATS.
Using two or more woofers in a low-frequency cabinet can give you a number of different advantages over single woofer designs. The three basic configurations are: standard, push/pull, and compound.

1.91 STANDARD CONFIGURATION.
This is defined as two or more identical woofers having the same enclosure and mount-

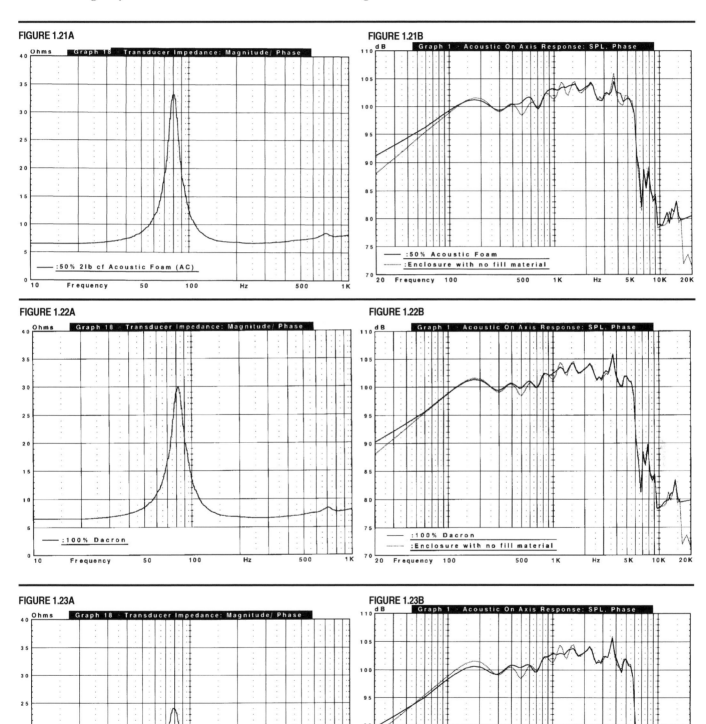

FIGURE 1.21A

FIGURE 1.21B

FIGURE 1.22A

FIGURE 1.22B

FIGURE 1.23A

FIGURE 1.23B

ed as close together as possible. For a two woofer situation the following applies:

A. F_s driver resonance for two drivers will be the same as that of a single driver.

B. Q_{ts} will be the same as that for a single driver.

C. V_{as} (and the associated box volume, V_b) will be twice that of a single driver.

D. Combined impedance will be half of a single unit when connected in parallel, and twice the value of a single unit when connected in series.

E. Sensitivity will increase +3dB for a parallel connection and −3dB for a series connection

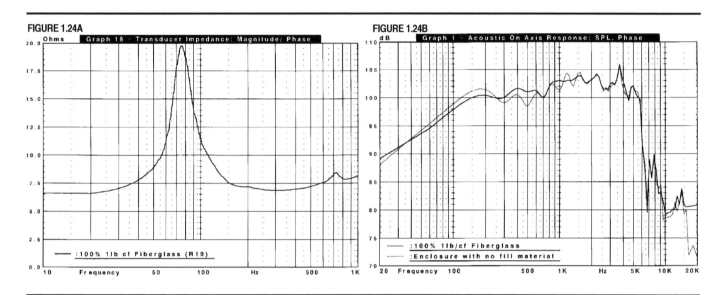

FIGURE 1.24A

FIGURE 1.24B

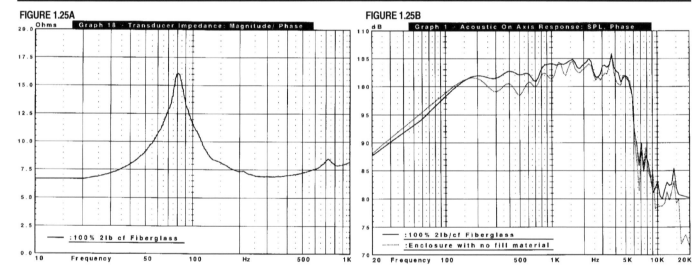

FIGURE 1.25A

FIGURE 1.25B

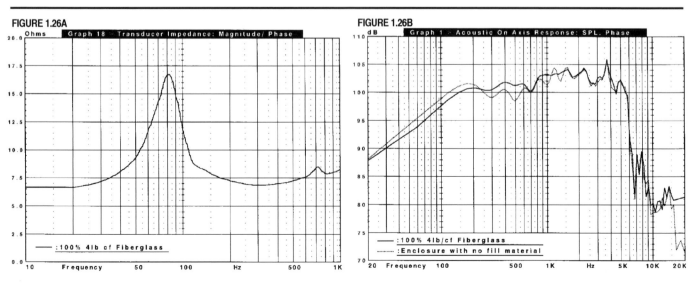

FIGURE 1.26A

FIGURE 1.26B

compared to a single driver.

F. Cone excursion will be half that of a single woofer enclosure.

Using four identical woofers in a series parallel configuration will also have 6dB efficiency gain over the single woofer, the same gain as the two-woofer combination. Doubling the cone area over the two-woofer design adds 3dB of acoustic efficiency, but putting two parallel sets of woofers in series decreases gain by −3dB, for a net change of zero when compared to the output of the two woofer box.

Figure 1.34 depicts the 1W SPL comparison of the computer simulation for single-

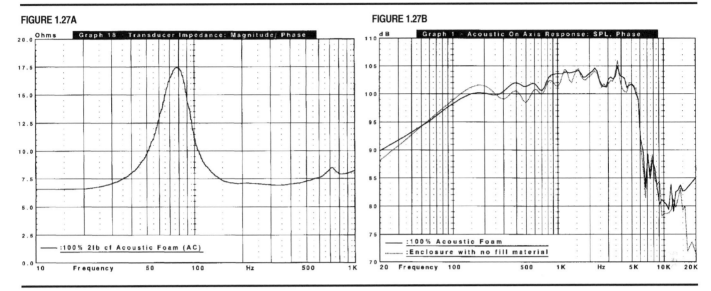

FIGURE 1.27A

FIGURE 1.27B

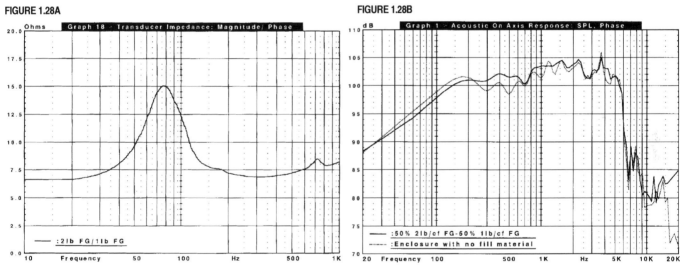

FIGURE 1.28A

FIGURE 1.28B

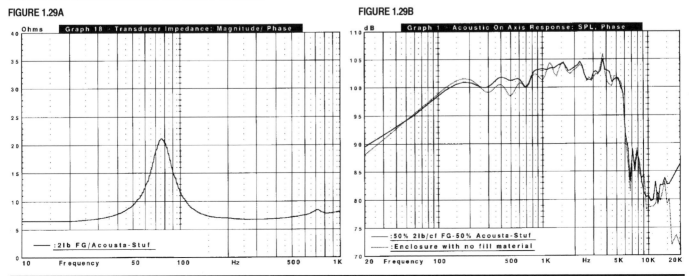

FIGURE 1.29A

FIGURE 1.29B

woofer, two-woofer, and four-woofer box designs. Notice the changes in the midband response of the four-driver combination over the two-driver one. When multiple drivers are closely grouped together, changes in the combined radiation impedance can cause substantial midband deviation. As expected, the gain for the two- and four-woofer designs is 6dB greater than the single driver. Looking at cone excursion and group delay curves in *Fig. 1.35*, the excursion of the two- and one-driver designs are approximately double that of the four-woofer design for the same drive voltage, while the group delay is identical for all three designs. The impedance curves in *Fig. 1.36* show the two- woofer box with half the impedance of one speaker. The single- and four-woofer boxes have similar impedance curves, except the impedance peak is lower for the four-woofer box, resulting from the difference in multiple driver radiation impedance. The advantage of multiple driver combinations is more obvious at higher power levels. *Figure 1.37* illustrates the SPL comparison for the same three box designs, one, two, and four drivers, but this time all

are at the same 100dB level. The voltage input to get this SPL is 12.68V for a single driver and 6.35V for the two- and four-driver combinations. The overall midband level of the single woofer has now dropped by about 1.5dB and the rolloff slope is somewhat steeper, indicating damping changes caused by increased resistance due to voice coil heating. The −3dB phase angle for all three designs was 90° at the 2.83V 1W level, but the single driver has now changed to 100° (an equivalent Q of 1.0), the two-driver box to 93°, and no change in the four-driver phase angle (still 90°). Looking at the group delay curves in *Fig. 1.38*, the shape of the curve is about the same for the two- and four-driver combinations, while the "knee" of the single-driver design is clearly sharper. Cone excursion curves are as expected. Excursion for the two-woofer design is half that of the single driver, and excursion for the four-woofer design is half that of the two-woofer design. The impedance curve in *Fig. 1.39* shows the voice coil heating in the single driver has caused an increase in resistance of about 1.7Ω across the bandwidth.

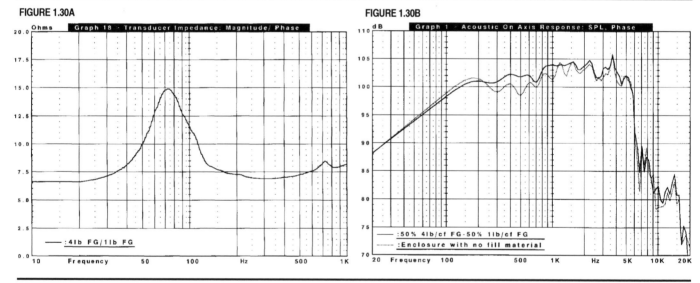

FIGURE 1.30A

FIGURE 1.30B

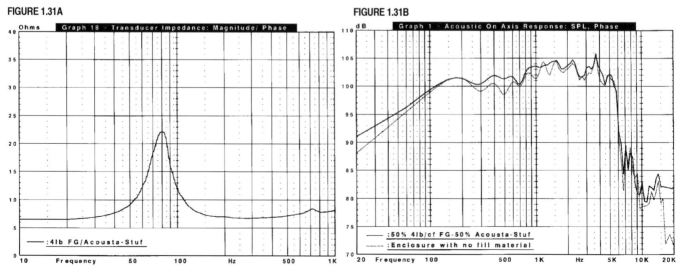

FIGURE 1.31A

FIGURE 1.31B

1.92 PUSH/PULL CONFIGURATION. This occurs if the two drivers are mounted on the box either back-to-back or front-to-front, and then driven electrically out of phase (*Fig. 1.40*). This type of setup will cause odd-order nonlinearities to cancel and result in substantially reduced driver distortion. All of the above-listed characteristics (A–E) apply to push/pull driver arrangements. In addition, the push/pull woofer arrangement can be used with vented, passive radiator systems, as well as with sealed boxes. Two construction articles which used this configuration appeared in *Speaker Builder*: "Tenth Row Center," by H. Hirsch, 2/84, p. 11; and "The Curvilinear Vertical Array," by S. Ellis, 2/85, p. 7.

1.93 COMPOUND WOOFER SYSTEMS. This design was first described by Olson in the early 1950s. The compound or Isobarik (constant pressure) system has several spectacular advantages over the other dual-woofer configurations. For the type of physical and electrical setup shown in *Fig. 1.41*, the following will apply.[F3]

A. Q_{ts} will be the same as a single driver.

B. F_s will be the same as a single driver.

C. V_{as} (and the associated box volume V_b) will be half that of a single driver.

D. Impedance will be half that of a single driver (assuming a parallel connection).

E. When configured in push/pull, the setup will have all the advantages listed in *Section 1.92.*

F. Sensitivity of the compound pair will be the same as that of a single driver (sensitivity will go up 3dB due to the 4Ω load, but will decrease by 3dB due to the doubling cone mass).

The major advantage of a constant pressure format is an enclosure volume that is half that of a single driver, making it an ideal choice for subwoofer applications.

The following construction details should be observed:

A. Your box size should be calculated as a closed-box system, using the single driver Q_{ts} and $V_{as}/2$.

B. Construct the short tunnel depicted in *Fig. 1.41*. It can be made square, out of lumber, or cylindrical, made from Sonotube (cardboard tubing used to build concrete pillars). Its length

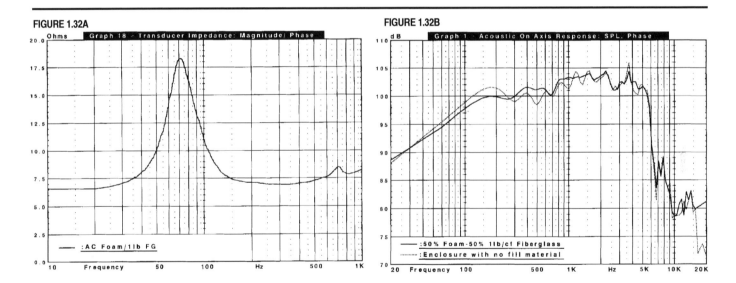

FIGURE 1.32A

Ohms
Graph 18 - Transducer Impedance: Magnitude/ Phase
:AC Foam/1lb FG
Frequency
Hz

FIGURE 1.32B

dB
Graph 1 - Acoustic On Axis Response: SPL, Phase
:50% Foam-50% 1lb/cf Fiberglass
:Enclosure with no fill material
Frequency
Hz

FIGURE 1.33A

Ohms
Graph 18 - Transducer Impedance: Magnitude/ Phase
:AC Foam/Acousta-Stuf
Frequency
Hz

FIGURE 1.33B

dB
Graph 1 - Acoustic On Axis Response: SPL, Phase
:50% Foam-50% Acousta-Stuf
:Enclosure with no fill material
Frequency
Hz

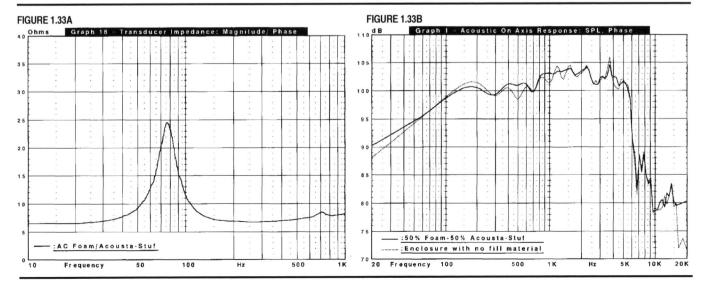

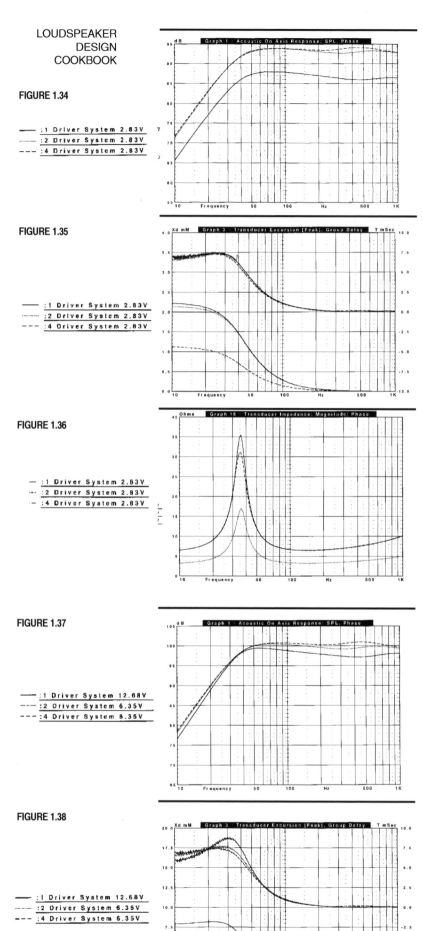

FIGURE 1.34

—— :1 Driver System 2.83V
······ :2 Driver System 2.83V
– – – :4 Driver System 2.83V

FIGURE 1.35

—— :1 Driver System 2.83V
······ :2 Driver System 2.83V
– – – :4 Driver System 2.83V

FIGURE 1.36

– :1 Driver System 2.83V
···· :2 Driver System 2.83V
–· :4 Driver System 2.83V

FIGURE 1.37

—— :1 Driver System 12.68V
······ :2 Driver System 6.35V
– – – :4 Driver System 6.35V

FIGURE 1.38

—— :1 Driver System 12.68V
······ :2 Driver System 6.35V
– – – :4 Driver System 6.35V

is not critical, but make sure the rear woofer cone never contacts the magnet structure of the front woofer. The tunnel should be airtight and, except for some sort of acoustical absorbent (such as felt) on the tunnel walls, should not contain damping material.

Again, you'll find two interesting construction articles printed in *Speaker Builder:* "Constructing a Sontek Subwoofer," by P. Todd, 2/85, p. 20; and "The Wonder of Symmetrical Isobarik," by Bill Schwefel, 5/90, p. 10.

1.100 ELECTRONICALLY ASSISTED CLOSED-BOX SYSTEMS.

A method sometimes suggested for altering the low end response of a closed-box system involves the use of some sort of active filter boost. Three published papers by Leach,[21, 22] Staggs,[23] and Greiner and Schoessow[24] describe different approaches to either lowering the cut-off frequency or enhancing the transient response of a loudspeaker. *Table 1.22* summarizes the parameters for the three filters.

Illustrating the problems created by boosting the low end output of a closed-box speaker can be accomplished using computer simulation. The computer simulation in this example duplicates the method described by Marshall Leach in his June 1981 *JAES* article, "Active Equalization of Closed Box Loudspeaker Systems." The box/driver combination had a Q_{tc} of 0.7 with an f_3 of 35Hz. The boost filter comprised a fourth-order response with about 8dB of gain at 24Hz. The transfer function of the filter is shown in *Fig. 1.42*. *Figure 1.43* gives the SPL comparison of the design with and without the filter at a 1W level. The closed box with the boost filter now has an f_3 of 24Hz, 11Hz lower than the unfiltered loudspeaker, but with a high order rolloff of 38.5dB/octave. This appears on the surface like a good choice until the cone excursion and group delay curves in *Fig. 1.44* are examined. Excursion (the lower set of curves) has more than doubled, while the damping change looks fairly radical with a very sharp "knee" in the group delay curve. The real story is told at a 20W power level, however. Although the SPL comparison in *Fig. 1.45* reveals only the changes expected at a higher power level, the cone excursion and group delay curves of the boosted design shown in *Fig. 1.46* are totally unacceptable. Excursion has increased to nearly 19mm, which is difficult since the driver has an X_{max} of only 7mm. While it can be argued that the peak is confined to a fairly narrow

TABLE 1.22
ACTIVE EQ FOR CLOSED BOXES

	Q_{tc}	new Q_{tc}	Boost	Result
A.	B_2–0.707	B_4–0.707	8dB	1 octave
B.	C_2–1.1	D_3–0.577	5dB	imp trans.
C.	B_2–0.707	B_2–0.707	10dB	1 octave

span at a frequency which has little program material, you could expect the cannon shots on the *1812* or some low synthesizer notes to send the voice coil completely out of the gap and likely crashing into the front plate on its return path causing catastrophic failure.

I have made no attempt to duplicate the design methodology presented for these filter systems. They are mentioned, however, as a possible alternative to previously discussed systems. The major drawback of the filters described above is the substantial increase in amplifier power required at lower frequencies. Such techniques for altering response, when used in conjunction with wide dynamic range sources such as CD players, become even more undesirable. If you want a lower cut-off frequency or improved transients, probably your best solution is to select an appropriate unassisted design in the first place.

1.200 MASS LOADING TECHNIQUES.

Modifying speakers by adding mass is an old technique which has been employed for years in the loudspeaker industry. This section describes two mass loaded configurations, simple mass loading and mass loading used to create a passively assisted speaker.

1.210 SIMPLE MASS LOADING MODIFICATIONS.

Adding mass to a speaker cone causes several of the operational parameters to change. *Table 1.23* summarizes the parameter modification in a 10″ woofer caused by increasing cone mass by 75%. Data is given for both woofer and enclosure operating parameter changes.

From the data in *Table 1.23*, it is apparent that as mass increases, Q_{ts} goes up, f_s goes down, and driver efficiency goes down. Likewise, as mass is added, the box Q increases as does the −3dB phase angle, and the f_3 and SPL go down. *Figure 1.47* illustrates the simulated SPL comparison of the woofer in the enclosure with and without the added mass. The cone excursion and group delay curves in *Fig. 1.48* confirm the decrease in damping, while the impedance curves in *Fig. 1.49* depict the lower f_c and the higher Q of the mass added combination. This tradeoff set seems fairly reasonable, even looking at the effects of increasing power to the 20W level. *Figure 1.50* shows the expected changes in SPL, but the increase in cone excursion shown in *Fig. 1.51* could be excessive. This particular driver has an X_{max} of 7mm. Figuring that third harmonic distortion increases to 3% at X_{max} +15%, or 8mm, the unchanged driver/box combination is still operating appropriately. The mass-loaded version is at about 9.5mm, which, though not really excessive, is starting to show more stress. A good solution would be to use two mass-loaded woofers instead of one, which would lower Q and excursion at higher operating levels and increase efficiency to an acceptable level. One caution, however: the additional mass can be sufficient to

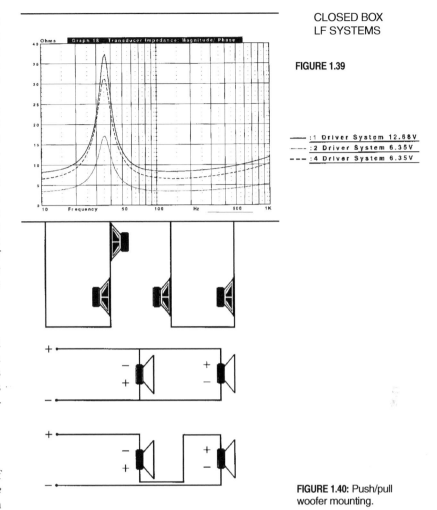

— :1 Driver System 12.88V
···· :2 Driver System 6.35V
- - - :4 Driver System 6.35V

FIGURE 1.40: Push/pull woofer mounting.

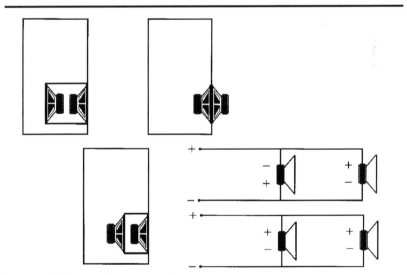

FIGURE 1.41: Compound woofer mounting.

cause cones to sag over time. If the sag is excessive, it can lead to misaligned voice coils and cause premature driver failure.

1.220 PASSIVELY ASSISTED MASS-LOADED SPEAKERS.

Passively assisted speakers do not necessarily require mass loading, but since that is one method of achieving the necessary parameters

FIGURE 1.42

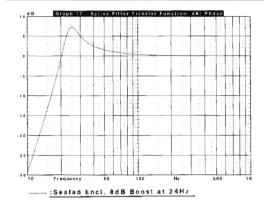

.... :Sealed Encl. 8dB Boost at 24Hz

FIGURE 1.43

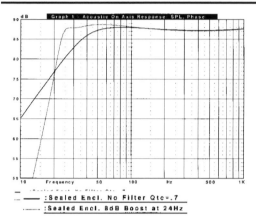

___ : Sealed Encl. No Filter Qtc=.7
.... :Sealed Encl. 8dB Boost at 24Hz

FIGURE 1.44

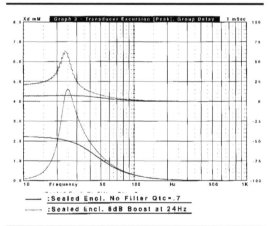

_____ :Sealed Encl. No Filter Qtc=.7
.... :Sealed Encl. 8dB Boost at 24Hz

FIGURE 1.45

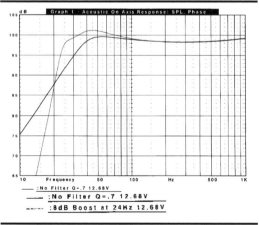

_____ :No Filter Q=.7 12.68V
_____ :No Filter Q=.7 12.68V
.... :8dB Boost at 24Hz 12.68V

TABLE 1.23

Q_{ts} AND Q_{tc} CHANGES WITH ADDED MASS

	Driver						Box		
Mmd.	Q_{ms}	Q_{es}	Q_{ts}	f_s Hz	SPL dB		Q_{tc}	f_3 Hz	SPL dB
47g	2.86	0.5	0.43	22.5	87		0.7	35	87.9
80g	3.67	0.65	0.55	17.5	82.7		1.0	25	84.9

sively assisted vented alignments. Benson[25] and Von Recklinghausen[26] first introduced the idea of extending the low end response of sealed enclosures by combining the transfer function of the box/driver combination with a passive high-pass filter. This was followed by Geddes' and Clark's presentation at the 79th AES Convention in 1985.[27] They suggested that such designs would be appealing for configurations where active equalization was not practical, such as with satellite speakers used with sub-woofers. It was also suggested that car audio applications would also likely be desirable because of the limited amplifier voltage available for active EQ (since active EQ plays a big role in car audio, the latter seems invalid). Bringing it into the practical world of real life speaker building, Tom Nousaine described an interesting application in *Speaker Builder* 1989.[28]

The basic criterion for this type of passively assisted design requires that the woofer have a Q_m between 7 and 10. The design described in Geddes' and Clark's paper used a JBL 128H 12″ woofer which has a high Q_m because it used a nonconductive voice coil former. Since most home hi-fi drivers use conductive formers and have Q_{ms} numbers in the vicinity of 3 to 5, the added mass is used to boost the Q_{ms} to an acceptable level.

To use this methodology, select a woofer with a fairly high Q_{ms} value of about 4.5 and add enough mass to increase the mechanical Q to about 7 or 8. The 12″ woofer used in the Nousaine article required increasing the mass by 80% (about the same as the previous mass loaded example). The unassisted box Q_{tc} should be set to about 1.1. As with the previous example, using a two driver combination will be more satisfactory in terms of excursion and efficiency.

The series high-pass filter is a single capacitor. The value of the capacitor is calculated by:

$$C = \frac{0.234}{R_E \times f_c}$$

Table 1.24 gives the driver and box values for the same driver, added mass and box dimensions used in Tom Nousaine's *SB* article (using a Precision TA305F 12″ woofer).

The capacitor required for this design was calculated to be about 1,750μF, which could be had from a single large motor starting capacitor or several paralleled nonpolar electrolytics.

The computer 1W SPL simulation for the driver in the enclosure without added mass,

for the methodology being described in this section, it is included as a type of mass loading technique. Passively assisted speakers grew out of the original Thiele work with fifth-order pas-

with added mass, and with both mass and the filter is depicted in *Fig. 1.52*. The assisted design lowers the f₃ by 10Hz from the mass only design. The response now, however, resembles a very sharp "knee" fourth-order vented response. This is confirmed by the group delay and cone excursion curves in *Fig. 1.53*. The shape of the group delay curve is very much like a poorly damped vented enclosure, but that is the tradeoff for the good low end extension. Cone excursion has also increased with the addition of the filter. Looking at the impedance curves in *Fig. 1.54*, the series high-pass filter does not alter the driver/box impedance.

Increasing the power input to 20W yields the SPL graphs in *Fig. 1.55*. The changes are what would be expected for an increase in voltage drive, and are within acceptable limits because of the multiple driver configuration. The group delay and cone excursion curves in *Fig. 1.56* also reflect the change in damping with power increase. The excursion requirements of the passively assisted design are reasonable at 7mm considering the X_max +15% is 6.9mm for this driver. About the only drawback is the poor group delay response which accompanies the fourth-order response shape and the possibility of voice coil alignment problems caused by the added mass. The sagging problem could likely be avoided by occasionally rotating the drivers 180° in the enclosures.

1.300 SEALED REAR CHAMBER BANDPASS ENCLOSURES.

1.310 DEFINITION.
A sealed rear chamber bandpass enclosure is basically a closed-box system with the addition of an acoustic filter in series with the front radiation of the driver. Because of the additional filter element, the possibilities for tradeoffs on bandwidth and efficiency are greater than with a simple sealed type enclosure.

1.315 ADDITIONAL TERMS.
f_b = tuning frequency of front chamber vent
f_L = f_3 of low frequency rolloff
f_H = f_3 of high frequency rolloff
L_v = vent length in inches (centimeters)
Q_{bp} = Q of bandpass rear chamber
R = vent radius in inches (centimeters)
S = passband ripple (variation in magnitude response)
V_f = volume of the front (acoustic filter) chamber
V_r = volume of the rear chamber

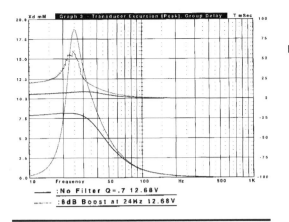

FIGURE 1.46

FIGURE 1.47

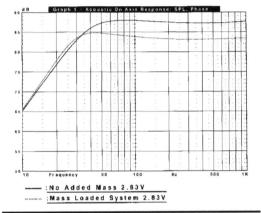

FIGURE 1.48

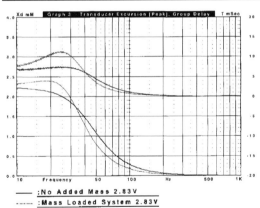

V_t = total volume of both chambers

1.320 HISTORY.
The bandpass enclosure, despite its current fad-like popularity, is not a new design concept. The original patent was filed in 1934 by Andre d'Alton[29] (patent no. 1,969,704) followed by another filed in 1952 by then MIT grad student

TABLE 1.24
PASSIVELY ASSISTED MASS LOADED DESIGN

	Mmd.	Q_{ms}	Q_{es}	Q_{ts}	SPL	f_s	Q_{tc}	Slope dB/oct.	SPL	f_3
w/o mass	72g	5.8	0.32	0.31	91dB	24Hz	0.7	11.03	95dB	54Hz
w/mass	129g	7.6	0.42	0.40	87dB	19Hz	1.1	12.34	90dB	38Hz
plus cap	—	—	—	—	—	—	4th	21.6	90dB	28Hz

FIGURE 1.49

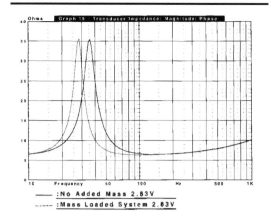

: No Added Mass 2.83V
...... : Mass Loaded System 2.83V

FIGURE 1.50

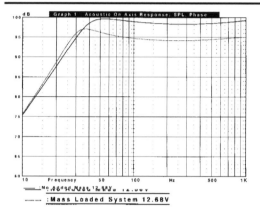

: No Added Mass 12.68V
...... : Mass Loaded System 12.68V

FIGURE 1.51

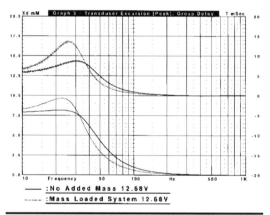

: No Added Mass 12.68V
...... : Mass Loaded System 12.68V

FIGURE 1.52

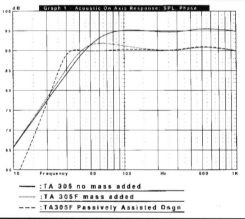

: TA 305 no mass added
...... : TA 305F mass added
- - - : TA305F Passively Assisted Dsgn

Henry Lang. Renewed interest probably was fired by Laurie Fincham's paper delivered at the 1979 63rd AES Convention titled "A Bandpass Loudspeaker Enclosure" (AES preprint no. 1512). In 1982 two French designers, Augris and

Santens, published a hand-calculator design system for sealed rear chamber bandpass speakers in the French publication *L'Audiophile*.[30] Not long after that, Bose was granted a patent on vented front and rear chamber bandpass enclosures in October, 1985 (patent no. 4,549,631), which was later to become the Acoustimass three piece speaker system. Earl Geddes kept things going with his presentation at the 81st AES Convention, November 1986, titled "Bandpass Loudspeaker Enclosure" (preprint no. 2383). The paper was revised and published in the *JAES* May, 1989. The real explosion of popularity among amateurs and manufacturers probably came, however, after Jean Margerand republished Augris and Santens methodology in *Speaker Builder* 6/88.

1.330 WOOFER SELECTION.
Woofer selection for sealed rear chamber bandpass enclosures is somewhat the same as for standard nonfiltered closed boxes. With bandpass enclosures, because of the increased tradeoff flexibility, the ratio of f_s/Q_{ts} is important. Low ratios, which mean the higher Qs usually employed in closed box systems, generally yield lower f_3 rolloffs.

1.340 BOX SIZE DETERMINATION.
The methodology described for box size determination is taken from the *SB* 6/88 article, but somewhat simplified. Three design tables are used for deriving enclosure parameters. Each table represents a different level of damping, "S," which is expressed as ripple within the passband. This ripple factor, S, describes the SPL variation in the box magnitude response between the two −3dB frequencies, f_L, the low frequency rolloff, and f_H, the high frequency rolloff. *Table 1.25* has S = 0.7 which indicates 0 ripple and the best transient performance. *Table 1.26* has S = 0.6 which allows for 0.35dB ripple (hardly significant) with somewhat degraded transient performance. *Table 1.27* has S = 0.5 which allows for 1.25dB ripple, and has transient performance somewhat degraded from S = 0.6. "S" is also a general indicator of overall bandwidth, which is widest at S = 0.5, and the most narrow at S = 0.7.

The recommended procedure is to start with *Table 1.25*, S = 0.7, and by trial and error determine whether or not the driver selected will provide the desired f_L and f_H frequencies at the sensitivity level required. The high and low frequency rolloff f_3 points are found by multiplying the driver f_s/Q_{ts} ratio by the f_L and f_H factors from the tables. This can be done for different levels of sensitivity, until a satisfactory compromise between f_L/f_H frequencies and sensitivity is determined.

V_f, the volume of the front enclosure, can be calculated once one of the tables has been selected, as it is independent of all factors, except "S." V_f is calculated by:

$$V_f = (2S \times Q_{ts})^2 \times V_{as}$$

As you scan the tables, be aware that the higher the Q_{bp} value, the higher the sensitivity and the nar-

TABLE 1.25
S = 0.7 Ripple = 0dB

Q_{bp}	f_L factor	f_H factor	Sensitivity
0.4507	0.2167	0.9373	−8dB
0.4774	0.2378	0.9584	−7dB
0.5057	0.2606	0.9812	−6dB
0.5356	0.2852	1.0058	−5dB
0.5674	0.3118	1.0324	−4dB
0.6010	0.3404	1.0610	−3dB
0.6366	0.3712	1.0918	−2dB
0.6743	0.4043	1.1248	−1dB
0.7143	0.4397	1.1603	0dB
0.7566	0.4777	1.1983	1dB
0.8014	0.5184	1.2390	2dB
0.8489	0.5619	1.2825	3dB
0.8772	0.6084	1.3290	4dB
0.9525	0.6581	1.3787	5dB
1.0090	0.7111	1.4317	6dB
1.0687	0.7675	1.4881	7dB
1.1321	0.8277	1.5483	8dB

TABLE 1.26
S = 0.6 Ripple = 0.35dB

Q_{bp}	f_L factor	f_H factor	Sensitivity
0.5258	0.2326	1.1886	−8dB
0.5570	0.2560	1.2119	−7dB
0.5900	0.2813	1.2373	−6dB
0.6249	0.3088	1.2648	−5dB
0.6619	0.3385	1.2945	−4dB
0.7012	0.3706	1.3266	−3dB
0.7427	0.4052	1.3612	−2dB
0.7867	0.4425	1.3986	−1dB
0.8333	0.4827	1.4387	0dB
0.8827	0.5258	1.4818	1dB
0.9350	0.5721	1.5281	2dB
0.9904	0.6217	1.5778	3dB
1.0491	0.6749	1.6309	4dB
1.1113	0.7317	1.6877	5dB
1.1771	0.7925	1.7485	6dB
1.2469	0.8573	1.8134	7dB
1.3207	0.9266	1.8826	8dB

TABLE 1.27
S = 0.5 Ripple = 1.25dB

Q_{bp}	f_L factor	f_H factor	Sensitivity
0.6310	0.2600	1.5312	−8dB
0.6683	0.2867	1.5579	−7dB
0.7079	0.3158	1.5870	−6dB
0.7499	0.3474	1.6186	−5dB
0.7943	0.3817	1.6528	−4dB
0.8414	0.4189	1.6900	−3dB
0.8913	0.4591	1.7302	−2dB
0.9441	0.5025	1.7736	−1dB
1.0000	0.5493	1.8204	0dB
1.0593	0.5997	1.8709	1dB
1.1220	0.6540	1.9251	2dB
1.1885	0.7122	1.9833	3dB
1.2589	0.7747	2.0458	4dB
1.3335	0.8417	2.1128	5dB
1.4125	0.9134	2.1845	6dB
1.4962	0.9901	2.2612	7dB
1.5849	1.0720	2.3431	8dB

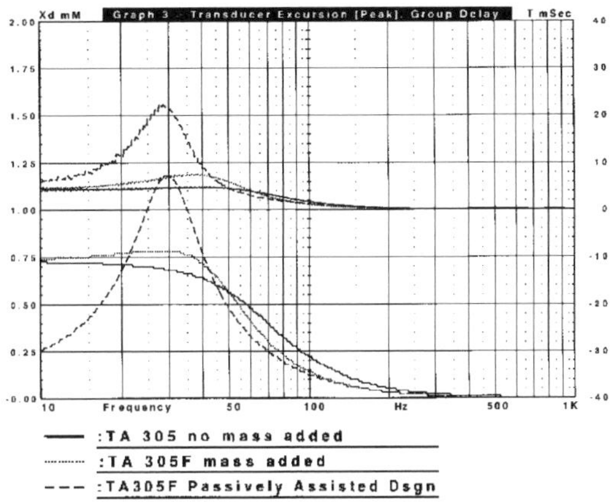

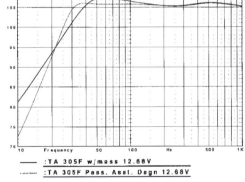

Xd mM
T mSec
Frequency
Hz
:TA 305 no mass added
:TA 305F mass added
:TA305F Passively Assisted Dsgn

FIGURE 1.53

Ohms
Frequency
Hz
:TA 305 no mass added
:TA 305F mass added
:TA305F Passively Assisted Dsgn

FIGURE 1.54

dB
Frequency
Hz
:TA 305F w/mass 12.68V
:TA 305F Pass. Asst. Dsgn 12.68V

FIGURE 1.55

rower the bandwidth. Conversely, the lower the Q_{bp} value, the lower the sensitivity and the wider the bandwidth, as you can see in *Fig. 1.57.*

Next, V_r can then be calculated from:

$$V_r = \frac{V_{as}}{(Q_{bp}/Q_{ts})^2 - 1}$$

The vent tuning frequency for the front chamber filter is found by:

$$f_b = Q_{bp} \times (f_s/Q_{ts})$$

Once the tuning frequency has been determined, the vent length in inches for V_{as} in cubic inches and vent radius R in inches is given by:

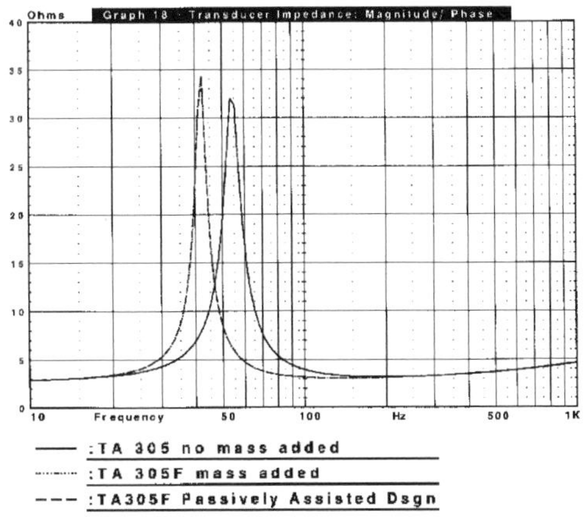

$$L_v = \frac{1.463 \times 10^7 R^2}{f_b{}^2 V_f} - 1.463R$$

For L_v in centimeters, with V_f in liters, and vent radius in centimeters use:

$$L_v = \frac{9.425 \times 10^4 R^2}{f_b{}^2 V_f} - 1.595R$$

Vent diameter should be made as large as practically possible for the depth of the enclosure to avoid excessive port nonlinearity. Shelf ports, discussed in *Chapter 2, Vented Box Low Frequency Systems*, generally will achieve the maximum port cross sectional area possible for a given enclosure depth. Longer tube vents require an elbow joint.

1.360 EXAMPLE DESIGN CHARTS and COMPUTER SIMULATION.

Table 1.28 gives an example of a sealed rear chamber bandpass design using an 8″ woofer. The data from the design tables was entered into LEAP 4.0 to generate a computer simulation. Compared to the calculator design charts, the modeling in the LEAP program is considerably more sophisticated. LEAP considers nonlinear BL, compliance, and port losses, as well as a variety of frequency dependent parameters. The general Thiele/Small model (which is the basis for all the calculator design charts in this book) relies upon generalized constants and does not account for the frequency dependent nature of losses or deal with driver nonlinearity. However, at a small signal (1W) level the calculator methods are usually fairly accurate and more than reliable enough for design work. The results shown in *Table 1.28* bear this out, since LEAP's predictions for f_L and f_H are within 1Hz of the predictions from the design tables.

The 1W SPL and phase is depicted in *Fig. 1.58*, showing the typical bandpass shape. The −3dB point is a respectable 42Hz, which is good compared to the roughly 55Hz −3dB point this woofer would produce in a 0.7ft³ sealed enclosure. The rolloff slope is about 15dB/octave on the low end, steep for a second-order rolloff, and 17.8dB/octave on the high end of the response. Much of the literature, however, seems to imply by omission that bandpass enclosures are good low-pass filters. This is not true. *Figure 1.59* shows the same enclosure from 20–20kHz with the LEAP port standing wave model turned on. This simulation was done without any front chamber damping to emphasize the effect. This is a good approximation of the response anomalies caused by the pipe resonances (pipe resonances are discussed in detail in *Chapter 2*) and standing waves transmitted out the port, but *Fig. 1.60* shows the real thing. This is the measure frequency response of a 10″ woofer in a sealed chamber bandpass enclosure (the front chamber of this design was lined with 1″ fiberglass). The measurement was taken near-field with the DRA MLSSA FFT analyzer and an ACO Pacific 7012 precision measurement microphone and the data then imported into LEAP 4.0. This problem is not trivial and it is difficult to use bandpass enclosures

FIGURE 1.56

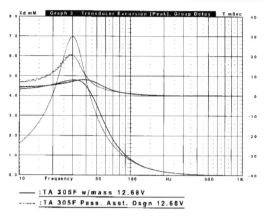

— : TA 305F w/mass 12.68V
······· : TA 305F Pass. Asst. Dsgn 12.68V

FIGURE 1.57

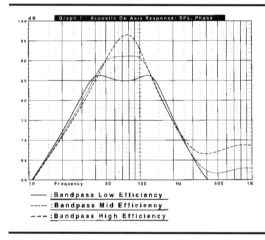

—— : Bandpass Low Efficiency
······· : Bandpass Mid Efficiency
– – – : Bandpass High Efficiency

FIGURE 1.58

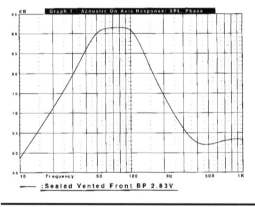

—— : Sealed Vented Front BP 2.83V

FIGURE 1.59

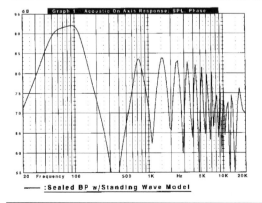

—— : Sealed BP w/Standing Wave Model

without additional filtering unless you accept the upper frequency output.

Several techniques can be employed to minimize the box standing wave modes being transmitted through the port. The first item is to line all inside walls in the front chamber with fiberglass or some other damping material. Stretching a length of material, such as the pressed cloth type used in automobiles, across the inside of the enclosure between the port and the driver can likewise be effective. Using an inside panel fastened at an angle between the port and driver to alter standing wave modes would also be somewhat effective. Pipe vents directly over the driver, as in many designs, are more offensive than shelf ports. Many designers do not try to use the enclosure without a filter and usually resort to an electronic filter or simple first-order filter and a conjugate circuit (impedance equalizer), which will require some retuning of the port frequency. Another method of adding first-order high-pass and low-pass filters to a bandpass enclosure accompanied by box retuning was described by Joe D'Appolito in a paper he presented at the 91st AES Convention.[31] This methodology was incorporated into the TopBox computer software and is readily available.[32,33] All of these methods will involve a certain amount of trial and error.

Figure 1.61 shows the group delay and cone excursion. The group delay indicates a damping picture much like a vented enclosure. Although the low end rolloff may be second-order like a closed box, the transient performance is degraded by comparison. The impedance is given in *Fig. 1.62* and the cone velocity in *Fig. 1.63*. The appearance of both graphs is similar to a vented enclosure with maximum driver acceleration happening at the resonance frequencies.

Increasing the power input to 20W, the magnitude response becomes somewhat peaked at upper end as seen in *Fig. 1.64*. Cone excursion, shown in *Fig. 1.65*, goes to about 6.4mm, which is not bad at this SPL (102dB), since X_{max} + 15% is 6mm. This design's total box volume is about 1.2ft³, which is large for a single 8″ speaker. Bandpass designs can be used with compound loading to create more compact variations.

1.370 BANDPASS ENCLOSURE VARIATIONS.

There are several possible variations of sealed rear chamber bandpass enclosures. The diagrams in *Fig. 1.66* depict five different possibilities: A. Single driver bandpass; B. Dual driver push/pull bandpass; C. Push/pull compound bandpass; D. Triple chamber bandpass; E. Push/pull triple chamber bandpass. The multiple driver configuration B uses the combined volumes of two drivers. The center chamber of the triple bandpass is equal to the combined

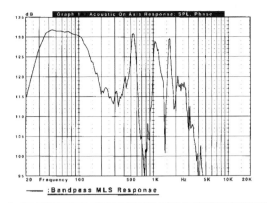

FIGURE 1.60

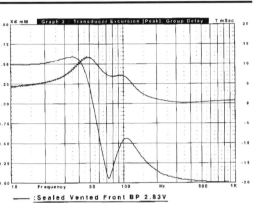

FIGURE 1.61

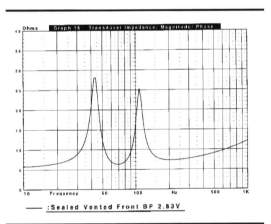

FIGURE 1.62

TABLE 1.28

8″ Woofer Bandpass Example S = 0.7

Driver

	f_s	Q_{ts}	V_{as}	f_s/Q_{ts}
	32	0.328	2.49ft³	0.99

Enclosure

	V_f	V_r	f_b	Q_{bp}	f_L	f_H	L_v	D
Calculated	0.53	0.66	71Hz	0.7142	43.5Hz	115Hz	9.75″	4″
Simulated	0.53	0.66	71Hz		42Hz	114Hz	9.25″	4″

FIGURE 1.63

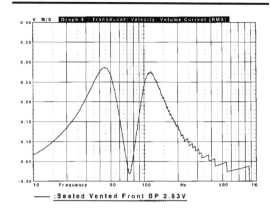

FIGURE 1.64

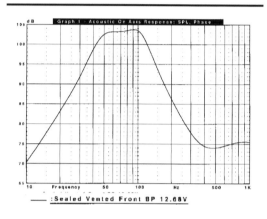

FIGURE 1.65

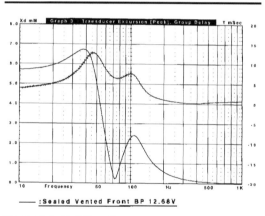

FIGURE 1.66: Bandpass enclosures.

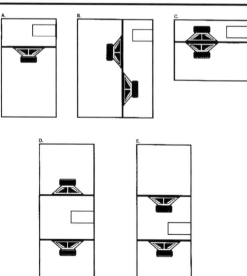

front volumes of each driver and tuned to that frequency. The drivers in B, C, and D are connected electrically out of phase for either a parallel or series wiring arrangement. A construction article describing a push/pull triple chamber bandpass enclosure for car audio is described in "Symmetrical Loading for Auto Subwoofers" by Matthew Honnert, *Speaker Builder* 6/90, p. 20.

1.400 APERIODIC CLOSED-BOX LOUDSPEAKERS.

The aperiodic speaker design is a variation of the closed box format, and is more like a special extension of the damping techniques described in *Section 1.82*. An aperiodic speaker is a closed box which uses a device known as a Variovent. The Variovent, sold by Scan Speak and Dynaudio, provides a flow resistant path out of the enclosure, converting the sealed box into a resistively leaky closed box. The device mounts in a 4″ diameter hole, resembling a vent, and consists of a plastic holder which sandwiches a 1″ thick piece of high density fiberglass material. The fiberglass provides the resistance leakage and does not function as a vent. The Variovent tends to damp the impedance of a closed box in the same way as adding 100% of high density fill material. Both techniques increase the effective size of the enclosure. The history of the aperiodic design goes back to the 1960s Dynaco A-25 series. The highly successful A-25 was produced by Scan Speak and used a Variovent.

The application of a Variovent is uncomplicated. You simply cut a mounting hole in the rear of the enclosure, and line the enclosure with fiberglass, leaving a path from the woofer to the Variovent. One Variovent is recommended for enclosures to 50 liters, two for volumes up to 80 liters, and three for volumes over 80 liters.

In order to place the enclosure damping capabilities of the Variovent in a useful perspective, a Dynaudio Variovent was measured and then compared to the damping effects of some damping materials described in *Section 1.82*. The same test procedures, woofer and test box were assembled using the Audio Precision System 1 to measure the impedance of the driver/box combination. Tests were done with the Variovent installed in the box without additional fiberglass and with the prescribed amount of fiberglass (approximately 50% fill).

The System 1 .DAT files were then imported into LEAP 4.0, converted to true impedance and graphed and displayed in *Fig. 1.67*. Three curves are depicted, the box without fiberglass or Variovent, the box with Variovent and no fiberglass, and the box with both Variovent and fiberglass.

The LEAP Automatic Speaker Parameter Measurement utility was used to calculate box parameters. *Table 1.29* compares the measured effects of the Variovent with and without fiberglass along with the data from *Table 1.21* for 50% fill of 1 lb/ft³ fiberglass (R19) and a 50/50 blend of 4 lb/ft³ fiberglass and Acousta-Stuf. The

TABLE 1.29

	Z_0	f_0	Q_m	Q_e	Q_{tc}	f_3
Variovent w/o Fiberglass	25.74	86.62	4.53	1.42	1.08	66.00
Variovent with Fiberglass	20.22	81.65	3.17	1.39	0.96	65.30
1lb/ft³ Fiberglass	27.77	79.51	4.93	1.40	1.09	60.37
4lb Fiberglass/Acousta-Stuf	21.96	79.11	3.49	1.33	0.96	63.40

effects of the Variovent are very nearly the same as the 50/50 4 lb/ft³ fiberglass/Acousta-Stuf combination. From this information, it is obvious that the Variovent would certainly be a viable alternative to some of the other stuffing formats. The frequency response was not measured, since it would be about the same as that of the 50% fill R19 example in *Section 1.82.*

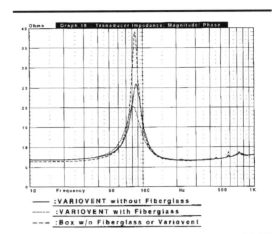

FIGURE 1.67

—— :VARIOVENT without Fiberglass
········· :VARIOVENT with Fiberglass
— — — :Box w/o Fiberglass or Variovent

REFERENCES

1. D. B. Weems, "Closed-Box Speaker System Design," *Popular Electronics,* June-July 1973.
2. E.M. Villchur, "Revolutionary Loudspeaker and Enclosure," *Audio,* October 1954.
3. R. Small, "Direct Radiator Loudspeaker System Analysis," *JAES,* June 1972.
4. R. Small, "Closed-Box Loudspeaker System, Part 1, 2" *JAES,* Jan./Feb. 1973.
5. V. Brociner, "Speaker Size and Performance in Small Cabinets," *Audio,* March 1970.
6. L. Beranek, *Acoustics,* McGraw-Hill, 1954, p. 226.
7. M. Colloms, *High Performance Loudspeakers,* Pentech Press, 1978, 1985.
8. Dr. Richard C. Cabot, "Audio Tests and Measurements," *Audio Engineering Handbook,* edited by K. Blair Benson, ©1988 by McGraw-Hill Inc.
9. J. Ashley and T. Saponas, "Wisdom and Witchcraft of Old Wives Tales and Woofer Baffles," *JAES,* October 1970.
10. H. D. Harwood, "Some Factors in Loudspeaker Quality," *Wireless World,* May 1975.
11. R. Small, "Suitability of Low-Frequency Drivers for Horn Loaded Loudspeaker Systems," AES preprint No. 1251.
12. H. J. J. Hoge, "Switched on Bass," *Audio,* August 1976.
13. M. Gander, "Moving Coil Loudspeaker Topology as an Indicator of Linear Excursion Capability," *JAES,* January 1981.
14. K. P. Zacharia, "On the Syntheses of Closed-Box Systems Using Available Drivers," *JAES,* November 1973.
15. D. B. Weems, "Ten Speaker Enclosure Fallacies," *Popular Electronics,* June 1976.
16. W. D'Ascenzo, "The AR-1 Rejuvenated," *Speaker Builder,* 2/82.
17. L. M. Chase, "The Thermo-Acoustic Properties of Fibrous Materials," *IEEE Transactions on Acoustics, Speech, and Signal Processing,* August 1974.
18. W. M. Leach, "Electroacoustic-Analogous Circuit Models for Filled Enclosures," *JAES,* July/August 1989.
19. A. Carrion-Isbert, "A New Method of Designing Closed Box Loudspeaker Systems Under the Influence of Enclosure Filling Material," 75th AES Convention, preprint No. 2058.
20. Small and Margolis, "Personal Calculator Program," *JAES,* June 1981.
21. W. M. Leach, "Active Equalizer of Closed-Box Loudspeaker Systems," *JAES,* June 1981.
22. W. M. Leach, "A Generalized Active Equalizer for Closed-Box Loudspeaker Systems, Part I—Isolated Filters Driving Second Order (Closed-Box) Systems," *JAES,* July/August 1979.
23. V. Staggs, "Transient-Response Equalization of Sealed-Box Loudspeakers," *JAES,* December 1982.
24. Greiner and Schoessow, "Electronic Equalization of Closed-Box Loudspeakers," *JAES,* March 1983.
25. J. E. Benson, "An Introduction to the Design of Filtered Loudspeaker Systems, Part I—Isolated Filters Driving Second Order (Closed-Box) Systems," *JAES,* July/August 1979.
26. D. R. Von Recklinghausen, "Low-Frequency Range Extension of Loudspeakers," *JAES,* June 1985.
27. Geddes and Clark, "Passively Assisted Loudspeakers," 79th AES Convention, preprint No. 2291.
28. Tom Nousaine, "A Passively Assisted Woofer System," *Speaker Builder,* 2/89, p. 16.
29. *Voice Coil,* August 1990.
30. Augris and Santens, "Optimization des enceintes a charge symetrique," *L'Audiophile,* No. 23, 1982.
31. J. A. D'Appolito, "Designing Symmetric Response Bandpass Loudspeakers," 91st AES Convention, preprint No. 3205.
32. V. Dickason, "New Design Software: TopBox," *Voice Coil,* June 1992.
33. V. Dickason, "TopBox Speaker Enclosure Program," *Audio,* June 1994.

FOOTNOTES

1. *Audio,* Oct. 1954, July 1955, Oct. 1957; also *JAES,* July 1957.
2. This criterion suggests optimal operating parameters. A closed box is, however, possible with any value of Q_{tc}.
3. The front-to-front variation is primarily suitable for subwoofer application with a low-pass crosspoint of 100Hz or less.
4. Oddly enough this paper focused on the use of passive radiators instead of vents. When KEF finally produced a bandpass speaker, the KEF 104-2 series, the speaker used vents and not radiators.

CHAPTER TWO

VENTED-BOX
LOW-FREQUENCY
SYSTEMS

2.10 DEFINITION.

A vented loudspeaker is analogous to a 24dB/octave cutoff high-pass filter, characterized by an enclosure having an open tunnel or port which allows the passage of air in and out of the box.[1] At low frequencies, the vent contributes substantially to the sound output of the system. It does so, however, by increasing the acoustic load at the rear of the cone, reducing cone motion and the output of the driver. As such, a vent only adds as much as it subtracts. As we will see later, vents also contribute other extraneous and unwanted sounds.

Compared to closed-box systems, vented enclosures possess several unique characteristics:

A. Lower cone excursion near the box resonance frequency, resulting in relatively higher power-handling and lower modulation distortion.[2] This attribute makes vented enclosures rather attractive for use in two-way loudspeakers. It is accompanied, however, by high excursion potential at frequencies substantially below resonance, making the design extremely sensitive to subsonic noise such as that caused by record warp. Fortunately, this problem can be easily overcome with a low frequency filter.

B. Lower cutoff using the same driver.

C. In theory, a +3dB higher efficiency for the same volume closed-box system. While this is not particularly significant in practice, driver design requirements of lower required cone mass and less required voice coil overhang do contribute significantly to increased efficiency for a given magnet assembly.

D. On the downside, vented enclosures are much more sensitive to misaligned parameters.[3] This factor makes the vented-box loudspeakers somewhat more difficult for the inexperienced builder.

2.20 HISTORY.

The original patent describing detailed driver and vent interaction was granted to A. C. Thuras in 1932. During the 1950s, papers by Locanthi, Beranek, van Leeuwen, de Boer, Lyon, and Novak developed in great detail the mathematical models analogous to high-pass filter synthesis. Working with a simplified model devised by Novak, A. N. Thiele published his landmark paper in 1961 (reprinted in the *JAES*, 1971).[4] Although Thiele's work was the most comprehensive and detailed in terms of practical realization, it did not include a systematic accounting of box losses.

Nomura's 1969 paper, "An Analysis of Design Conditions of a Bass-Reflex Loudspeaker Enclosure for Flat Frequency Response," described in some detail the effect of box losses on response deviations. In 1973, Richard Small published his series of vented loudspeaker system papers in the *JAES*.[1] Also noteworthy is Robert Bullock's resynthesis of Small's design graphs into more accurate and easy to read design tables.[5] At any rate, when we hear the frequently used words "Thiele/Small" in regard to vented loudspeakers, it is good to remember that although the contribution of these two individuals is extremely important to the field, their theory depended on the tremendous amount of work by others before them.

2.30 DRIVER "Q" and
ENCLOSURE RESPONSE.

As with closed-box systems, the low frequency response characteristics of vented designs can be predicted and controlled by adjusting the total box-speaker Q. The primary difference is the manner in which your design is carried out. With a closed-box, you may choose a value of Q_{tc} (*Fig. 1.1*), and proceed to derive a box size which will achieve this response. Vented enclosures, however, are usually discussed in terms of specific alignments,[4] which adjust all parameters to achieve a more or less flat response (Q = 1.0) with a particular f_3.

In other words, you cannot adjust the parameters of a vented system to give response curves such as Q_{tc} of 0.7 or 1.5. Variations which cause increases or decreases in the bass region are referred to as "misalignments." For instance, changing the Q_{ts} of the driver by ±20% will produce SPL deviations at f_3 of ±2 to 4dB.[6] This is not the same as varying the Q_{tc} of a sealed enclosure between 0.7 and 1.5. Vented systems have a much steeper cutoff and, if severely misaligned, tend to be troubled by audible transient ringing. In fact, the vented system's previously maligned reputation for being "boom boxes" undoubtedly was earned by severely misaligned loudspeakers.

2.40 WOOFER SELECTION.

Compared to woofers which are maximized for closed-box designs, woofers for vented enclosures tend to have less cone mass, less voice coil outside the magnetic gap (because of less excursion requirements), and lower overall Q_{ts}. As with closed boxes, they can use almost any value of Q_{ts}, however, only a Q_{ts} between 0.2–0.5 generally provides satisfactory responses. Small's EBP

FIGURE 2.1

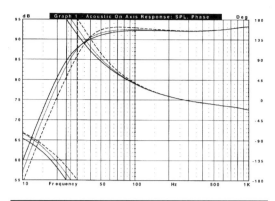

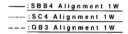

FIGURE 2.2

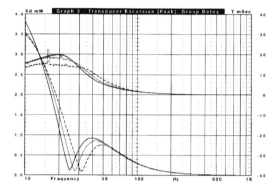

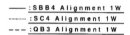

FIGURE 2.3

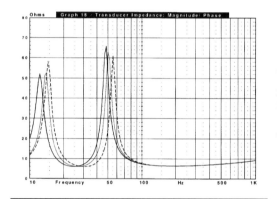

dealing with. "Alignment" is not an especially difficult concept to grasp. An alignment is a classification of a particular box size and tuning combination, producing a realizable frequency response which is more or less useful, i.e., flat. There are, at this writing, at least 15 well-defined alignment categories.

The two basic types are assisted and unassisted. Assisted alignments, first described by Thiele,[4] require some degree of active electronic filter equalization to achieve a defined response (*Section 2.110*). Unassisted alignments do not require the electronic equalization to achieve a predicted response, and are the type most often utilized by manufacturers.

For unassisted alignments, there are two basic domains: flat and nonflat. The flat response domain generally requires values of Q_{ts} less than 0.4, and is represented by six categories:

A. SBB_4. The Super Fourth-Order Boom Box is a lower required Q_{ts} extension of Hoge's BB_4. First described by Bullock,[5] the SBB_4 is characterized by a large box, low tuning (longer vent length), and good transient response. In this case the term "Boom Box" is a bit of a misnomer.

B. SC_4. The Fourth-Order Sub-Chebyshev is a lower required Q_{ts} extension of the Chebyshev (C_4) response. This response is roughly the same size and f_3 as the SBB_4, but has different tuning. Compared with the SBB_4, the SC_4 transient response is somewhat degraded.

C. QB_3. The Quasi Third-Order alignment is the most commonly used vented alignment because it yields a smaller box and lower f_3 for a given driver Q_{ts}. Its transient response, however, is not quite as good as the SBB_4 or the SC_4.

D. Discrete Alignments. There are three discrete alignments:

B_4 (Fourth-Order Butterworth)
BE_4 (Fourth-Order Bessel)
IB_4 (Butterworth Inter-Order)

These alignments are designated discrete because they exist for only one single value of Q_{ts}. Since the box losses affect the value of a discrete alignment, they are difficult, though not impossible, to obtain. Of the three, the BE_4 has the best transient response.

The nonflat domains are usually generated using the higher values of Q_{ts}, and have inferior

(*Section 1.40*) suggests values around 100 will work well for vented enclosures.

Although losses can play a critical role in determining box size and tuning, apparent leakage problems from some types of woofers should probably be ignored and the offending drivers dealt with in the manner they were designed. Porous dust caps do provide a loss path for air in the box, but are usually installed by the woofer designer for voice coil cooling purposes, and sealing them may well create as many problems as it solves. Lossy pleated cloth surrounds can also provide high leakage losses. As long as driver parameters and losses are appropriately measured, however, there is no reason to avoid this type of product unless the performance is not adequate.

2.50 ALIGNMENTS.

To determine a box size, start by choosing an appropriate alignment to satisfy whatever design criteria or driver limitation you may be

TABLE 2.0

Alignment Type	V_b	f_b	f_3	Slope dB/oct
Flat Alignments				
SBB_4	2.7	25Hz	36Hz	18
SC_4	2.4	27Hz	36Hz	19
QB_3	2.0	31Hz	36Hz	20
Nonflat Alignments				
SQB_3	7.6	30Hz	34Hz	27
BB_4	2.8	37Hz	30Hz	30
C_4	5.3	30Hz	27Hz	30

transient and frequency response characteristics. As such, they have somewhat limited use in high-quality audio applications. If their negative attributes can be tolerated for a particular situation, however, they will yield lower values of f_3, for a given driver:

A. C_4. The Chebyshev equal ripple alignment can be useful for low values of ripple less than 1dB. Originally described by van Leeuwen in 1956 (see [1] p. 364 for reference).

B. BB_4. The Fourth-Order Boom Box was originally described by Hoge.[7] The name comes from the peak in response close to rolloff, which, if large enough, has the same undesirable characteristics as a high Q_{tc} (1.2+) closed box.

C. SQB_3. The Super Third-Order Quasi-Butterworth is a high value of Q_{ts} extension of the QB_3 alignment, and was described by R. Bullock in *Speaker Builder* 3/81.

Computer simulation allows us to compare the six alignment types. Two woofers were used: a 12″ woofer with a Q_{ts} of 0.3 for the SBB_4, SC_4, and QB_3 flat alignments, and a 10″ woofer with a Q_{ts} of 0.5 for the SQB_4, BB_4, and C_4 nonflat alignments (discrete alignments, because of their selective nature, are not represented). Although the simulations are representations of the performance of particular drivers, generalized conclusions can be made about the class of responses. The various box parameters are compared in *Table 2.0* (phase angle at the −3dB frequency will not be used as an indicator for vented designs, as it does not readily relate to alignment type as it does to Q_{tc} with sealed enclosures).

The data in *Table 2.0* for the flat alignment group shows rolloff of all three alignment types to be about the same (at least for this woofer). The difference lies in the somewhat better transient performance reflected in the slightly more shallow rolloff of the SBB_4 alignment. However, the tradeoff is in enclosure size. The graphs of the flat alignment simulation are shown in *Figs. 2.1–2.5*. The response and phase curves depicted in *Fig. 2.1* are very closely grouped, showing a typically steep slope in the vicinity of 18–20dB/octave. Looking at the group delay curves in *Fig. 2.2*, the shape of all three curves is similar to that of a sealed box Q_{tc} of 0.9 but with much higher absolute delay. Compared with sealed enclosures, the transient performance of the best vented enclosure is worse than the best sealed box woofer, and the differences are certainly audible.[6] However, it remains true that there have been examples of well reviewed, well liked, and financially successful vented designs. Of the three alignments in the flat group, the QB_3 is probably the best choice since it produces nearly the same f_3 and similar transient performance of any in the group and has the most compact enclosure.[8]

The cone excursion curves in *Fig. 2.2* illustrate the main advantage of vented enclosures, low excursion.

Above the tuning frequency, cone excursion demands are substantially less than a sealed

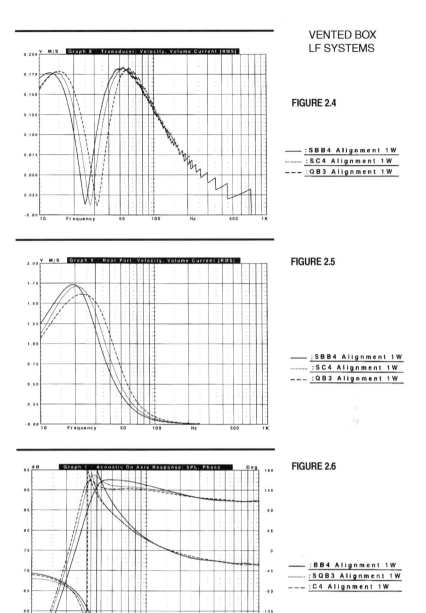

FIGURE 2.4

——— : SBB4 Alignment 1W
·········· : SC4 Alignment 1W
– – – : QB3 Alignment 1W

FIGURE 2.5

——— : SBB4 Alignment 1W
·········· : SC4 Alignment 1W
– – – : QB3 Alignment 1W

FIGURE 2.6

——— : BB4 Alignment 1W
·········· : SQB3 Alignment 1W
– – – : C4 Alignment 1W

enclosure, which also means lower distortion. This advantage is likewise one of the main problems, because the excursion rate rapidly increases below the tuning frequency f_b, and continues to increase into the subsonic regions. This causes excursion problems with ultra-low frequency program material and subsonic noises, such as rumble from warped records (*Section 2.120*).

Figure 2.3 compares the impedance curves, which are similar for all three alignments. Transducer velocity, given in *Fig. 2.4*, shows the peaks in driver acceleration to be roughly coincident with the impedance peaks, and minimum at the impedance minimum. Port velocity, the speed of the air moving in the port, is depicted in *Fig. 2.5*. The velocity peaks in the vicinity of f_b.

The nonflat alignments for higher Q drivers presented in *Figs. 2.6–2.10* portray a somewhat different performance. Looking at the magnitude responses in *Fig. 2.6*, we see variation in

FIGURE 2.7

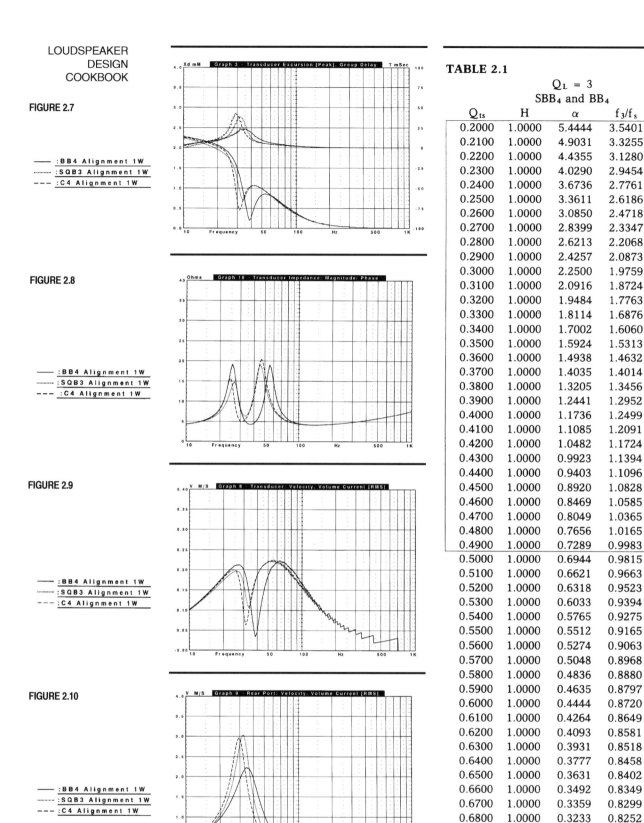

— : BB4 Alignment 1W
········· : SQB3 Alignment 1W
– – – : C4 Alignment 1W

FIGURE 2.8

— : BB4 Alignment 1W
········· : SQB3 Alignment 1W
– – – : C4 Alignment 1W

FIGURE 2.9

— : BB4 Alignment 1W
········· : SQB3 Alignment 1W
– – – : C4 Alignment 1W

FIGURE 2.10

— : BB4 Alignment 1W
········· : SQB3 Alignment 1W
– – – : C4 Alignment 1W

TABLE 2.1

$Q_L = 3$
SBB_4 and BB_4

Q_{ts}	H	α	f_3/f_s	Peak–dB
0.2000	1.0000	5.4444	3.5401	0
0.2100	1.0000	4.9031	3.3255	0
0.2200	1.0000	4.4355	3.1280	0
0.2300	1.0000	4.0290	2.9454	0
0.2400	1.0000	3.6736	2.7761	0
0.2500	1.0000	3.3611	2.6186	0
0.2600	1.0000	3.0850	2.4718	0
0.2700	1.0000	2.8399	2.3347	0
0.2800	1.0000	2.6213	2.2068	0
0.2900	1.0000	2.4257	2.0873	0
0.3000	1.0000	2.2500	1.9759	0
0.3100	1.0000	2.0916	1.8724	0
0.3200	1.0000	1.9484	1.7763	0
0.3300	1.0000	1.8114	1.6876	0
0.3400	1.0000	1.7002	1.6060	0
0.3500	1.0000	1.5924	1.5313	0
0.3600	1.0000	1.4938	1.4632	0
0.3700	1.0000	1.4035	1.4014	0
0.3800	1.0000	1.3205	1.3456	0
0.3900	1.0000	1.2441	1.2952	0
0.4000	1.0000	1.1736	1.2499	0
0.4100	1.0000	1.1085	1.2091	0.01
0.4200	1.0000	1.0482	1.1724	0.05
0.4300	1.0000	0.9923	1.1394	0.12
0.4400	1.0000	0.9403	1.1096	0.20
0.4500	1.0000	0.8920	1.0828	0.30
0.4600	1.0000	0.8469	1.0585	0.41
0.4700	1.0000	0.8049	1.0365	0.53
0.4800	1.0000	0.7656	1.0165	0.66
0.4900	1.0000	0.7289	0.9983	0.79
0.5000	1.0000	0.6944	0.9815	0.93
0.5100	1.0000	0.6621	0.9663	1.08
0.5200	1.0000	0.6318	0.9523	1.23
0.5300	1.0000	0.6033	0.9394	1.38
0.5400	1.0000	0.5765	0.9275	1.54
0.5500	1.0000	0.5512	0.9165	1.70
0.5600	1.0000	0.5274	0.9063	1.86
0.5700	1.0000	0.5048	0.8968	2.02
0.5800	1.0000	0.4836	0.8880	2.18
0.5900	1.0000	0.4635	0.8797	2.34
0.6000	1.0000	0.4444	0.8720	2.50
0.6100	1.0000	0.4264	0.8649	2.66
0.6200	1.0000	0.4093	0.8581	2.82
0.6300	1.0000	0.3931	0.8518	2.98
0.6400	1.0000	0.3777	0.8458	3.14
0.6500	1.0000	0.3631	0.8402	3.30
0.6600	1.0000	0.3492	0.8349	3.46
0.6700	1.0000	0.3359	0.8299	3.61
0.6800	1.0000	0.3233	0.8252	3.77
0.6900	1.0000	0.3113	0.8207	3.92
0.7000	1.0000	0.2999	0.8165	4.08

the overall response shape among the different alignment types. The BB_4 alignment is similar to the flat alignment group in cabinet volume and f_3, but is less damped and has a much higher rolloff rate. Although higher Q drivers will work in vented enclosures, their performance is less than spectacular in their transient capability. The other alignment types, C_4 and SQB_3, will yield impressive low f_3s, but also call for rather large enclosures for this woofer. The group delay curves shown in *Fig. 2.7* confirm the degraded transient performance of this alignment family using high Q drivers. Cone excursion, shown on the same graph, gives about the same picture as that for the flat alignments, good above f_b, highly sensitive below f_b.

TABLE 2.2

$Q_L = 7$
SBB$_4$ and BB$_4$

Q_{ts}	H	α	f_3/f_s	Peak–dB
0.2000	1.0000	5.8980	3.3686	0
0.2100	1.0000	5.3339	3.1518	0
0.2200	1.0000	4.8457	2.9521	0
0.2300	1.0000	4.4204	2.7674	0
0.2400	1.0000	4.0478	2.5960	0
0.2500	1.0000	3.7114	2.4366	0
0.2600	1.0000	3.4286	2.2883	0
0.2700	1.0000	3.1699	2.1503	0
0.2800	1.0000	2.9388	2.0220	0
0.2900	1.0000	2.7315	1.9031	0
0.3000	1.0000	2.5448	1.7932	0
0.3100	1.0000	2.3761	1.6922	0
0.3200	1.0000	2.2233	1.6000	0
0.3300	1.0000	2.0843	1.5162	0
0.3400	1.0000	1.9576	1.4406	0
0.3500	1.0000	1.8419	1.3728	0
0.3600	1.0000	1.7357	1.3122	0
0.3700	1.0000	1.6392	1.2583	0
0.3800	1.0000	1.5484	1.2104	0.01
0.3900	1.0000	1.4656	1.1679	0.06
0.4000	1.0000	1.3890	1.1302	0.14
0.4100	1.0000	1.3181	1.0966	0.24
0.4200	1.0000	1.2523	1.0667	0.37
0.4300	1.0000	1.1911	1.0399	0.51
0.4400	1.0000	1.1341	1.0160	0.66
0.4500	1.0000	1.0809	0.9944	0.82
0.4600	1.0000	1.0313	0.9750	1.00
0.4700	1.0000	0.9849	0.9574	1.17
0.4800	1.0000	0.9414	0.9415	1.36
0.4900	1.0000	0.9006	0.9270	1.55
0.5000	1.0000	0.8622	0.9137	1.74
0.5100	1.0000	0.8262	0.9015	1.93
0.5200	1.0000	0.7923	0.8904	2.13
0.5300	1.0000	0.7603	0.8801	2.33
0.5400	1.0000	0.7302	0.8706	2.53
0.5500	1.0000	0.7017	0.8619	2.73
0.5600	1.0000	0.6747	0.8537	2.93
0.5700	1.0000	0.6493	0.8462	3.13
0.5800	1.0000	0.6251	0.8391	3.33
0.5900	1.0000	0.6022	0.8325	3.53
0.6000	1.0000	0.5805	0.8264	3.73
0.6100	1.0000	0.5599	0.8206	3.93
0.6200	1.0000	0.5403	0.8152	4.12
0.6300	1.0000	0.5216	0.8102	4.32
0.6400	1.0000	0.5038	0.8054	4.51
0.6500	1.0000	0.4869	0.8009	4.70
0.6600	1.0000	0.4708	0.7967	4.90
0.6700	1.0000	0.4554	0.7926	5.09
0.6800	1.0000	0.4407	0.7889	5.27
0.6900	1.0000	0.4267	0.7853	5.46
0.7000	1.0000	0.4133	0.7819	5.65

TABLE 2.3

$Q_L = 15$
SBB$_4$ and BB$_4$

Q_{ts}	H	α	f_3/f_s	Peak–dB
0.2000	1.0000	6.0844	3.2996	0
0.2100	1.0000	5.5113	3.0818	0
0.2200	1.0000	5.0149	2.8811	0
0.2300	1.0000	4.5821	2.6955	0
0.2400	1.0000	4.2025	2.5233	0
0.2500	1.0000	3.8678	2.3633	0
0.2600	1.0000	3.5711	2.2146	0
0.2700	1.0000	3.3070	2.0764	0
0.2800	1.0000	3.0708	1.9483	0
0.2900	1.0000	2.8588	1.8301	0
0.3000	1.0000	2.6678	1.7214	0
0.3100	1.0000	2.4650	1.6222	0
0.3200	1.0000	2.3384	1.5323	0
0.3300	1.0000	2.1958	1.4514	0
0.3400	1.0000	2.0357	1.3790	0
0.3500	1.0000	1.9467	1.3146	0
0.3600	1.0000	1.8375	1.2576	0
0.3700	1.0000	1.7372	1.2071	0.01
0.3800	1.0000	1.6447	1.1626	0.07
0.3900	1.0000	1.5593	1.1233	0.16
0.4000	1.0000	1.4803	1.0886	0.27
0.4100	1.0000	1.4070	1.0577	0.41
0.4200	1.0000	1.3390	1.0303	0.57
0.4300	1.0000	1.2757	1.0059	0.73
0.4400	1.0000	1.2167	0.9840	0.91
0.4500	1.0000	1.1616	0.9643	1.10
0.4600	1.0000	1.1101	0.9466	1.30
0.4700	1.0000	1.0619	0.9305	1.50
0.4800	1.0000	1.0167	0.9160	1.71
0.4900	1.0000	0.9743	0.9027	1.91
0.5000	1.0000	0.9344	0.8906	2.13
0.5100	1.0000	0.8969	0.8795	2.34
0.5200	1.0000	0.8616	0.8696	2.56
0.5300	1.0000	0.8282	0.8599	2.78
0.5400	1.0000	0.7967	0.8513	2.99
0.5500	1.0000	0.7670	0.8432	3.21
0.5600	1.0000	0.7388	0.8358	3.43
0.5700	1.0000	0.7121	0.8289	3.65
0.5800	1.0000	0.6868	0.8224	3.86
0.5900	1.0000	0.6628	0.8164	4.08
0.6000	1.0000	0.6400	0.8108	4.29
0.6100	1.0000	0.6183	0.8055	4.51
0.6200	1.0000	0.5977	0.8006	4.72
0.6300	1.0000	0.5781	0.7959	4.93
0.6400	1.0000	0.5594	0.7916	5.14
0.6500	1.0000	0.5415	0.7874	5.35
0.6600	1.0000	0.5245	0.7836	5.55
0.6700	1.0000	0.5083	0.7799	5.76
0.6800	1.0000	0.4927	0.7764	5.96
0.6900	1.0000	0.4779	0.7731	6.16
0.7000	1.0000	0.4637	0.7700	6.36

Figure 2.8 shows the comparison of the impedance curves. The minimum impedance height at f_b is greater for the less damped C_4 and SQB$_3$ alignments. *Figure 2.9* shows the cone velocity curves peaking at approximately the same frequencies as the impedance maximums, f_L and f_H (*Fig. 2.42*), and minimum at about the same frequency as the impedance minimum.

Port velocity, presented in *Fig. 2.10*, peaks in the vicinity of f_b.

2.60 BOX SIZE DETERMINATION and RELEVANT PARAMETERS.

Determining box size is somewhat more involved than for closed-box loudspeakers. Begin by obtaining the following woofer parameters:

TABLE 2.4

$Q_L = 3$
QB$_3$ and SQB$_3$

Q_{ts}	H	α	f_3/f_s	Peak–dB
0.1000	4.3303	31.2904	5.6709	0
0.1100	3.9371	25.6824	5.1456	0
0.1200	3.6096	21.4169	4.7069	0
0.1300	3.3325	18.0974	4.3348	0
0.1400	3.0950	15.4635	4.0150	0
0.1500	2.8892	13.3386	3.7371	0
0.1600	2.0792	11.5994	3.4932	0
0.1700	2.5504	10.1581	3.2772	0
0.1800	2.4092	08.9502	3.0844	0
0.1900	2.2830	07.9280	2.9113	0
0.2000	2.1694	07.0552	2.7548	0
0.2100	2.0666	06.3041	2.6125	0
0.2200	1.9733	05.6531	2.4824	0
0.2300	1.8881	05.0851	2.3630	0
0.2400	1.8100	04.5866	2.2528	0
0.2500	1.7381	04.1467	2.1508	0
0.2600	1.6719	03.7566	2.0559	0
0.2700	1.6105	03.4090	1.9674	0
0.2800	1.5536	03.0980	1.8845	0
0.2900	1.5006	02.8186	1.8065	0
0.3000	1.4512	02.5666	1.7331	0
0.3100	1.4050	02.3386	1.6636	0
0.3200	1.3617	02.1317	1.5978	0
0.3300	1.3210	01.9432	1.5351	0
0.3400	1.2828	01.7712	1.4754	0
0.3500	1.2467	01.6136	1.4183	0
0.3600	1.2127	01.4690	1.3636	0
0.3700	1.1806	01.3360	1.3110	0
0.3800	1.1501	01.2133	1.2605	0
0.3900	1.1213	01.0999	1.2118	0
0.4000	1.0939	0.9949	1.1649	0
0.4100	1.0679	0.8974	1.1198	0
0.4200	1.0431	0.8069	1.0763	0
0.4300	1.0195	0.7225	1.0346	0
0.4400	0.9970	0.6439	0.9947	0
0.4500	0.9755	0.5704	0.9568	0
0.4600	0.9550	0.5016	0.9210	0.02
0.4700	0.9354	0.4672	0.8875	0.06
0.4800	0.9166	0.3767	0.8563	0.14
0.4900	0.8986	0.3199	0.8276	0.27
0.5000	0.8813	0.2665	0.8014	0.45
0.5100	0.8647	0.2161	0.7775	0.70
0.5200	0.8488	0.1686	0.7558	1.00
0.5300	0.8336	0.1238	0.7363	1.36
0.5400	0.8189	0.0814	0.7186	1.77
0.5500	0.8047	0.0413	0.7027	2.25
0.5600	0.7911	0.0033	0.6883	2.78

TABLE 2.5

$Q_L = 7$
QB$_3$ and SQB$_3$

Q_{ts}	H	α	f_3/f_s	Peak–dB
0.1000	3.8416	34.3925	5.2233	0
0.1100	3.4947	28.2341	4.7386	0
0.1200	3.2058	23.5499	4.3337	0
0.1300	2.9615	19.9046	3.9902	0
0.1400	2.7525	17.0150	3.6949	0
0.1500	2.5712	14.6784	3.4381	0
0.1600	2.4129	12.7685	3.2126	0
0.1700	2.2743	11.1855	3.0128	0
0.1800	2.1495	9.8589	2.8345	0
0.1900	2.0388	8.7361	2.6741	0
0.2000	1.9393	7.7775	2.5289	0
0.2100	1.8494	6.9524	2.3968	0
0.2200	1.7678	6.2372	2.2759	0
0.2300	1.6935	5.6132	2.1647	0
0.2400	1.6254	5.0655	2.0620	0
0.2500	1.5629	4.5822	1.9667	0
0.2600	1.5054	4.1535	1.8778	0
0.2700	1.4522	3.7714	1.7946	0
0.2800	1.4029	3.4295	1.7165	0
0.2900	1.3571	3.1223	1.6429	0
0.3000	1.3145	2.8421	1.5732	0
0.3100	1.2748	2.5944	1.5070	0
0.3200	1.2376	2.3667	1.4439	0
0.3300	1.2028	2.1594	1.3836	0
0.3400	1.1702	1.9699	1.3258	0
0.3500	1.1395	1.7964	1.2702	0
0.3600	1.1106	1.6371	1.2167	0
0.3700	1.0834	1.4905	1.1651	0
0.3800	1.0578	1.3552	1.1153	0
0.3900	1.0335	1.2300	1.0674	0
0.4000	1.0106	1.1141	1.0214	0
0.4100	0.9889	1.0065	0.9776	0
0.4200	0.9683	0.9064	0.9362	0.01
0.4300	0.9488	0.8131	0.8975	0.05
0.4400	0.9303	0.7260	0.8618	0.14
0.4500	0.9128	0.6445	0.8294	0.31
0.4600	0.8961	0.5682	0.8001	0.56
0.4700	0.8802	0.4966	0.7741	0.90
0.4800	0.8651	0.4294	0.7510	1.32
0.4900	0.8507	0.3661	0.7307	1.85
0.5000	0.8370	0.3065	0.7129	2.46
0.5100	0.8240	0.2503	0.6972	3.18
0.5200	0.8116	0.1971	0.6835	4.01
0.5300	0.7998	0.1468	0.6715	4.97
0.5400	0.7886	0.0992	0.6610	6.08
0.5500	0.7779	0.0540	0.6518	7.36
0.5600	0.7677	0.0111	0.6438	8.87

(1) f_s, the free air resonance
(2) Q_{ts}, total Q of the driver, including all series resistances
(3) V_{as}, volume of air equal to driver compliance
(4) X_{max}, amount of voice coil overhang in mm
(5) S_d, effective driver radiating area in square meters
(6) V_d, displacement volume $= S_d \times X_{max}$ in cubic meters

Parameters 4, 5, and 6 can be obtained from the driver manufacturer. Refer to *Chapter 8,*

Loudspeaker Testing, for the proper procedure in calculating 1, 2, and 3. Even if you decide to use the published data for f_s, Q_{ts}, and V_{as}, be certain to include the series resistances you expect in the Q_{ts} value (this can be done if Q_{ts} is given along with Q_{es} and Q_{ms}).

The best way to proceed is to generate a design table, as was done for closed-box systems. In this case, assemble all the data for the three flat alignments. Looking at the design *Tables 2.1–2.10,*[5] you will notice three sets of three tables plus one table for the discrete

TABLE 2.6

$Q_L = 15$
QB_3 and SQB_3

Q_{ts}	H	α	f_3/f_s	Peak-dB
0.1000	3.6841	35.4793	5.0715	0
0.1100	3.3494	29.1286	4.6004	0
0.1200	3.0732	24.2984	4.2069	0
0.1300	2.8398	20.5392	3.8730	0
0.1400	2.6400	17.5563	3.5859	0
0.1500	2.4670	15.1498	3.3362	0
0.1600	2.3158	13.1802	3.1169	0
0.1700	2.1826	11.5478	2.9225	0
0.1800	2.0644	10.1797	2.7488	0
0.1900	1.9589	9.0218	2.5926	0
0.2000	1.8640	8.0331	2.4512	0
0.2100	1.7784	7.1822	2.3225	0
0.2200	1.7007	6.4446	2.2045	0
0.2300	1.6299	5.8010	2.0960	0
0.2400	1.5652	5.2361	1.9956	0
0.2500	1.5058	4.7375	1.9023	0
0.2600	1.4512	4.2952	1.8153	0
0.2700	1.4007	3.9011	1.7338	0
0.2800	1.3540	3.5484	1.6571	0
0.2900	1.3106	3.2314	1.5846	0
0.3000	1.2703	2.9455	1.5159	0
0.3100	1.2327	2.6867	1.4504	0
0.3200	1.1976	2.4517	1.3880	0
0.3300	1.1648	2.2376	1.3281	0
0.3400	1.1341	2.0420	1.2705	0
0.3500	1.1052	1.8629	1.1251	0
0.3600	1.0781	1.6983	1.1615	0
0.3700	1.0526	1.5468	1.1099	0
0.3800	1.0286	1.4070	1.0602	0
0.3900	1.0059	1.2777	1.0125	0
0.4000	0.9845	1.1579	0.9672	0
0.4100	0.9643	1.0466	0.9245	0.02
0.4200	0.9452	0.9430	0.8849	0.08
0.4300	0.9272	0.8464	0.8488	0.21
0.4400	0.9101	0.7562	0.8162	0.43
0.4500	0.8939	0.6719	0.7872	0.76
0.4600	0.8786	0.5928	0.7618	1.18
0.4700	0.8641	0.5185	0.7395	1.72
0.4800	0.8503	0.4488	0.7202	2.36
0.4900	0.8373	0.3830	0.7034	3.13
0.5000	0.8249	0.3211	0.6889	4.04
0.5100	0.8132	0.2625	0.6764	5.09
0.5200	0.8021	0.2072	0.6656	6.33
0.5300	0.7916	0.1547	0.6563	7.79
0.5400	0.7817	0.1050	0.6483	9.56
0.5500	0.7723	0.0577	0.6416	11.80
0.5600	0.7635	0.0128	0.6359	14.70

TABLE 2.7

$Q_L = 3$
SC_4 and C_4

Q_{ts}	H	α	f_3/f_s	Ripple-dB
0.2500	1.0093	3.4080	2.6083	0
0.2600	1.0322	3.2301	2.4391	0
0.2700	1.0529	3.0516	2.2860	0
0.2800	1.0703	2.8731	2.1473	0
0.2900	1.0871	2.6952	2.0217	0
0.3000	1.1004	2.5188	1.9078	0
0.3100	1.1109	2.3447	1.8042	0
0.3200	1.1187	2.1738	1.7097	0
0.3300	1.1236	2.0069	1.6232	0
0.3400	1.1255	1.8448	1.5437	0
0.3500	1.1244	1.6885	1.4703	0
0.3600	1.1203	1.5387	1.4023	0
0.3700	1.1133	1.3961	1.3390	0
0.3800	1.1034	1.2616	1.2798	0
0.3900	1.0909	1.1356	1.2244	0
0.4000	1.0758	1.0187	1.1723	0
0.4100	1.0586	0.9110	1.1236	0
0.4200	1.0394	0.8128	1.0778	0
0.4300	1.0188	0.7238	1.0348	0
0.4400	0.9770	0.6439	0.9947	0
0.4500	0.9744	0.5726	0.9572	0
0.4600	0.9515	0.5093	0.9222	0
0.4700	0.9286	0.4533	0.8898	0
0.4800	0.9059	0.4040	0.8597	0
0.4900	0.8837	0.3605	0.8318	0
0.5000	0.8621	0.3223	0.8060	0.01
0.5100	0.8412	0.2885	0.7822	0.02
0.5200	0.8212	0.2586	0.7601	0.02
0.5300	0.8021	0.2321	0.7397	0.03
0.5400	0.7838	0.2084	0.7208	0.05
0.5500	0.7664	0.1872	0.7033	0.06
0.5600	0.7499	0.1681	0.6871	0.08
0.5700	0.7341	0.1508	0.6720	0.10
0.5800	0.7192	0.1350	0.6579	0.12
0.5900	0.7049	0.1205	0.6447	0.14
0.6000	0.6913	0.1072	0.6324	0.17
0.6100	0.6784	0.0984	0.6209	0.20
0.6200	0.6661	0.0832	0.6101	0.23
0.6300	0.6543	0.0723	0.5999	0.26
0.6400	0.6430	0.0630	0.5906	0.29
0.6500	0.6322	0.0524	0.5812	0.32
0.6600	0.6218	0.0431	0.5726	0.35
0.6700	0.6118	0.0343	0.5644	0.39
0.6800	0.6022	0.0258	0.5567	0.42
0.6900	0.5929	0.0175	0.5493	0.46
0.7000	0.5840	0.0096	0.5423	0.50

alignments. They are: *Tables 2.1–2.3*, SBB_4 and BB_4; *Tables 2.4–2.6*, QB_3 and SQB_3; *Tables 2.7–2.9*, SC_4 and C_4; *Table 2.10*, Discrete Alignments. The three tables for each alignment pair (one flat and one nonflat) correspond to the different box losses or Q_L.

2.61 BOX LOSSES.

Three types of losses can affect the ultimate box volume and tuning: leakage (Q_L), absorption (from damping material) (Q_A), and vent losses (Q_P). Total losses (Q_B) for any vented enclosure is the sum of these separate losses, and is given as:

$$\frac{1}{Q_B} = \frac{1}{Q_L} + \frac{1}{Q_A} + \frac{1}{Q_P}$$

In practice, Q_A and Q_P tend to be so low that they are not very significant. This is assuming the vents are unobstructed, and a minimal amount of damping material (1″) lines the enclosure walls. Since leakage losses dominate, they are the only ones considered in the differ-

TABLE 2.8

$Q_L = 7$

SC_4 and C_4

Q_{ts}	H	α	f_3/f_s	Ripple–dB
0.2500	1.0338	3.8961	2.3949	0
0.2600	1.0534	3.6755	2.2282	0
0.2700	1.0703	3.4551	2.0784	0
0.2800	1.0842	3.2360	1.9439	0
0.2900	1.0951	3.0193	1.8229	0
0.3000	1.1028	2.8062	1.7137	0
0.3100	1.1073	2.5977	1.6149	0
0.3200	1.1086	2.3952	1.5251	0
0.3300	1.1065	2.1997	1.4431	0
0.3400	1.1012	2.0125	1.3679	0
0.3500	1.0926	1.8347	1.2986	0
0.3600	1.0810	1.6672	1.2345	0
0.3700	1.0667	1.5109	1.1751	0
0.3800	1.0498	1.3665	1.1200	0
0.3900	1.0309	1.2343	1.0689	0
0.4000	1.0103	1.1146	1.0215	0
0.4100	0.9886	1.0070	0.9777	0
0.4200	0.9662	0.9113	0.9373	0
0.4300	0.9436	0.8266	0.9001	0
0.4400	0.9212	0.7521	0.8660	0
0.4500	0.8992	0.6868	0.8348	0.01
0.4600	0.8780	0.6297	0.8064	0.01
0.4700	0.8578	0.5798	0.7804	0.02
0.4800	0.8385	0.5361	0.7567	0.03
0.4900	0.8203	0.4978	0.7351	0.05
0.5000	0.8031	0.4642	0.7155	0.07
0.5100	0.7870	0.4345	0.6975	0.09
0.5200	0.7719	0.4083	0.6810	0.12
0.5300	0.7578	0.3849	0.6659	0.15
0.5400	0.7445	0.3640	0.6520	0.19
0.5500	0.7321	0.3453	0.6393	0.23
0.5600	0.7205	0.3284	0.6275	0.27
0.5700	0.7096	0.3131	0.6166	0.31
0.5800	0.6993	0.2992	0.6065	0.36
0.5900	0.6896	0.2865	0.5971	0.41
0.6000	0.6805	0.2749	0.5883	0.46
0.6100	0.6719	0.2641	0.5802	0.51
0.6200	0.6638	0.2542	0.5726	0.57
0.6300	0.6561	0.2449	0.5654	0.63
0.6400	0.6488	0.2363	0.5587	0.68
0.6500	0.6418	0.2283	0.5524	0.74
0.6600	0.6353	0.2208	0.5465	0.80
0.6700	0.6289	0.2136	0.5409	0.89
0.6800	0.6229	0.2069	0.5355	0.92
0.6900	0.6171	0.2006	0.5305	0.98
0.7000	0.6116	0.1946	0.5258	1.05

TABLE 2.9

$Q_L = 15$

SC_4 and C_4

Q_{ts}	H	α	f_3/f_s	Ripple–dB
0.2500	1.0420	4.0890	2.3097	0
0.2600	1.0601	3.8500	2.1477	0
0.2700	1.0751	3.6119	1.9970	0
0.2800	1.0871	3.3757	1.8647	0
0.2900	1.0958	3.1429	1.7460	0
0.3000	1.1011	2.9147	1.6391	0
0.3100	1.1031	2.6924	1.5426	0
0.3200	1.1016	2.4774	1.4549	0
0.3300	1.0966	2.2711	1.3749	0
0.3400	1.0884	2.0748	1.3016	0
0.3500	1.0769	1.8896	1.2342	0
0.3600	1.0626	1.7166	1.1720	0
0.3700	1.0456	1.5567	1.1146	0
0.3800	1.0265	1.4101	1.0615	0
0.3900	1.0058	1.2779	1.0125	0
0.4000	0.9840	1.1591	0.9675	0
0.4100	0.9615	1.0535	0.9262	0
0.4200	0.9390	0.9604	0.8884	0
0.4300	0.9167	0.8787	0.8539	0
0.4400	0.8951	0.8074	0.8226	0.01
0.4500	0.8744	0.7453	0.7942	0.02
0.4600	0.8547	0.6911	0.7684	0.03
0.4700	0.8361	0.6439	0.7451	0.05
0.4800	0.8187	0.6027	0.7239	0.07
0.4900	0.8025	0.5666	0.7047	0.09
0.5000	0.7873	0.5348	0.6873	0.12
0.5100	0.7732	0.5068	0.6714	0.16
0.5200	0.7601	0.4820	0.6569	0.20
0.5300	0.7479	0.4599	0.6437	0.24
0.5400	0.7366	0.4402	0.6315	0.29
0.5500	0.7260	0.4225	0.6204	0.34
0.5600	0.7162	0.4065	0.6101	0.39
0.5700	0.7070	0.3921	0.6006	0.45
0.5800	0.6984	0.3789	0.5919	0.51
0.5900	0.6903	0.3670	0.5838	0.57
0.6000	0.6828	0.3560	0.5762	0.63
0.6100	0.6757	0.3459	0.5692	0.70
0.6200	0.6690	0.3366	0.5626	0.77
0.6300	0.6627	0.3281	0.5565	0.83
0.6400	0.6567	0.3201	0.5508	0.90
0.6500	0.6511	0.3127	0.5454	0.97
0.6600	0.6458	0.3058	0.5403	1.00
0.6700	0.6408	0.2994	0.5355	1.12
0.6800	0.6360	0.2933	0.5311	1.19
0.6900	0.6314	0.2876	0.5268	1.26
0.7000	0.6271	0.2823	0.5228	1.33

ent design tables. The effect these different loss levels have on the frequency response of a given driver and alignment are illustrated in *Fig. 2.11.* Unfortunately, these losses are altogether unpredictable and must be measured with a working enclosure.

To correct for losses, begin by assuming the "typical" loss figure of $Q_L = 7$, construct the appropriate-sized enclosure, tune it to the indicated tuning frequency, and then measure the new enclosure for exact losses. If the measured loss is close to the target $Q_L = 7$, no additional

changes are necessary. If, however, the measured loss figure is less or more than $Q_L = 7$, you must recalculate and alter the enclosure size and tuning.

The diagram in *Fig. 2.12* depicts the box-size relationship for the range of possible Q_Ls. It is not unusual for the measured value of Q_L to be lower than 7. As a result, you will have to increase the box size. If you are accustomed to woodworking, quickly knocking out a particleboard box will help you determine losses. If this seems like too much of a burden, you can always start by over-voluming the

TABLE 2.10

Discrete Alignments

Q_L	Q_{ts}	H	α	f_3/f_s
		Butterworth		
3	0.4386	1.0000	0.6543	1.0000
7	0.4048	1.0000	1.0613	1.0000
15	0.3937	1.0000	1.2444	1.0000
		Bessel		
3	0.3535	0.9696	1.4036	1.4911
7	0.3312	0.9735	1.9076	1.4941
15	0.3230	0.9749	2.1296	1.4951
		Butterworth Inter-Order		
3	0.3835	1.1397	1.1722	1.2432
7	0.3572	1.1184	1.6802	1.2315
15	0.3477	1.1117	1.9030	1.2278

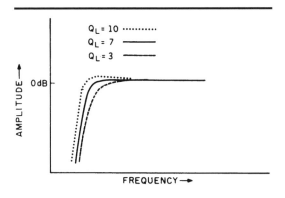

FIGURE 2.11: Effects of enclosure loss on box response.

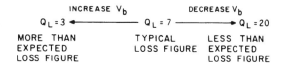

FIGURE 2.12: QL and relative box size.

FIGURE 2.13

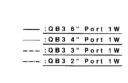

box by about 25% and adjusting the volume with a removable solid filter.

If the Q_L turns out to be lower, you can remove the required amount of filler. If Q_L is higher, add more solid filler. Be sure to also over-volume the box for anything that will detract from the required volume (*Section 1.70*). The procedure for actual measurement of Q_L will be given in *Section 2.100*.

2.62 USING DESIGN TABLES 2.1–2.10.
Once you decide on an alignment, you can use the values of H (the tuning ratio), α (the system compliance or box volume ratio), and f_3/f_s (the f_3 ratio) to find the box tuning frequency (f_B), the box volume (V_B), and (f_3), respectively.

$$V_B = \frac{V_{as}}{\alpha} \qquad\qquad f_B = H \times f_s$$

$$f_3 = (f_3/f_s) \times f_s$$

2.70 CALCULATING VENT DIMENSIONS.
PVC pipe used in house plumbing is virtually the best, easiest to fabricate, and most readily available material for constructing speaker vents. It comes in a number of useful diameters (½, ¾, 1, 1.5, 2, 3, and 4″) and can be easily cut for tuning. While you can construct rectangular vents out of wood, changing vent length for tuning is time-consuming. For that reason, we will discuss only tube-type vents.

For a tubular vent flush-mounted on a speaker baffle, calculate the length by:

$$L_v = \frac{1.463 \times 10^7 R^2}{f_B{}^2 V_B} - 1.463R$$

where: L_v = length in inches
  f_B = tuning frequency in Hz
  V_B = box volume in cubic inches
  R = radius of the vent in inches

Because virtually all the acoustic power is radiated by the vent at f_B, a minimum volume displacement is required in order to prevent power compression. You can find an approximate figure for the minimum possible diameter from:[9]

$$1. \qquad d_v \geq 39.37 \left(\frac{411.25 V_d}{\sqrt{f_B}}\right)^{1/2}$$

where:

$d_v \geq$ minimum diameter of vent in inches
f_B = tuning frequency in Hz
V_d = cone displacement volume in cubic meters

As a general guideline[1] Small offers a somewhat more conservative formula (for the same units as above):

$$2.$$
$$d_v \geq 39.37 (f_B V_d)^{1/2}$$

For a 10″ woofer in a box tuned to 33.5Hz, the minimum vent diameter would be 3.57″ in the first case, and 2.45″ in the second. Since these figures are approximations, the formulas suggest a 3–4″ port would be adequate. However,

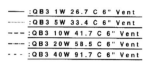

FIGURE 2.14

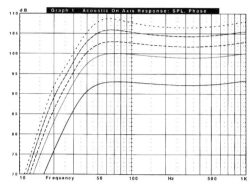

```
—— :QB3 1W 26.7 C 6" Vent
······ :QB3 5W 33.4 C 6" Vent
––– :QB3 10W 41.7 C 6" Vent
–·– :QB3 20W 58.5 C 6" Vent
– – – :QB3 40W 91.7 C 6" Vent
```

FIGURE 2.15

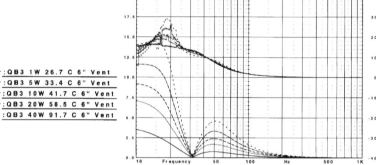

```
—— :QB3 1W 26.7 C 6" Vent
······ :QB3 5W 33.4 C 6" Vent
––– :QB3 10W 41.7 C 6" Vent
–·– :QB3 20W 58.5 C 6" Vent
– – – :QB3 40W 91.7 C 6" Vent
```

FIGURE 2.16

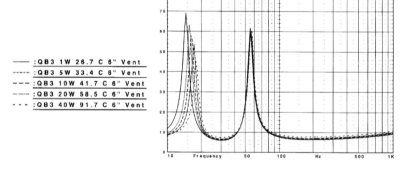

```
—— :QB3 1W 26.7 C 6" Vent
······ :QB3 5W 33.4 C 6" Vent
––– :QB3 10W 41.7 C 6" Vent
–·– :QB3 20W 58.5 C 6" Vent
– – – :QB3 40W 91.7 C 6" Vent
```

FIGURE 2.17

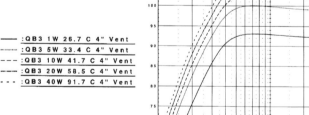

```
—— :QB3 1W 26.7 C 4" Vent
······ :QB3 5W 33.4 C 4" Vent
––– :QB3 10W 41.7 C 4" Vent
–·– :QB3 20W 58.5 C 4" Vent
– – – :QB3 40W 91.7 C 4" Vent
```

computer simulation and empirical testing have shown that almost any practical port diameter is nonlinear.

The impedance curves in *Fig. 2.13* display a computer simulation comparison of four tubular vent diameters measured at a 1W level with the same enclosure, driver, and tuning frequency (the

same 12″ driver and QB$_3$ enclosure used in *Section 2.50*). Diameter and port lengths are as follows:

Dia.	Length
6″	30.0″
4″	12.3″
3″	6.2″
2″	2.3″

Obviously, the 6″ diameter port presents a port length which is not practical but will act as a reference to understand the dynamics of port nonlinearity. The various impedance curves in *Fig. 2.13* make it apparent that even at 1W the 4″, 3″, and 2″ port lengths are operating in a nonlinear fashion, which is to say that they are unable to move sufficient quantities of air at the velocity required. To get an even better perspective on the implications of port function nonlinearity, a power series set of simulations was done for 6″, 4″, and 2″ diameter ports. The series included running simulations for five power levels, 1, 5, 10, 20, and 40W with appropriately increasing voice coil temperatures. *Figures 2.14–2.16* are for the 6″ diameter tubular vent, *Figs. 2.17–2.19* are for the 4″ diameter vent, and *Figs. 2.20–2.22* are for the 2″ diameter vent. The 6″ diameter port SPL graphs of *Fig. 2.14* show not only the changes expected with increasing power input but indications of port nonlinearity. f_3 goes from 40.6Hz at 1W to 42.3 at 40W, which would be expected as power increases. The slope changes moderately from 18.4dB/octave at 1W to 17.5dB/octave at 40W. However, at higher power levels a discontinuity takes place at about 20Hz. This change is due to port nonlinearity at higher power levels raising the port resonance frequency and causing the response to alter at 20Hz to a higher rolloff rate. The group delay curves in *Fig. 2.15* get progressively worse as power increases, with the group delay at 40W being the worst case. Excursion is less than 5mm at the highest power level, which is still 0.5mm less than the 6mm X_{max} of this 12″ woofer (as mentioned, excursion for vented speakers is very good above f_b). The impedance shown in *Fig 2.16* shows the same pattern of nonlinearity as was seen in the 1W impedance of different diameter port tubes in *Fig. 2.13*. The height of the lower peak, f_L, is 24Ω lower (69Ω at 1W and 45Ω at 40W) in magnitude and 4Hz higher in frequency at the 40W level than at 1W.

The 4″ diameter port SPL curves, *Fig. 2.17*, show about the same SPL change in the response as the 6″ vent example. The change in f_3 from 38Hz at 1W to 42Hz at 40W is mostly attributed to dynamic changes. The slope undergoes a slight change from 17.7dB/octave to 17.4dB/octave, again less than expected due to port nonlinearity. The slope at higher power levels develops a similar discontinuity at 20Hz as the 6″ vent. The group delay, *Fig. 2.18*, shows the development of a sharper knee at the 40W level as transient performance degrades because of dynamic changes. The impedance curves in *Fig.*

2.19 show the peak at f_L to be gradually disappearing with the magnitude decreasing 33Ω from 1W to 40W (58Ω-25Ω), shifting upward in frequency by 3Hz and showing more compression overall than the 6″ port.

The 2″ diameter port SPL curves given in *Fig. 2.20* illustrate the severe nonlinearity of this port size. The tail end of the curve is starting to alter toward a second-order response below 20Hz. The f_3 changes from 40Hz at 1W to 45Hz at 40W. Slope rate changes from 19dB/octave at 1W to 14.4dB/octave at 40W, reflecting the change in response at 20Hz. The group delay picture, *Fig. 2.21*, indicates more radical changes in the transient performance as power increases, obviously due mostly to the port nonlinearity. Note that the minimum excursion magnitude is no longer at zero at all power levels, but increases in minimum level for increasing power. The impedance curves, *Fig. 2.22*, show the port practically non-existent at levels above 5W. The differences between a fairly linear port, like the 6″ diameter example, and a very nonlinear port such as the 2″ example are compared at a 5W input level (100dB SPL) in *Figs. 2.23–2.25* and require no additional comment.

The conclusion is fairly obvious from the data presented. Although not widely understood, almost no realizable tubular vent operates in a completely linear fashion, and practically all realizable vents compromise performance at higher power levels. Minimum vent size goes beyond either of the formulas given. While these calculations may suggest an absolute minimum vent diameter, the recommended port size will still yield substantially nonlinear performance. Larger cross-sectional areas will always produce better linearity in any given situation. For high powered applications, such as the speakers designed for stage performance, it is desirable to use vent areas as nearly equal to the driver area as possible. The downside of using larger vent diameters is the pipe organ resonance caused by the increased length. While the pipe resonances produced by longer vents can cause minor response anomalies in the range of 1–2dB, the response anomalies are probably less worrisome than the severe nonlinearity of undersized ports.

Vent nonlinearity has always been a problem, but one which most designers were either blissfully unaware of, or not very concerned about. For home systems at moderate volume levels, the effects are not particularly obtrusive, especially if you stay on the high side of the minimum vent diameter recommendations. As long as the ratio of vent area to driver area is at least a minimum of 9/1, the nonlinear effects are not severe. Ratios of 4/1 and up generally yield reasonably good port linearity. In terms of standard plastic tube diameters, 1″ diameter ports are good for only 4″ diameter drivers; 2″ diameter ports are good for 4″ and 5″ speakers, and usable for 6″ speakers; 3″ diameter ports are good for 6″ speakers, and minimally usable for 8″ speakers; 4″ diameter ports are good for 8″ and 10″ speakers, and minimally usable for 12″ and 15″; 6″ ports are good for 12″ and 15″ types (and any smaller diameter driver with a box large enough to accommodate the length).

While tubular vents are enormously conve-

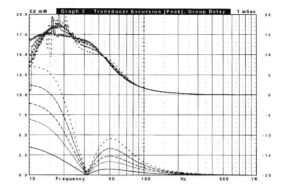

FIGURE 2.18

——— : QB3 1W 26.7 C 4″ Vent
·········· : QB3 5W 33.4 C 4″ Vent
— — — : QB3 10W 41.7 C 4″ Vent
—·—· : QB3 20W 58.5 C 4″ Vent
- - - - : QB3 40W 91.7 C 4″ Vent

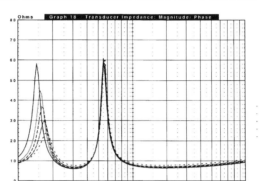

FIGURE 2.19

——— : QB3 1W 26.7 C 4″ Vent
·········· : QB3 5W 33.4 C 4″ Vent
— — — : QB3 10W 41.7 C 4″ Vent
—·—· : QB3 20W 58.5 C 4″ Vent
- - - - : QB3 40W 91.7 C 4″ Vent

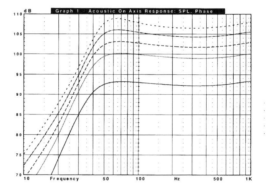

FIGURE 2.20

——— : QB3 1W 26.7 C 2″ Vent
·········· : QB3 5W 33.4 C 2″ Vent
— — — : QB3 10W 41.7 C 2″ Vent
—·—· : QB3 20W 58.5 C 2″ Vent
- - - - : QB3 40W 91.7 C 2″ Vent

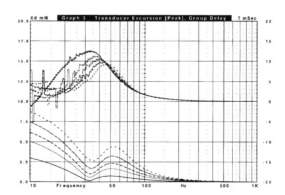

FIGURE 2.21

——— : QB3 1W 26.7 C 2″ Vent
·········· : QB3 5W 33.4 C 2″ Vent
— — — : QB3 10W 41.7 C 2″ Vent
—·—· : QB3 20W 58.5 C 2″ Vent
- - - - : QB3 40W 91.7 C 2″ Vent

FIGURE 2.22

----- : QB3 1W 26.7 C 2" Vent
.... : QB3 5W 33.4 C 2" Vent
- - - : QB3 10W 41.7 C 2" Vent
-- -- : QB3 20W 58.5 C 2" Vent
- - - : QB3 40W 91.7 C 2" Vent

FIGURE 2.23

⎯ : QB3 5W 6" Vent
.... : QB3 5W 2" Vent

FIGURE 2.24

⎯ : QB3 5W 6" Vent
.... : QB3 5W 2" Vent

FIGURE 2.25

⎯ : QB3 5W 6" Vent
.... : QB3 5W 2" Vent

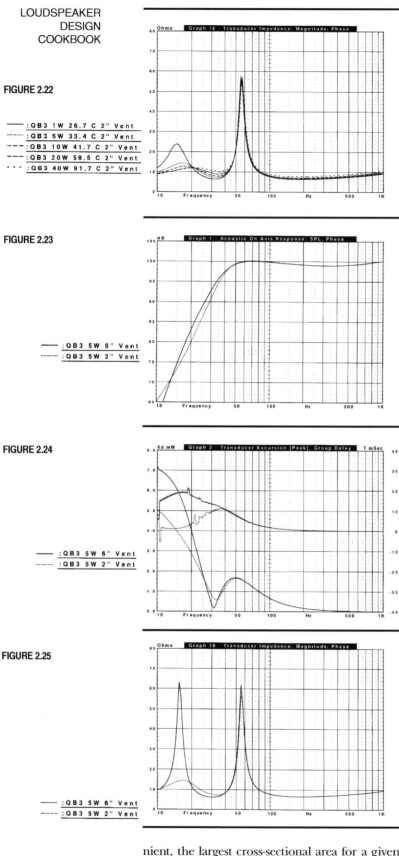

nient, the largest cross-sectional area for a given cabinet depth (without using an elbow in the vent) can be had from a shelf port, as described by Thiele in his first paper (reprinted in the AES *Loudspeaker Anthology*, Volume 1). Shelf ports are not particularly easy to tune, can require successive prototyping, and for that reason are not discussed here.

For large diameter woofers, where the values

of $R \geq 2''$, multiple vents are practical. The combination of two vent tubes, d_1 and d_2, is given by:

$$d_t = (d_1^2 + d_2^2)^{1/2}$$

Using two 4″ diameter vents together would yield a combined diameter of 5.7″.

If the vent length is such that the distance from the inside rear enclosure wall (assuming a front baffle-mounted vent tube) to the end of the vent tube is less than 3″, use a 90 PVC elbow joint (*Fig. 2.26*). Small cautions against such long vents, suggesting they create excessive noises. Others, including Weems and Bullock, do not consider this a problem.

In addition to making the port as large a diameter as possible to increase linearity (which includes techniques such as multiple ports or using an elbow to fit a long larger diameter port into an enclosure), another methodology has surfaced in the last five years that is now a fairly common practice—flared ports. Unfortunately, this technology is a little hard to implement without fabricating the vent out of injection molded plastic, but there is no question of the increased linearity these devices achieve.

Flared ports are not a really new idea. Patents began surfacing as early as 1980[10] that specified a rounded over surface on the entrance and exit of the port as a method decreasing port noise and increasing performance. Since then numerous patents from companies like Bose[11], Polk Audio[12], and Philips[13] have described various flared port configurations to the point that the majority of manufactured vented speakers are now being fitted with flared ports of one kind or another. Not only does the inclusion of a flared vent improve performance; it also gives the speaker a more interesting cosmetic appearance.

Research into port non-linearity and port noise due to turbulence have pretty much verified the information in the previous paragraphs of this section, that all ports are non-linear given sufficient SPL,[14,15,16,17] and further concluded that adding almost any round-over on the entrance and exit of a port will help the situation[14].

I published a short study on flared ports in 1996[18] that described the increased linearity to be had from a flared port. In this case, the port was supplied by Lightning Audio who sells a kind of modular flared port system. What they did was come up with flared inlet (inside the box) and exhaust flared ends that had a mounting flange that would fit onto a straight pipe. Once you fit the two flared ends onto the tube (that can be cut to length), the result is a vent like the one pictured in *Fig. 2.27*.

The test procedure included measuring impedance and SPL at multiple voltage levels for three different types of vents, a straight vent, a straight vent with a flared exhaust (no inlet flare), and a straight vent with a flared exhaust and a flared inlet. Using a LinearX VIBox (*see Chapter 8, Section 8.31*) I measured impedance at

1V, 2.83V, and 10V for each vent type (straight, single flare, and double flare) and also performed groundplane SPL measurements for each vent type at 2.83V and 10V. The results certainly confirmed what other research has shown, that flared ports which decrease the turbulence that occurs at the ends of a port are significantly more linear at higher voltage levels. *Figure 2.28* shows the impedance comparison of all three vent types at 10V. The type of non-linear impedance behavior simulated earlier in this section shows up in these pictures in the same manner as a type of compression. Note that the lower impedance peak is decreasing in level and the trough between the two peaks increases in magnitude as compared to the same measurement taken at 1V (see *Fig. 2.29*). However, although all three vent types tested showed this behavior, the double flared vent was the best of the three and shows less disturbance to the system. This is further illustrated for each type vent's impedance examined at 1V, 2.83V, and 10V in *Figs. 2.30–2.32*. Note the increase in the port tuning frequency as voltage increases. This is likely due to the effective shortening of the vent length as turbulence increases at the vent ends.

This change in linearity was also examined by performing groundplane type SPL measurements at both 1V (see *Fig. 2.33*) and at 10V (see *Fig. 2.34*). At frequencies below the roll-off of the woofer being used (a Vifa M26WR 10″ woofer), output with the double flared vent increased 1.5–2dB, due to the increased linearity of a double flared port. Again this is the result of decreased end turbulence.

The most definitive study to date was done by three Harman International engineers from JBL Professsional and Infinity,[19] and drew a number of interesting conclusions regarding flared ports as follows:

1. Flared ports exhibit less power compression than do straight ports.
2. Port tuning of flared ports is only slightly dependent on flare and primarily dependent on the minimum diameter and the total length of the port.
3. Resonance increase as ports compress is likely attributed to the effective length of the port decreasing as turbulence increases.
4. Extreme flared ports show more compression and distortion at maximum SPL levels than more moderate flared ports, suggesting that a compromise between the two is desirable.
5. Roughness of the interior vent surface appears to not be relevant in port functioning.
6. There is no one port profile that is optimum and the effect is had by a near infinite variety of port flares.
7. Asymmetrical inlet and exhaust ports with a shallow radius on the inlet end and a larger radius on the exhaust end (see *Fig. 2.35*) provide the best balance of lowering distortion and compression.

These results pretty much confirmed those of the study done in Philips Research Labs in The

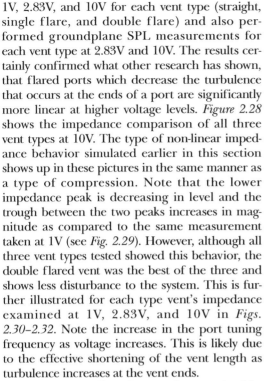

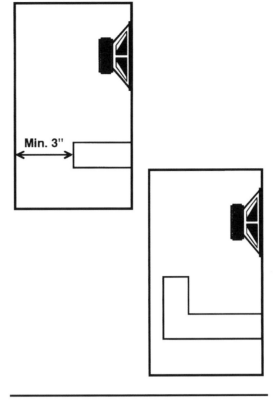

FIGURE 2.26: Vent configuration.

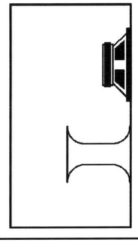

FIGURE 2.27: Example of dual flared port.

Netherlands[17] that some type of middle ground radius produced better results. The port profile that produced the lowest port noise (lowest distortion) in this study was a port that had a shallow 6° flare starting at the center of the port outward, and moderate small radius flare at both ends, as shown in *Fig. 2.36*. This port was reported to produce 5.5dB less blowing noise at a 95dB SPL level, which is certainly significant. Although some practitioners regard this noise to be effectively masked by program material, there is no doubt that controlling its effect and increasing the port linearity provide benefit.

2.71 VENT RESONANCE and MUTUAL COUPLING.
Vents can produce a variety of unwanted response variations that are the result of the acoustic coupling between the vent and the dri-

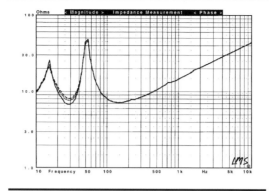

FIGURE 2.28: 10V impedance comparison (solid = dual-flare vent, dot = single-flare vent, dash = straight vent).

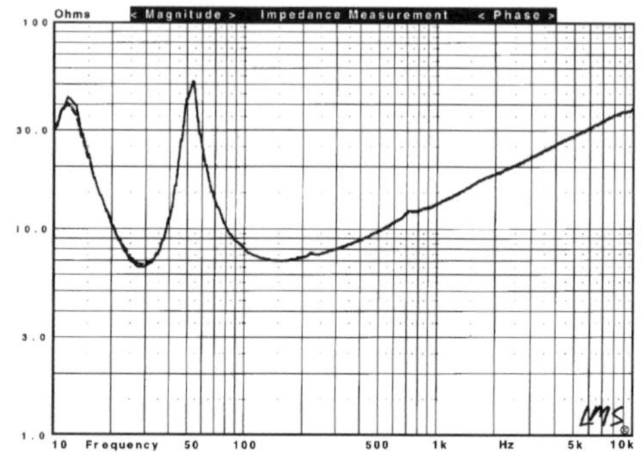

FIGURE 2.29: 1V impedance comparison (solid = dual-flare vent, dot = single-flare vent, dash = straight vent).

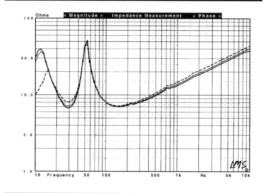

FIGURE 2.30: Impedance of straight vent at multiple input levels (solid = 1V, dot = 2.83V, dash = 10V).

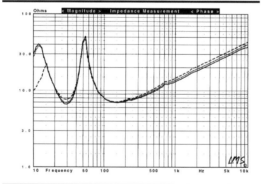

FIGURE 2.31: Impedance of single-flare vent at multiple input levels (solid = 1V, dot = 2.83V, dash = 10V).

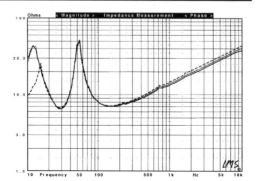

FIGURE 2.32: Impedance of dual-flare vent at multiple input levels (solid = 1V, dot = 2.83V, dash = 10V).

ver and between multiple vents. Besides its contribution to the total low-end sound output of the speaker, the vent also contributes wind noise, resonance disturbances, and colorations caused by the transmission of standing wave modes from within the enclosure.

The "pipe organ" resonances in speaker vents are functions of vent diameter and length. Generally, tubular vents which are not much longer than their diameter do not suffer from the resonance problems of longer vents. The computer simulation in *Fig. 2.37* illustrates the vent resonance problems of the vents used in the simulation in *Section 2.70*, with the addition of a multiple vent example. The curves have been deliberately displaced by even decibel increments to make them easier to read. The port diameters, lengths, and the center-to-center distances from the driver to the port are shown below:

Diameter	Length	Spk.-Port	Port-Port
6″	30.25″	9.75″	——
4″	12.3″	8.5″	——
3″	6.25″	8.0″	——
2″ × 3″	15.0″	9.75″	6.5″

All of the vents in the simulation are located immediately adjacent to the driver for a beginning reference. The magnitudes in these particular simulations are exaggerated by a factor of at least 10, and are presented only to give you an idea of the relative location and level of pipe resonances in various sized vents. As mentioned in *Section 2.70*, pipe resonances usually only produce minor changes in driver response, and are typically difficult to separate from other driver response anomalies. Pipe resonances are extremely unpredictable and are generally dependent on their location on the baffle, vent proximity to enclosure walls, and the location of damping material in the immediate vicinity of the vent. The rear of the cabinet is an alternative vent location. This will cause some marginal change in low-end output, depending upon where the speaker is located in the room, but will tend to subdue the subjective importance of vent-related noise problems.

The mutual coupling of the vent and driver can change depending on the location and diameter of the vent. No consistent rules for a guideline exist, except that computer simulations can give you at least a clue as to what to expect. The examples in *Figs. 2.38* (6″ vent), *2.39* (4″ vent), *2.40* (3″ vent), and *2.41* (two 3″ vents) show the changes in the influence of port resonances depending on the distance the vent is located from the driver (again, the magnitude of the pipe resonances is exaggerated by at least 10 times). The three curves in *Figs. 2.38, 2.39* and *2.40* represent the vent located immediately adjacent to the speaker, and 13″ and 16″ center-to-center distance from the 12″ cone. The two 3″ vents were located the same distance from the driver in both examples, but

FIGURE 2.33: SPL comparison at 2.83V (solid = dual flare vent, dot = single flare vent, dash = straight vent).

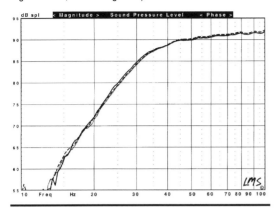

FIGURE 2.34: SPL comparison at 10V (solid = dual-flare vent, dot = single-flare vent, dash = straight vent).

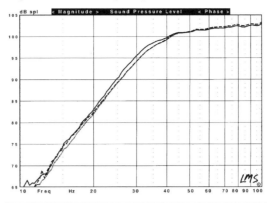

adjacent to each other in curve 1 and spaced some distance apart in curve 2. In the case of the 6″ and 4″ diameter vents, very close proximity to the driver seems to cause the least amount of disturbance. The 3″ vent showed fewer problems when located at a distance from the driver. The dual 3″ ports, *Fig. 2.41*, produce fewer problems when the two ports are separated by a reasonable distance.

2.72 BOX TUNING.

Once you select a vent size, you can accomplish final tuning. Generate an impedance curve by using a signal generator and a voltmeter (*Chapter 8*), and measure the frequency of f_B, which will occur at the trough between the twin impedance peaks. If the measured value of f_B is lower or higher than the target value, adjust the vent length until you obtain the correct value (*Fig. 2.42*).

Since f_B is subject to variation, depending on both box losses and the voice coil inductance, you may wish to use a more accurate tuning technique. Use a sound level meter placed close to the woofer diaphragm and, with the aid of a signal generator, adjust for minimum output. This will be the true value of f_B. With either method, your measurements should be made with the enclosure filled and crossover connected. If you have difficulty locating the center frequency, f_B, find the two areas which show increasing response and divide them by 2. Damping material can be removed and the crossover disconnected to make the frequency easier to read, but if a large quantity of fill material above 30% of the total volume is being used, the location of f_B will be affected and the material should be included in the measurement.

2.80 ADDITIONAL PARAMETERS.

As with sealed box enclosures, three additional parameters are useful in the evaluation of driver performance: reference efficiency, displacement limited acoustic power output, and the required electrical input.

2.81 REFERENCE EFFICIENCY (η_0).

For vented enclosures, η_0 (Eta) can be considered the free air reference efficiency:

$$\eta_0 = \frac{K(f_s{}^3 V_{as})}{Q_{es}}$$

Where K = 9.64×10^{-10} V_{as} in liters
9.64×10^{-7} V_{as} in cubic m.
2.70×10^{-8} V_{as} in cubic ft.
1.56×10^{-11} V_{as} in cubic in.

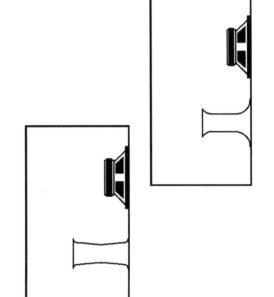

FIGURE 2.35: Example of an asymmetrically flared port.

FIGURE 2.36: Example of a compromise flared port design for lower noise and distortion.

FIGURE 2.37

```
——— : 6" PORT
········· : 4" PORT
— — — : 3" PORT
—·—·— : 2" PORT
- - - - : 2 3" PORTS
```

To convert η_0 to:

Percentage $\qquad \% = \eta_0 \times 100$

SPL, 1W/1m dB $= 112 + 10\log_{10} \eta_0$

2.82 DISPLACEMENT LIMITED ACOUSTIC OUTPUT (P_{ar}).

P_{ar}, for program material primarily within the system frequency response range, can be calculated by:

$$P_{ar} = 3.0\, f_3{}^4 V_d{}^2$$

where:

P_{ar} is in watts
V_d is the cone displacement volume in cubic meters

Expressed as an SPL:

$$dB = 112 + 10\log_{10} P_{ar}$$

2.83 ELECTRICAL INPUT MAXIMUM (P_{er}).

P_{er} is related to P_{ar} by the reference efficiency of the system, as is given by:

$$P_{er} = \frac{P_{ar}}{\eta_0}$$

where P_{ar} is in watts and η_0 is a decimal equivalent. You can then compare P_{er} to the manufacturer's thermal rating (in watts) to see whether the output required by your proposed box design will exceed the power capacity of the driver. As a rule, these figures do not absolutely determine the acceptability of an alignment. Also take into account the type of program material you're using, acceptable levels of distortion, and your average anticipated listening levels. In other words, if you listen to acoustic jazz played at around 85–90dB, it might not be very important if you exceed the thermal rating. If you like heavy metal rock music at 115dB+, then a more conservative posture makes sense.

FIGURE 2.38

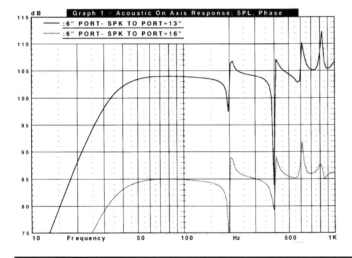

FIGURE 2.39

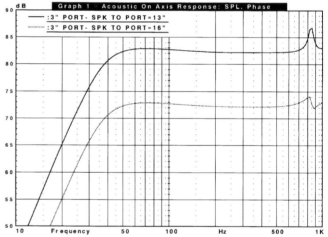

FIGURE 2.40

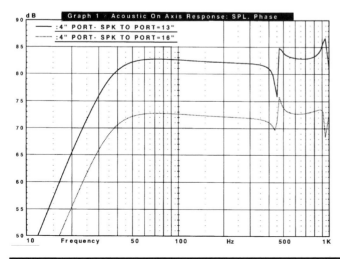

FIGURE 2.41

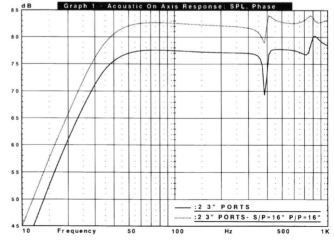

2.90 EXAMPLE DESIGN CHARTS.

Tables 2.11 and *2.12* give examples of two drivers suitable for vented-box construction. When calculating your total box size, make sure you over-volume the enclosure to compensate for anything which will detract from its target volume (*Section 1.70*). It is also important that you account for all series resistances when you calculate final Q_{ts} (*Chapter 8*).

For further examples of vented-box calculation and construction, you'll find the following in *Speaker Builder* magazine to be quite helpful:

1. P. Stamler, "How to Improve That Small Cheap Speaker," 1/80.

2. R. Saffran, "Build a Mini-Pipe Speaker," 3/81.

3. M. Lampton, "A Three-Way Corner Loudspeaker System," 4/82.

4. R. Parker, "A Thiele/Small Aligned Satellite/Subwoofer System," 1/84.

5. H. Hirsch, "Tenth Row, Center," 2/84.

6. D. Baldwin, "A Beginner Builds His First Speaker," 3/84.

7. S. Ellis, "The Curvilinear Vertical Array," 2/85.

8. W. Marshall Leach, Jr., "The Audio Laboratory Loudspeaker System," 2/89.

9. Bill Schwefel, "The Beer Budget Window Rattler," 3/90.

10. Thomas Nousaine, "Four Eight by Twos," 6/90.

11. M. Rumreich, "Box Design and Woofer Selection: A New Approach," 1/92, p. 9.

12. G. R. Koonce and R. O. Wright, Jr., "Alignment Jamming," 4/92, p. 14.

13. P. E. Rahnefeld, "Non-Optimum Vented-Box Spreadsheet Documentation," 5/92, p. 16.

14. R. Gonzalez, "Quasi-Monotonic Vented

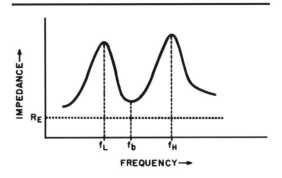

FIGURE 2.42: Impedance plot used for vent tuning.

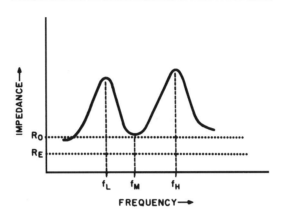

FIGURE 2.43: Impedance plot used for finding box loss.

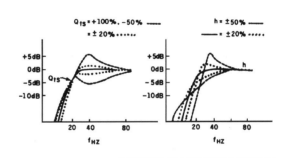

FIGURE 2.44: Variations in response for different Q_{ts} and tuning (h).

TABLE 2.11
8″ WOOFER EXAMPLE

$Q_L = 7$

Q_{ts}	Q_{es}	Q_{ms}	f_s	X_{max}	S_d (m²)		V_d (m³)		V_{as}
0.26	0.28	4.0	32	3.5mm	2.15×10^{-2}		7.53×10^{-5}		2.76 ft³

AL	V_B ft³	f_3 Hz	f_B Hz	L_v in	d_v in	$P_{ar(p)}$ watts	SPL dB	η_0 %	SPL dB	P_{er} watts
QB₃	0.67	61	49	2	1.5	0.235	106	0.908	92	26
SC₄	0.75	72	34	4	1.5	0.457	109	0.908	92	50
SBB₄	0.81	74	32	8.7	2	0.509	109	0.908	92	56

TABLE 2.12
10″ WOOFER EXAMPLE

$Q_L = 7$

Q_{ts}	Q_{es}	Q_{ms}	f_s	X_{max}	S_d (m²)		V_d (m³)		V_{as}
0.25	0.29	2.1	22	3.5mm	3.3×10^{-2}		1.16×10^{-4}		7.51 ft³

AL	V_B ft³	f_3 Hz	f_B Hz	L_v in	d_v in	$P_{ar(p)}$ watts	SPL dB	η_0 %	SPL dB	P_{er} watts
QB₃	1.7	42	34	3	2	0.126	103	0.724	91	17.4
SC₄	2.0	51	23	12	2.5	0.267	106	0.724	91	36.9
SBB₄	2.1	52	22	12	2.5	0.288	106	0.724	91	39.8

Alignments," 1/93, p. 24.

15. M. Redhill, "Stalking f_3," 2/93, p. 24.

2.100 Q_L MEASUREMENT.

As discussed previously, once you have determined box size, and have built it using $Q_L = 7$ as a target loss figure, you must recheck and readjust the system for actual losses. Start by measuring the impedance of the new system and record values f_L, f_M, f_H, and R_0 (*Fig. 2.43*). R_0 equals the calibrated impedance at f_M.

If you have no box filling material, and your crossover is by-passed, f_B will generally be the same frequency as the measured minimum

FIGURE 2.45

FIGURE 2.46

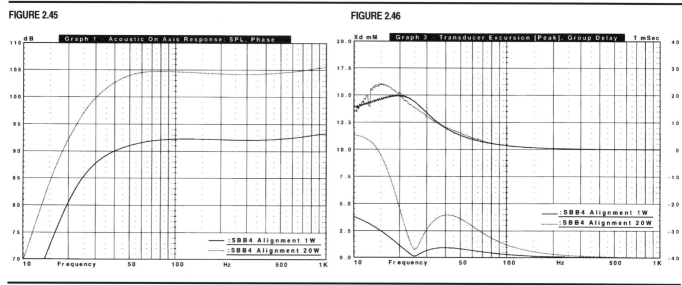

FIGURE 2.47

FIGURE 2.48

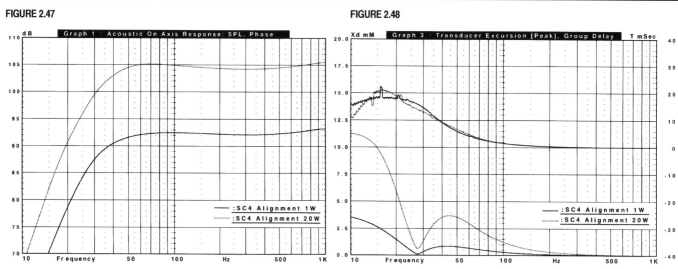

FIGURE 2.49

FIGURE 2.50

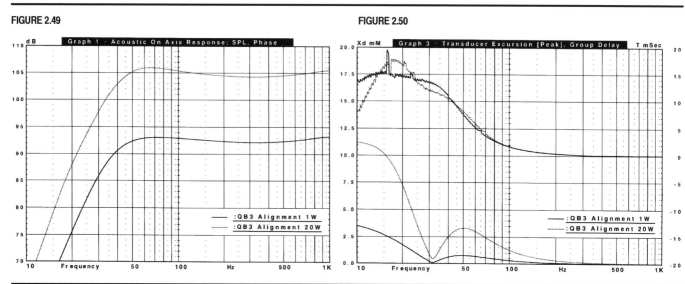

impedance, f_M ($f_M = f_B$). Phase shift imposed by large inductance voice coils can, however, cause the two frequencies to be somewhat different. To find a better approximation of f_B, simply cover the vent and measure the closed-box resonance (f_c) and then:

$$f_B = (f_L{}^2 + f_H{}^2 - f_c{}^2)^{1/2}$$

The following equation sequence will give you the value of measured Q_L:

$$f_{sb} = \frac{f_L f_H}{f_B}$$

$$r_m = \frac{R_0}{R_E}$$

where R_E is the driver DC resistance.

$$Q_{msb} = \left(\frac{f_s}{f_{sb}}\right) Q_{ms}$$

$$Q_{esb} = \left(\frac{f_s}{f_{sb}}\right) Q_{es}$$

$$Q_{tsb} = \left(\frac{f_s}{f_{sb}}\right) Q_{ts}$$

$$h_a = \frac{f_B}{f_{sb}}$$

$$\alpha' = \frac{(f_H{}^2 - f_B{}^2)(f_B{}^2 - f_L{}^2)}{(f_H{}^2 f_L{}^2)}$$

$$Q_L = \frac{h_a}{\alpha'} \left(\frac{1}{Q_{esb}(r_M - 1)} - \frac{1}{Q_{msb}} \right)$$

If your measured Q_L is very different from the target $Q_L = 7$ used to construct the enclosure, select the new design table closest to the new value and recalculate all parameters using Q_{tsb}. If you measured Q_{ts} on a baffle whose dimensions are similar to the enclosure front baffle, its value will be close to Q_{tsb}.

The accuracy of measured Q_L can be checked by:

$$\frac{f_B}{f_M} = 1 \simeq \left(\frac{\alpha' Q_L{}^2 - h_a{}^2}{\alpha' Q_L{}^2 - 1} \right)^{1/2}$$

FIGURE 2.51

FIGURE 2.52

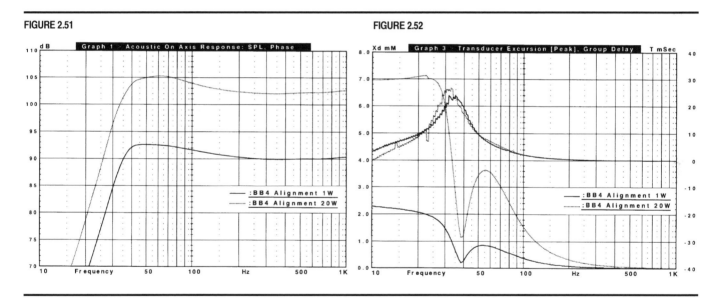

FIGURE 2.53

FIGURE 2.54

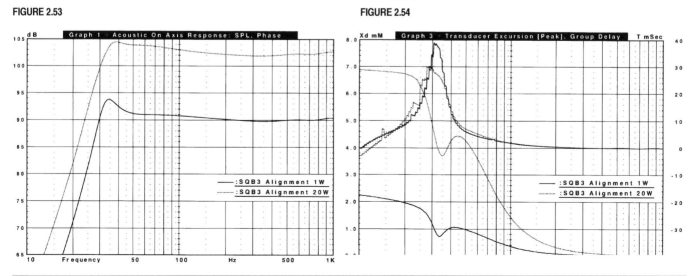

If your calculation is reasonably close to 1, you can assume $f_B = f_M$, and the procedure was accurate.

TABLE 2.13
MISALIGNMENT RESPONSE VARIATIONS

Q_{ts}		H	
% target	dB	% target	dB
+ 100	+ 7	+ 50	+ 7
+ 20	+ 2	+ 20	+ 2
− 20	− 3	− 20	− 2
− 50	− 5	− 50	− 4

2.110 FREQUENCY RESPONSE VARIATIONS CAUSED BY MISALIGNMENT.

As mentioned previously, vented design requires some degree of precision, mostly because there is a moderately severe penalty for misaligned parameters. *Table 2.13* summarizes the response variations which will occur at the corner frequency (just prior to rolloff) for situations where Q_{ts} was incorrectly measured, and where the box was incorrectly tuned (*Fig. 2.44*).[1,3]

2.115 DYNAMIC CHANGES IN FREQUENCY RESPONSE OF DIFFERENT ALIGNMENTS.

The changes in frequency response and other factors caused by increasing power input and voice coil temperature are not as easily seen with vented enclosures as with sealed enclosures because of port nonlinearity. This was apparent in the discussion in *Section 2.70*. Computer simulation of the six flat and nonflat alignments will still help clarify the dynamic changes due to increased power input. The frequency response and cone excursion/group-delay simulations are essentially the same as those shown at 1W in *Section 2.50* with the addition of the 20W simulation added for comparison. The illustrations for the various alignments are as follows:

Alignment Type	Figure
SBB_4	2.35–2.36
SC_4	2.37–2.38
QB_3	2.39–2.40
BB_4	2.41–2.42
SQB_4	2.43–2.44
C_4	2.45–2.46

The conclusion from looking at the various alignments at 1W and 20W of input power are similar to those of the sealed enclosure in *Chapter 1*. Damping tends to decrease with power increase, accompanied by a slight increase in f_3. Since damping is generally decreasing, the flat alignments develop a peak at higher power levels. If computer modeling is not available, you can compensate for the changes in response at higher power levels by opting for a slightly lower tuning than recommended by the Thiele/Small models from the design charts. The Thiele/Small model is a small signal prediction and designs flat at low input signal level. Tuning to a 10-20% lower f_B will yield a more underdamped response at low power levels, and a more nearly flat response at higher power levels. This can be difficult, however, since using a sufficiently large diameter vent already calls for a relatively long length. Tuning to a lower f_B makes the length even longer and possibly impractical for the particular enclosure dimensions.

2.120 SUBSONIC FILTERING.

As discussed in *Section 2.10*, vented enclosures are highly sensitive to subsonic information from such sources as warped records. Woofer excursion will greatly exceed X_{max} and create large amounts of distortion under these circumstances. It is essential, therefore, that you use either active or passive low-frequency filtering with any vented loudspeaker. A 12dB/octave or 18dB/octave active filter, such as the Old Colony model KF-6, is relatively inexpensive, effective, and available in kit form.*

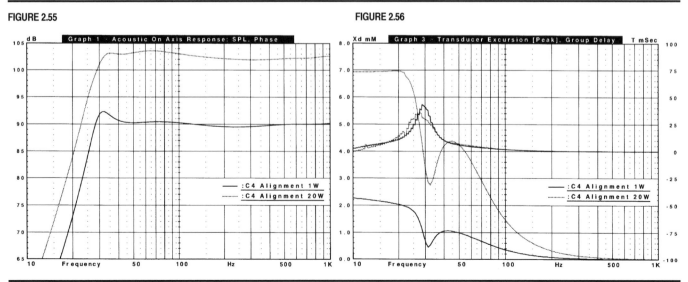

FIGURE 2.55

FIGURE 2.56

As an alternative, you can construct a simple passive CR filter, such as the one illustrated in *Fig. 2.57*. One of the characteristics of this type of filter is f_3's dependence on the output impedance of your amplifier. The values given will work with most solid-state amplifiers, although the cut-off frequency will vary by ±6Hz or so. While this modest circuit is fairly effective in suppressing subsonic cone motion, it is still more desirable to build a filter with an IC buffer.[20]

2.130 BOX DAMPING.

"Traditional" enclosure damping, to suppress standing waves in a vented-box system, consists of lining one of each opposite side with 1"–2" of fiberglass. It is recommended, however, that you cover all surfaces directly behind, and adjacent to, the woofer. Colloms recommends that such damping material be placed within the volume or open area, not on the box walls.

The effects of damping can be observed by computer simulation. Using the same 12" woofer and QB_3 enclosure from the *Section 2.50* simulation, three enclosures were built with 0%, 10%, and 50% fill of standard fiberglass (R19). The 10% was made by lining one of each opposite sides with 1" fiberglass. The 50% sample would be equivalent to lining all four sides and the rear wall with 3" thick material. The computer-generated graphs shown in *Figs. 2.58–2.60* give the results. The SPL curves in *Fig. 2.58* show minor response changes, while the damping changes seen in the group delay in *Fig. 2.59* also are slight. The impedance curves in *Fig. 2.60* likewise indicate only minor changes. This being the case, the primary benefit would be from decreased response changes due to box standing wave modes, making the 50% fill an attractive choice. Be sure you do not obstruct the vent with fill.

2.140 DUAL WOOFER FORMATS.

Everything discussed in the closed-box section (*Section 1.90*) applies when you are using a vented enclosure. The only exception is the compound woofer setup. In some circumstances, the smaller enclosure could require an impracticably long vent tube.

The following *Speaker Builder* articles provide construction examples of compound vented enclosures:

1. John Cockroft, "The Demonstrator: A Vented, Compound Speaker System," 2/87.

2. Chris Edmondson, "A Thunderbird Isobarik," 3/89 (a compound speaker for car audio).

2.150 RESISTIVE and DISTRIBUTED VENTS.

According to Thiele's analysis[4], resistive vents (series resistance in the form of fibrous stuffing or a tight cloth placed over the vent) and distributed vents (parallel resistance using several small holes clustered together instead of one larger tube) result in the following changes in normal

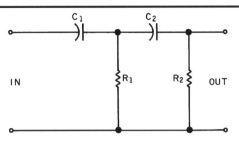

C_1 = 0.47μF POLYPROPYLENE CAP
R_1 = 120K, 5%, 1/4–1/2 W
C_2 = 0.68μF POLYPROPYLENE CAP
R_2 = 100K, 5%, 1/2 W

FIGURE 2.57: Passive low-frequency filter.

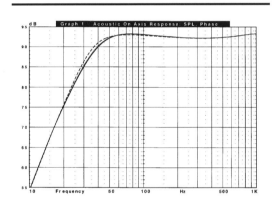

FIGURE 2.58

—— :QB3 0% fill 1W
········· :QB3 10% fill 1W
– – – :QB3 50% fill 1W

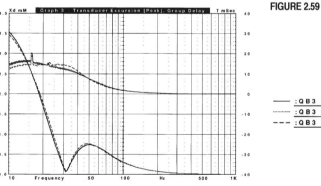

FIGURE 2.59

—— :QB3 0% fill 1W
········· :QB3 10% fill 1W
– – – :QB3 50% fill 1W

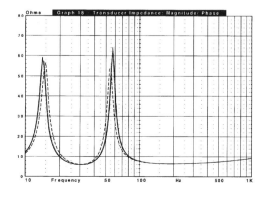

FIGURE 2.60

—— :QB3 0% fill 1W
········· :QB3 10% fill 1W
– – – :QB3 50% fill 1W

TABLE 2.14

		Assisted (Q_L = 7)		Unassisted (Q_L = 7)	
Q_{ts}	Class	Lift-dB	f_3/f_s	Q_{ts}	f_3/f_s
.3	I	5.1	1.098	.31(QB_3)	1.573
.3	II	1.7	1.990	.30(QB_3)	1.573
.55	III	6.3	1.009	.49(C_4)	.735

vent-box operation:

1. increase in f_3
2. decrease in output (efficiency)
3. increased cone excursion near cutoff
4. lowered Q_{ts}

Thiele's point is that if a regular alignment with no resistance has a lower cutoff and higher efficiency than the same size box with resistive loading, why bother? As with closed-box systems, however, being able to tweak an already built enclosure can sometimes be advantageous.

The following is a list of the various techniques which you can apply. Besides the penalties mentioned above, your successful application of any of these methods will require successive measurement and a bit of trial and error:

A. *Basket Damping*. One of the most effective techniques to lower driver Q is to stretch acoustically-resistive cloth over the holes at the rear of the driver frame.[1,21] Fabric adhesive works well for this purpose. You may need to apply successive layers until a target Q is reached. This method of adjusting Q is preferred to adding port resistance.

B. *Vent Damping*. You can use Dacron, wool, fiberglass, or foam to add resistance to a vent. A tight, porous cloth stretched over the vent end will also have the same effect.

C. *Enclosure Stuffing*. If you fill the enclosure with fiberglass or Dacron, you will achieve results similar to B, except that the stuffing will yield a better midrange quality from the driver.[21]

D. *Straw-Filled Vent*. This technique, a variation on the distributed vent idea, works rather well. Simply fill the vent tube with a bundle of plastic straws or rolled, decorative corrugated cardboard. This provides better control than the series resistance methods. By trial and error, you can change the length of the port to produce the same tuning frequency as an unfilled vent tube. Measuring Q_L will give you a clue as to how radical your change is.

2.160 ELECTRONICALLY ASSISTED VENTED DESIGNS.

One of the notable details of Thiele's first papers was his suggestion of taking a QB_3 or C_4 alignment, lowering the tuning, and providing a certain amount of boost (as opposed to the extensive contouring for closed-box electronic assists) accompanied by subsonic filtering to lower f_3 by about ½-octave. Since Thiele's initial paper, several authors have expanded upon the idea,[22,23,24] but none as thoroughly as Robert Bullock in his article on sixth-order alignments in *Speaker Builder* 1/82. *Table 2.14* compares each of the three classes of assisted alignments to its unassisted counterpart.

Looking at *Table 2.14*, we can see that, with only a moderate amount of boost, the Class I system is capable of about a half-octave extension. Although Class II and III systems actually produce a higher f_3, they tend to produce higher undistorted SPLs than their unassisted counterparts. While the wide dynamic range sources, such as compact digital discs, make any boost somewhat questionable at higher SPLs, the sixth-order class of alignments is a far more attractive means of altering a low-end response than the circuits on closed-box assisted enclosures (*Section 1.10*). For details, and an easy method of using a modified Old Colony KF-6 filter kit, see the above-referenced article in *SB*.

Although electronically assisted vented boxes are less likely to produce the excursion catastrophes possible with sealed enclosures having

FIGURE 2.61

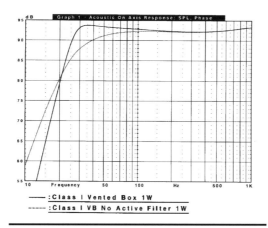

FIGURE 2.62

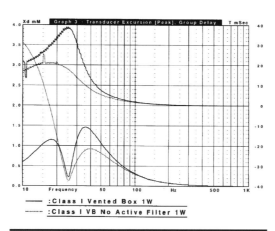

FIGURE 2.63

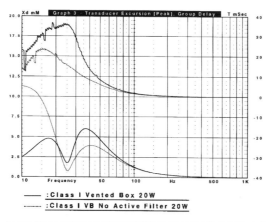

added boost, there are tradeoffs. Using the methodology described in Bullock's 1/82 *Speaker Builder* article (pp. 20–24), a computer simulation was produced to show the dynamic consequences of a Class I sixth-order system. Again the same 12″ woofer parameters were used for a QB_3 enclosure. V_B of the box was 2.7ft^3, tuned to f_B 24.8Hz. The active filter tuning frequency was 27.14Hz, with a Q of 1.77, which gives a bass lift of 5.33dB. *Figure 2.61* shows the 1W SPL comparison of the response with and without the added electronic filter. The f_3 and slope for the two responses were as follows:

Type	f_3	Slope dB/oct.
no filter	36Hz	20
with filter	26Hz	34

The extension is the predicted ½-octave, but with considerable increase in rolloff rate. Looking at the 1W group delay curve in *Fig. 2.62*, the curve with the filter has developed a very sharp "knee" and has increased group delay by a factor of 3. *Figure 2.63* shows the cone excursion and group delay at a 20W input level. Excursion of only 5.8mm maximum for the speaker/filter combination, plus the typical continuing increase in excursion rate below f_B, has been attenuated to a low level. Given the 6mm X_{max} of the driver, the Class I system should provide good high SPL performance. The only drawback, or tradeoff, is the rather severe change in damping which accompanies the use of the filter. If the system were being used in a subwoofer application, rolled off on the high end at 24dB/octave at 75Hz, the transient performance loss would not be as noticeable.

2.170 VENTED REAR CHAMBER BANDPASS ENCLOSURES.

Vented rear chamber bandpass enclosures are substantially more complicated mathematically than the sealed rear chamber type described in *Chapter 1*. Although they possess a great deal more design flexibility, no one has published a hand-calculator methodology for designing this enclosure variation. It is likely that many of the manufactured units made available over the last several years were designed by trial and error, although several computer programs can now be used to create simulations of these designs. The three programs capable of this, at the time of publishing, are Low Frequency Designer by SpeakEasy, Speak by DLC Design, and LEAP 4.0 by LinearX Systems.

LEAP 4.0 was used to create a simulation using the same 12″ woofer used in most of the previous simulations in this chapter. The rear volume was 5.65ft^3, tuned to 21Hz, and the front volume was 1.25ft^3 tuned to 48Hz. This produces a fairly large enclosure with a total volume of nearly 7ft^3. *Figure 2.64* depicts the frequency response at both 1W and 20W input levels. f_{3L} is at 15Hz with a 16dB/octave slope rate, and f_{3H}

is at 78Hz with a 15dB/octave slope rate. Efficiency is about the same as the other alignments described for this woofer, 90dB, so the primary tradeoff is enclosure size for low-end extension.

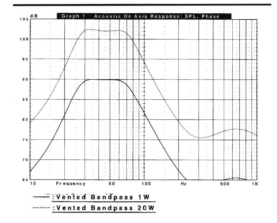

FIGURE 2.64

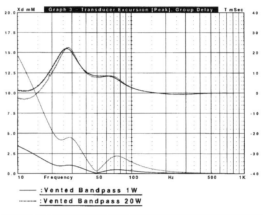

FIGURE 2.65

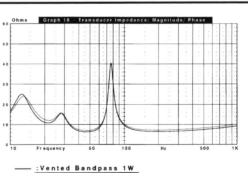

FIGURE 2.66

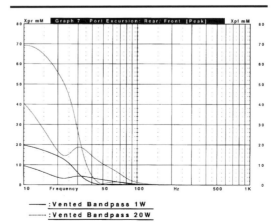

FIGURE 2.67

Figure 2.65 shows the group delay and cone excursion. Excursion, as with any vented enclosure, increases below the low tuning frequency, but even at 20W is only 4.4mm maximum above the tuning frequency compared to the driver X_{max} of 6mm. Group delay is high compared to the other alignments, with an absolute value of 20ms for this particular woofer/box combination. The impedance curves, shown in *Fig. 2.66*, illustrate the typical three-peak impedance curve characteristic of vented rear chamber bandpass enclosures. *Figures 2.67–2.70* show the rear and front port excursion, transducer cone velocity, rear port volume velocity, and front port volume velocity.

FIGURE 2.68

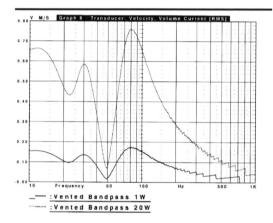

FIGURE 2.69

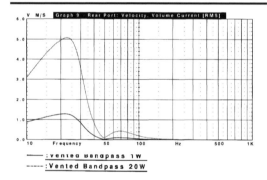

FIGURE 2.70

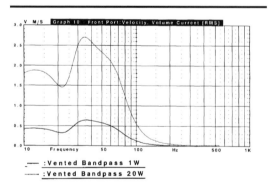

REFERENCES

1. R. Small, "Vented-Box Loudspeaker Systems," *JAES*, June through October 1973.

2. M. Lampton and L. Chase, "Fundamentals of Loudspeaker Design," *Audio*, December 1973.

3. D. B. Keele, Jr., "Sensitivity of Thiele's Vented Loudspeaker Enclosure Alignments to Parameter Variations," *JAES*, May 1973.

4. A. N. Thiele, "Loudspeakers in Vented-Boxes," *JAES*, May–June 1971.

5. R. Bullock III, "Thiele, Small and Vented Loudspeaker Design," *Speaker Builder* 3/81.

6. Gunter J. Krauss, "Low Frequency Transient Response Problems in Vented Boxes," 88th AES Convention, March 1990, Preprint #2895.

7. W. J. J. Hoge, "A New Set of Vented Loudspeaker Alignments," *JAES*, Vol. 25 1977.

8. G. R. Koonce, "The QB3 Vented Box is Best," *Speaker Builder* 5/88, p. 22.

9. M. E. Engebretson, "Low Frequency Sound Reproduction," *JAES*, May 1984.

10. R. Laupman, US Patent No. 4,213,515, "Speaker System," awarded July 1980.

11. B. Gawronski, US Patent No. 5,714,721, "Porting," assigned to Bose Corporation, awarded Feb. 3, 1998.

12. M. Polk and C. Campbell, US Patent No. 5,717,573, "Ported Loudspeaker System and Method of Reduced Air Temperature," assigned to Polk Audio, awarded May 14, 1996.

13. Roosen, Nicolaas, Vael, Jozef, Nieuwendijk, and Joris, US Patent No. 5,892,183, "Loudspeaker System Having a Bass-Reflex Port," assigned to Philips Corporation, awarded April 6, 1999.

14. Juha Backman, "The Nonlinear Behavior of Reflex Ports," 98th AES Convention, February 1995, preprint no. 3999.

15. John Vanderkooy, "Loudspeaker Ports," 103rd AES Convention, September 1997, preprint no. 4523.

16. John Vanderkooy, "Nonlinearities in Loudspeaker Ports," 104th AES Convention, May 1998, preprint no. 4748.

17. Roosen, Vael, and Nieuwendijk, "Reduction of Bass-Reflex Port Nonlinearities by Optimizing the Port Geometry," 104th AES Convention, May 1998, preprint no. 4661.

18. Vance Dickason, "Aeroport Low-Distortion Ports," *Voice Coil*, September 1996.

19. Button, Devantier, and Salvatti, "Maximizing Performance from Loudspeaker Ports," 105th AES Convention, September 1998, preprint no. 4855.

20. W. Jung, "A L.F. Garbage Filter," *Audio Amateur*, April 1975.

21. J. Graver, "Acoustic Resistance Damping for Loudspeakers," *Audio*, March 1965.

22. D. B. Keele, Jr., "A New Set of Sixth-Order Vented-Box Loudspeaker System Alignments," *JAES*, June 1975.

23. R. Bywater and H. Wiebell, "Alignment of Filter Assisted Vented-Box Loudspeaker Systems," *JAES*, May 1982.

24. R. Normandin, "Extended Low Frequency Performance of Existing Loudspeaker Systems," *JAES*, Jan./Feb. 1984.

*Information on Old Colony kits can be requested by writing: Old Colony Sound Lab, Box 876, Peterborough, NH 03458-0243, FAX (603) 924-9467, E-mail custserv@audioXpress.com.

PASSIVE-RADIATOR LOW-FREQUENCY SYSTEMS

3.10 DEFINITION.

Passive-radiators (PR) are a type of vent substitute, and closely follow vented loudspeaker design methodology and performance characteristics. "Drone cones," as they are sometimes called, have two important advantages over vents. First, they eliminate vent colorations (such as resonant pipe sounds), wind noises, and the internal high frequency sound reflected out of the vent. Second, they are practical for small enclosures which call for vent lengths in excess of internal box dimensions. PRs are also simpler to deal with, having fewer alignments and less concern with loss calculation. On the downside, PRs have a steeper cutoff (and less transient stability), a slightly higher cut-off frequency and greater overall losses (Q_L) than vented designs.

A popular misconception about PRs is that they operate in low-frequency regions, mechanically cross over to the driver at a higher frequency and extend the bass of that driver. Actually, the PR operates in conjunction with the driver at low frequencies, sharing the acoustic load and reducing driver excursion. Working as a vent, PRs only add as much as they subtract. This implies they have the same positive attributes as a vent, such as higher power handling and lower distortion.

3.20 HISTORY.

Passive-radiators were first described by Harry Olson in his patent "Loudspeakers and Method of Propagating Sound," issued January 1935. Except for an article by Olson in 1954,[1] very little was published about PRs until Nomura and Kitamura in their IEEE paper in October 1973,[2] and Small's *JAES* paper in October 1974.[3] Polk Audio is one of the largest and most successful commercial producers of PRs in the US.

3.30 DRIVER "Q" and ENCLOSURE RESPONSE.

PRs exhibit about the same Q/box relationship as a vented-box, and are treated similarly. *Figure 3.1* shows the amplitude response comparison for closed, vented and PR systems with identical drivers. The characteristic which sets PRs apart from vented systems is a dip or notch at the PR resonance frequency (f_p). Though located below system cutoff, the notch increases the slope of the driver low-frequency rolloff, and degrades transient performance. As will be shown in a separate section, Clarke's augmented passive-radiator system substantially solves this problem.

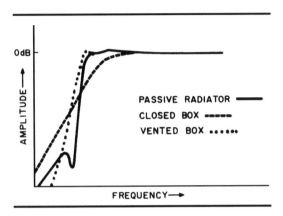

FIGURE 3.1: Response comparison for vented, closed box, and passive radiator systems.

3.35 WOOFER SELECTION.

Follow the same procedures described in *Section 2.40*, for vented driver selection.

3.40 ALIGNMENTS.

As mentioned above, alignments for PRs are generally restricted to QB_3, B_4 and C_4 types.[3] Q_{ts}s above 0.5 are not useable because they produce C_4 responses with excessive frequency response ripple. This narrows the choice of Q_{ts} for a flat response to values of 0.2 to about 0.35 QB_3 alignments. There are also no described electronically assisted alignments.

Transient behavior for low alpha alignments is rather poor and has a high ripple C_4 response (alpha equal to 1–2, corresponding to a Q_{ts} of 0.44–0.35). This is true for Small's analysis, which assumes the radiator compliance ratio (δ or delta) is equal to alpha (α), the system compliance ratio. In other words, the passive-radiator is constructed from the same exact cone, surround, spider and basket as the driver ($\delta = \alpha$). If, however, the radiator is made more compliant than the woofer, even low alpha (high Q_{ts}) alignments will be more like their vented counterparts and have acceptable transient response.

The frequency response and phase graph in *Fig. 3.2* is a computer simulation of a $\delta = \alpha$ passive radiator QB_3 alignment using the same 12″ woofer as in the vented simulations in *Chapter 2*. This response is at a 1W power level and illustrates the typical shape of the bottom end response of a passive radiator enclosure. The enclosure volume is the same as the QB_3 vented enclosure, 2ft³, but the radiator is tuned to 18Hz, compared to the vent tuning frequency of 31Hz. With the $\delta = \alpha$ alignment, the radiator has the same cone area and compliance as the driver. 131.9 grams of mass is required to tune

FIGURE 3.2

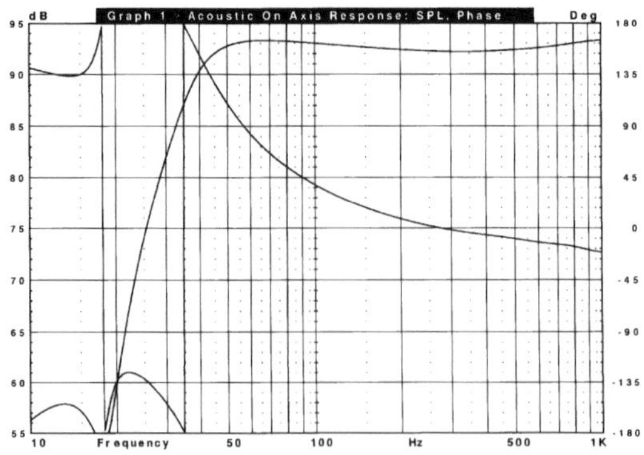

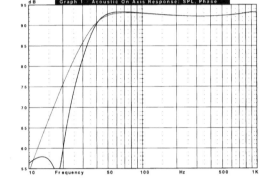

—— : PR 12"SPK/12"D 1W

FIGURE 3.3

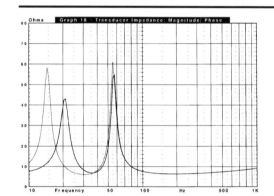

—— : PR 12"SPK/12"D 1W
········ : QB3 12" SPK

FIGURE 3.4

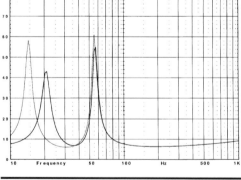

—— : PR 12"SPK/12"D 1W
········ : QB3 12" SPK

FIGURE 3.5

—— : PR 12"SPK/12"D 1W
········ : QB3 12" SPK

the 12″ radiator to the required frequency. The f_3 of the PR box is at 41Hz with a rolloff slope of 29dB/octave. *Figure 3.3* compares the QB$_3$ passive radiator 1W response with the QB$_3$ vented enclosure 1W response. In this example the vented enclosure will have more effective low end response than the PR enclosure. Despite the somewhat decreased low end content for the same driver, the PR's primary advantage is that it doesn't suffer from the pipe resonance and standing wave transmission problems found in

vents. PRs also tend to have a distinctive subjective sound quality which is different than that of a vented speaker. *Figure 3.4* depicts the impedance curves for both enclosures. *Figure 3.5* shows the dynamic changes in the response with the input power increased to 20W and the SPL output raised to about 105dB for both enclosure types. The response indicates the expected decreased damping accompanied a slight increase in the −3dB frequency to 42Hz. The group delay and cone excursion curves are in *Fig. 3.6*. The excursion, as with the vented enclosure, is moderate with a maximum of 3mm above the frequency f_m center frequency between the impedance maxima (as opposed to the radiator tuning frequency). Since the driver X_{max} is 6mm, the distortion level is low. However, like vented enclosures, excursion increases rapidly at low frequencies, and a 20Hz high-pass filter is recommended. Group delay, shown on the same graph, is at about 10mS for both types of enclosures. However, the phase change at low frequencies in the vicinity of the notch results in the negative group delay. *Figure 3.7* shows the radiator excursion, which is at maximum at the radiator tuning frequency. *Figure 3.8* gives the speaker cone velocity with a shape similar to the vented QB$_3$, having maximum velocity at the impedance maxima, and a minimum velocity at the impedance minima. Radiator velocity is maximum at the driver minimum velocity, as shown in *Fig. 3.9*.

The procedures for designing passive radiators and the design example above are limited to the use of a radiator with the same cone and compliance as the driver. Using different sized radiators with different compliance values requires either trial and error, or the use of one of the available computer box design programs. Although not repeated here, another guideline for using different passive radiators can be found in a method described in AES preprint no. 2539, "Generalized Design Method of Lossy Passive-Radiator Loudspeaker Systems" by Carrion-Isbert.

A complete analysis of passive radiator systems including the various types of losses and the damping effects caused by fill materials plus the mathematics needed to describe this is found in a "Complete Response Function and System Parameters for a Loudspeaker with Passive Radiator," by Douglas Hurlburt[4].

3.45 BOX SIZE DETERMINATION and RELEVANT PARAMETERS.

Except for the extra radiator parameters, you can determine the box size and tuning for passive-radiator systems the same way you determine them for vented enclosures. Begin by assembling the following driver parameters:

1. f_s calculated on a baffle simulating final enclosure loading
2. Q_{ts} driver Q includes all series resistances
3. V_{as}
4. X_{max}
5. S_d
6. V_d

7. S_{dp} area of passive-radiator
8. δ delta, the compliance ratio of the PR

You can obtain parameters 1–6 in the same manner as 2.6. Calculate S_{dp} using the formula for driver S_{dp} found in *Chapter 1*. Delta requires a separate test procedure, described in *Section 3.50*.

3.50 FINDING DELTA FOR PRS.
The technique described here is a variation on one for tuning vented enclosures, originally suggested by Weems[5] and applied to PRs by G. R. Koonce.[6]

Definition of Additional Terms:

1. V_T —test volume equal to V_b, given in cubic meters
2. C_{ab} —the acoustic compliance of the enclosure
3. C_{mp} —the mechanical compliance of the PR in meters/newton
4. C_{ap} —acoustic compliance of the PR = $C_{mp} \times S_{dp}^2$
5. S_{dp} —area of PR in square meters
6. V_{ap} —the volume of air equal to the radiator compliance

The following procedures will work for any type of PR, including PRs made from a cone, surround, spider and basket, or PRs made from flat cardboard and foam with an attached surround.

A. To calculate V_{ap}, find the free-air resonance of the passive radiator, f_p. Mount the radiator on a baffle (the same size and shape as that of the test box used to measure V_{ap}) and then drive it by holding a small driver directly behind the radiator to couple the motion of the driver to the radiator. Use a sine wave oscillator to drive the speaker and vary the frequency until the passive radiator's maximum excursion is found. Determine the point by watching the radiator's motion.

B. Using the test setup in *Fig. 3.10*, vary the frequency of the oscillator until the box resonance frequency f_c is found. You can then find V_{ap} by:

$$V_{ap} \approx V_T \left[(f_c/f_p)^2 - 1 \right]$$

Then:

$$C_{ap} = \frac{V_{ap}}{1.42 \times 10^5}$$

C. C_{ab} of the design enclosure is calculated by:

$$C_{ab} = \frac{V_{ab}}{1.42 \times 10^5}$$

D. Finally, calculate delta (δ), the PR compliance ratio, by:

$$\delta = \frac{C_{ap}}{C_{ab}}$$

As an alternative method that does not require

drilling and filling a hole in the enclosure, you can calculate alpha using the final enclosure. First, run an impedance curve (as in *2.10*). Then, assuming $f_M = f_B^2$:

$$f_{SB} = \frac{f_L f_H}{f_B}$$

Then:

$$\alpha = \left(\frac{f_H^2 + f_L^2 - f_B^2}{f_{SB}^2} \right) - 1$$

To find δ, solve the equation:

$$\frac{\alpha\delta}{\alpha + \delta + 1} = \frac{(f_H^2 - f_B^2)(f_B^2 - f_L^2)}{f_L^2 f_H^2}$$

3.51 USING DESIGN TABLE 3.1.
Because PRs have very consistent losses usually equal to a $Q_L = 7$, and because only one alignment "family" will be explored, you will need only one design table (*Table 3.1*). Use this table in the same manner as with vented boxes, but note the additional ratio of V_{pr}/V_d, which is the ratio of the

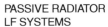

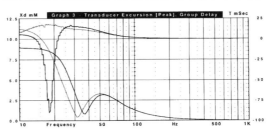

FIGURE 3.6

——— :PR 12"SPK/12"D 20W
··········· :QB3 12" SPK 20W

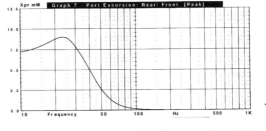

FIGURE 3.7

——— :PR 12"SPK/12"D 20W

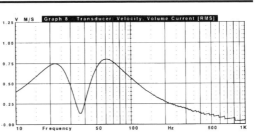

FIGURE 3.8

——— :PR 12"SPK/12"D 20W

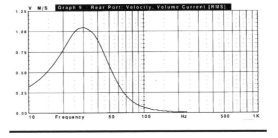

FIGURE 3.9

——— :PR 12"SPK/12"D 20W

volume displacement of the PR to the volume displacement of the driver.

Box volume $\qquad V_b = \dfrac{V_{as}}{\alpha}$

−3dB Point $\qquad f_3 = (f_3/f_s) \times f_s$

Tuning Frequency $\qquad f_B = H(f_s)$

PR Displacement $\qquad V_{pr} = (V_{pr}/V_d) \times V_d$

3.52 PR LOCATION and MUTUAL COUPLING.

Because of the large (compared to vented designs) center-to-center distance from a driver to a PR, response problems caused by the proximity of the two are negligible.

3.53 BOX TUNING.

Tune a PR system the same way you would a vented enclosure (*Section 2.72*). The only difference will be your frequency adjustment method. With a vented-box you can change the length of the vent tube, but with a passive-radiator, you must add or subtract weight (clay or a metal dust cap) from the radiator until you reach f_B.

3.54 ADDITIONAL PARAMETERS.

Calculate η_0, $P_{ar(p)}$, and P_{er} the same as you would with vented enclosures (*Section 2.8*).

3.55 EXAMPLE DESIGN CHARTS.

The same two woofers used in the design charts in *Section 2.90* were used to generate *Tables 3.2* and *3.3*.

Since Small's design methodology specifies $\delta = \alpha$, find the PR's required excursion from:

$$\frac{V_{pr}}{S_d} = 7.5mm$$

This is more than twice the excursion of the driver, but is feasible for most high compliance surrounds. If the PR were made from a 10″ cone ($S_d = 3.3 \times 10^{-2}$ m^2), the travel would decrease to 4.7mm.

As with the 8″ example, your PR excursion is 7.5mm. If you use a 12″ PR, required excursion is 5.3mm. For systems where $\delta = \alpha$, the required mass to tune the enclosure to f_B is twice that of the driver.

3.56 Q_L MEASUREMENT.

As previously stated, passive-radiator system losses are typically $Q_L = 7$, so no other design chart is required. To make a good check on a finished design, however, recalculate Q_L using the same procedures described in *Section 2.10*. If Q_L is much less than 5 or 6, you probably have a box air leak or possibly a driver, PR seal, or dust cap leak.

TABLE 3.1

$Q_L = 7$
Passive-Radiator
QB$_3$, B$_4$, and C$_4$
$\delta = \alpha$

Q_{ts}	H	a	f_3/f_s	V_{pr}/V_d
0.2000	2.10	8.21	2.65	1.81
0.2100	2.02	7.26	2.51	1.84
0.2200	1.94	6.38	2.36	1.88
0.2300	1.88	5.76	2.26	1.92
0.2400	1.82	5.20	2.16	1.98
0.2500	1.77	4.76	2.06	2.02
0.2600	1.73	4.33	1.98	2.07
0.2700	1.68	4.01	1.90	2.10
0.2800	1.64	3.65	1.82	2.15
0.2900	1.59	3.34	1.74	2.20
0.3000	1.56	3.08	1.67	2.24
0.3100	1.51	2.78	1.59	2.35
0.3200	1.48	2.58	1.53	2.44
0.3300	1.45	2.38	1.49	2.53
0.3400	1.42	2.20	1.44	2.61
0.3500	1.39	2.06	1.38	2.67
0.3600	1.35	1.91	1.33	2.76
0.3700	1.33	1.80	1.30	2.84
0.3800	1.30	1.66	1.27	2.94
0.3900	1.26	1.53	1.23	3.09
0.4000	1.23	1.41	1.19	3.11
0.4100	1.21	1.30	1.17	3.19
0.4200	1.19	1.22	1.14	3.25
0.4300	1.16	1.12	1.11	3.32
0.4400	1.13	1.03	1.08	3.38
0.4500	1.10	0.96	1.05	—
0.4600	1.06	0.87	1.01	—
0.4700	1.03	0.80	0.98	—
0.4800	1.00	0.73	0.95	—
0.4900	0.98	0.69	0.92	—
0.5000	0.95	0.65	0.90	—
0.5100	0.92	0.60	0.87	—
0.5200	0.90	0.55	0.84	—
0.5300	0.87	0.52	0.82	—
0.5400	0.84	0.48	0.79	—
0.5500	0.81	0.44	0.76	—
0.5600	0.78	0.39	0.72	—
0.5700	0.75	0.37	0.69	—
0.5800	0.72	0.33	0.67	—
0.5900	0.70	0.31	0.65	—
0.6000	0.68	0.28	0.62	—

Note: V_{pr}/V_d is not given for values above $Q_{ts}=0.44$ because the data provided by Small's computer simulation did not record past that point. As already mentioned, however, high Q_{ts} alignments with alpha close to 1 have definite problems with transient stability (again, for $\delta = \alpha$), so avoid them anyway.

3.57 FREQUENCY RESPONSE VARIATIONS.
Same as for vented-boxes (*Section 2.110*).

3.58 SUBSONIC FILTERING.
Same as for vented-boxes (*Section 2.120*).

3.59 DAMPING.
Same as for vented-boxes (*Section 2.130*).

TABLE 3.2
8″ WOOFER EXAMPLE

Q_{ts}	Q_{es}	Q_{ms}	f_s	X_{max}	$S_d(m^2)$	$V_d(m^3)$		$V_{as}(ft^3)$
0.26	0.28	3.95	32	3.5mm	2.15×10^{-2}	7.53×10^{-5}		2.764
AL	V_b	f_3	f_B	V_{pr}	$P_{ar(p)}$	η_o		P_{er}
	ft^3	Hz	Hz	m^3	watts	SPL	%	SPL	watts
QB_3	0.64	64	56	1.6×10^{-4}	0.285	107dB	0.91	91.6	31

TABLE 3.3
10″ WOOFER EXAMPLE

Q_{ts}	Q_{es}	Q_{ms}	f_s	X_{max}	$S_d(m^2)$	$V_d(m^3)$		$V_{as}(ft^3)$
0.25	0.29	2.05	22	3.5mm	3.3×10^{-2}	1.16×10^{-4}		7.51
AL	V_b	f_3	f_B	V_{pr}	$P_{ar(p)}$	η_o		P_{er}
	ft^3	Hz	Hz	m^3	watts	SPL	%	SPL	watts
QB_3	1.64	44	38	2.4×10^{-2}	0.151	104dB	0.72	91dB	20.9

3.60 PASSIVE RADIATOR BANDPASS ENCLOSURES.

Like vented bandpass enclosures, passive radiator bandpass enclosures do not lend themselves to hand-calculator methods. Besides the tedious cut-and-try methodology, the only other way to design this type of enclosure is through computer simulation. At the time this edition was written, only two speaker design programs were capable of simulating rear and front chamber passive radiator bandpass enclosures: Speak by DLC Design and LEAP 4.0 by LinearX Systems.

Although there has been some work with passive radiator bandpass loudspeakers, such as Laurie Fincham's paper presented at the 63rd AES Convention in 1979 ("A Bandpass Enclosure," AES preprint no. 1512), most commercial designs have been of the vented variety.

One of the main disadvantages of vented bandpass enclosures is the front chamber pipe resonances and standing waves transmitted through the front port (illustrated in *Chapter 1*). Passive radiators used in the front chamber, however, do not suffer from the same malady, or at least not to the same degree. The graph in *Fig. 3.11* compares the response of a passive radiator rear chamber/passive radiator front chamber bandpass, to a passive radiator rear chamber/vented front chamber bandpass enclosure. The simulation is the same as that described in *Chapter 2*, with the same 12″ woofer, and a 5.65ft³ rear chamber and a 2.12ft³ front chamber. Both rear chambers were designed with a 12″ passive radiator ($\delta = \alpha$), mass loaded to 104.4 grams, which tuned the chamber to 20Hz. The front chamber tuning is quite different for the two devices. The front radiator is tuned very close to the rear

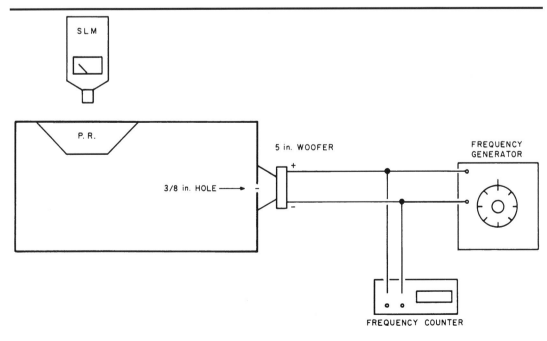

FIGURE 3.10: Test setup for finding driver parameters.

FIGURE 3.11

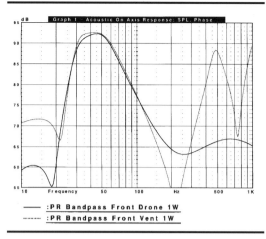

— :PR Bandpass Front Drone 1W
········· :PR Bandpass Front Vent 1W

FIGURE 3.12

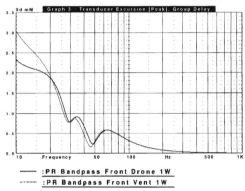

— :PR Bandpass Front Drone 1W
········· :PR Bandpass Front Vent 1W

FIGURE 3.13

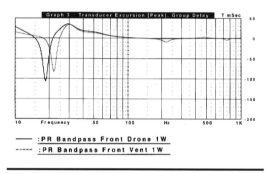

— :PR Bandpass Front Drone 1W
········· :PR Bandpass Front Vent 1W

FIGURE 3.14

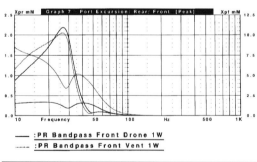

— :PR Bandpass Front Drone 1W
········· :PR Bandpass Front Vent 1W

chamber at 24Hz with 68.6 grams. The front vent in the example is tuned to 45Hz with a 6″ diameter vent 10″ long. The response shape is similar, but the radiator version lacks the standing wave and pipe resonance anomalies.

Figure 3.12 compares the cone excursion curves for designs, showing the excursion for both bandpass enclosures to be about the same. *Figure 3.13* gives the group delay curves. Comparing the front and rear chamber drone/vent excursion, shown in *Fig. 3.14*, the

front vent excursion is greater than the front radiator excursion, which simply reflects the difference in surface area between the 6″ vent and 12″ radiator. Driver cone velocity, shown in *Fig. 3.15*, is somewhat greater at low frequencies for the drone/vent combination than for the drone/drone combination.

3.70 AUGMENTED PASSIVE RADIATORS.

3.71 DEFINITION.
The augmented passive-radiator (APR) is a twin cavity variation of the normal drone cone concept. It is capable of most of the vented and APR systems' alignment variations. For a given driver (Q_{ts}), the APR will have a higher power output rating, and a 15–25% lower cutoff (up to ½ -octave extension). The tradeoff for this substantial increase in low frequency range is a 20% increase in total enclosure volume. Compared to normal PRs, the APR has better transient performance and lower cutoff frequencies. This occurs because the notch frequency is lowered to an out-of-band location.[2,7]

3.72 HISTORY.
The APR is a relatively recent device, originally patented in November 1973 by E. Hossbach. Another variation was patented by Thomas Clarke in February 1978. The design procedures presented here are primarily based on Clarke's *JAES* article (June, July/August 1981). (*See also* Speaker Builder *2/86, p. 20. –Ed.*)

3.73 CONFIGURATION.
Figure 3.16 illustrates the unusual layout for the APR. It consists of two unequal area PRs, connected back-to-back, with the front baffle joined to the inner dividing baffle. The radiators are made from surround-cone combinations (usually available from professional reconing stations) directly attached to the enclosure without the use of metal baskets. V_1 is from 33% to 75% of the total volume, depending on the alignment type.

3.74 WOOFER SELECTION.
Same as for vented and PR systems.

3.75 ALIGNMENTS.
While a variety of alignments can be generated for the APR format, only the flat response QB_3 and the low ripple C_4 will be dealt with in this text.

3.76 BOX SIZE DETERMINATION.
To determine box size for an APR, follow the procedure for vented and PR systems. Start by obtaining the following standard driver parameters:

1. f_s free-air resonance
2. Q_{ts} total Q of the driver, including series resistances

3. V_{as} volume of air equal to driver compliance
4. X_{max} amount of voice coil overhang in mm
5. S_d effective radiating area in square meters
6. V_d displacement volume of driver in cubic meters

As with the other operating systems described in this book, the best way to proceed is to generate a design chart showing all the various calculated parameters for whatever alignment you choose. You will notice design *Table 3.4* is more or less the same as the vented and PR design tables already presented. There are, however, several new parameters you need to consider.

Definition of Additional Terms:

1. α (alpha) equals the compliance ratio of both V_1 and V_2.
2. Γ (gamma) is the ratio of the area for the two PR cones.
   For $\Gamma = 1.67$, the following cone combinations (in inches) will be appropriate:

$$8 : 6.5$$
$$12 : 8$$
$$15 : 10$$

The larger cone loads to the outside, while the smaller cone connects the internal volumes.

3. Σ (Epsilon) is the ratio of the driver enclosure volume to total volume:

$$\Sigma = \frac{V_1}{V_b}$$

3.77 USING DESIGN TABLE 3.4.
Unlike vented and PR systems, design *Table 3.4* includes no compensation for leakage losses. To make things a little easier, the following equations have been adjusted to give an approximate $Q_L = 7$.

Box Volume $\quad V_b = V_1 + V_2 = \dfrac{V_{as}}{.95\alpha}$

$$V_1 = V_b \times \Sigma$$
$$V_2 = V_b - V_1$$
Tuning Frequency $\quad f_B = 1.1H \times f_s$
−3dB Point $\quad f_3 = 1.09(f_3/f_s) \times f_s$

3.78 MUTUAL COUPLING OF DRIVER and APR.
Same as for standard PRs.

3.79 BOX TUNING PROCEDURES.
Same as for PRs (*Section 3.53*).

3.80 ADDITIONAL PARAMETERS.
Calculate η_0, $P_{ar(p)}$, and P_{er} the same as for vented and PR systems. Note that $P_{ar(p)}$ is a much more conservative figure than that for the other types.

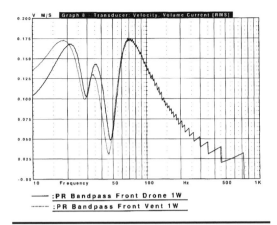

FIGURE 3.15

—— :PR Bandpass Front Drone 1W
········ :PR Bandpass Front Vent 1W

3.81 V_{pr} FOR APR SYSTEMS.
APR volume displacement is generally twice the driver displacement volume. Since the area of an APR is the sum of the two radiator cone areas, the amount of required excursion should not be a problem with most surrounds. Do not be concerned with the radiators' compliance because they are chiefly determined by the V_2 air-spring.

3.82 EXAMPLE DESIGN CHARTS.
The same two drivers used in *Section 2.90* were used to generate design charts in *Tables 3.5* and *3.6*.

For $\Gamma = 1.67$ (8″ woofer) and a given $V_{pr}/V_d = 2$, an APR composed of one 8″ and one 6.5″ cone would be adequate. If P_{er} and P_{ar} seem low, Clarke maintains the figures are conservative by a factor of 2 or 3 times (20–30W P_{er}).

TABLE 3.4

$Q_L = \inf.$
Augmented Passive-Radiator
QB_3 and C_4
$\Gamma = 1.67 \quad E = 0.40$

Q_{ts}	α	H	f_3/f_s
0.2000	3.20	1.75	1.85
0.2100	3.01	1.70	1.80
0.2200	2.81	1.65	1.75
0.2300	2.71	1.60	1.65
0.2400	2.42	1.55	1.55
0.2500	2.25	1.50	1.50
0.2600	2.03	1.49	1.49
0.2700	1.84	1.40	1.40
0.2800	1.64	1.35	1.38
0.2900	1.45	1.30	1.25
0.3000	1.25	1.20	1.15
0.3100	1.18	1.19	1.13
0.3200	1.10	1.18	1.11
0.3300	1.03	1.17	1.10
0.3400	0.96	1.15	1.05
0.3500	0.89	1.10	0.99
0.3600	0.81	1.02	0.95
0.3700	0.74	1.00	0.90
0.3800	0.67	0.95	0.85
0.3900	0.59	0.87	0.80
0.4000	0.52	0.80	0.75

TABLE 3.5
8″ WOOFER EXAMPLE

Q_{ts}	Q_{es}	Q_{ms}	f_s	X_{max}	$S_d(m^2)$	$V_d(m^3)$	$V_{as}(ft^3)$	Γ	E
0.26	0.28	3.95	32	3.5mm	2.15×10^{-2}	7.5×10^{-5}	2.76	1.67	.40

AL	V_b ft³	V_1 ft³	V_2 ft³	f_3 Hz	f_B Hz	$P_{ar(p)}$ watts	SPL	η_0 %	SPL	P_{er} watts
QB₃	1.43	.57	0.86	52	53	0.09	102dB	0.91	92dB	9.9

TABLE 3.6
10″ WOOFER EXAMPLE

Q_{ts}	Q_{es}	Q_{ms}	f_s	X_{max}	$S_d(m^2)$	$V_d(m^3)$	$V_{as}(ft^3)$	Γ	E
0.25	0.29	2.05	22	3.5mm	3.3×10^{-2}	1.16×10^{-4}	7.51	1.67	.40

AL	V_b ft³	V_1 ft³	V_2 ft³	f_3 Hz	f_B Hz	$P_{ar(p)}$ watts	SPL	η_0 %	SPL	P_{er} Watts
QB₃	3.67	1.47	2.2	36	36	0.048	99dB	.72	90.6dB	6.6

For $\Gamma = 1.67$ (10″ woofer) and a given $V_{pr}/V_d = 2$, an APR composed of one 12″ and one 8″ cone is satisfactory. Again, P_{er} and P_{ar} are quite conservative by a factor of 2–3 (13–20W P_{er}).

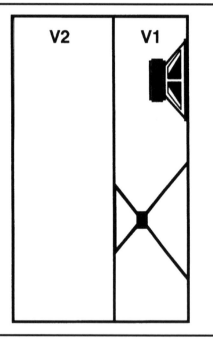

FIGURE 3.16: Augmented passive radiator (APR) diagram.

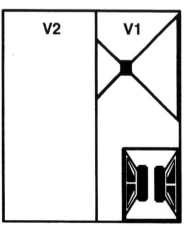

FIGURE 3.17: Compound APR diagram.

3.83 MISALIGNMENT RESPONSE VARIATIONS.
Same as for vented and PR systems.

3.84 SUBSONIC FILTERING.
Same as *Section 2.120*.

3.85 DAMPING.
You may apply minimum damping material to V_1 for standing wave suppression. V_2 requires no damping, but must be air tight (as V_1).

3.86 DUAL WOOFER FORMATS.
Although dual driver formats used in conjunction with the APR would be complex, all the applicatons discussed in *Section 1.90* apply. Of particular interest is the use of the compound dual woofer setup with an APR. The combination of the two yields a normal sized enclosure with a superior f_3. The normal box size occurs because the compound V_{as} is half, because the APR alpha is roughly half, and (in terms of "normal"-sized vented and PR systems) because their synthesis cancels out their mutual effects. *Figure 3.17* depicts the physical layout for a compound APR. Depending on your driver (Q) and box dimensions, you may need to construct a short Sonotube tunnel on the inside PR.

3.90 ACOUSTIC LEVERS.
1999 saw the invention of a new class of passive radiator bandpass loudspeaker[8], the Acoustic Lever[9], thanks to the creative work of Dr. Earl Geddes (see the references to Dr. Geddes' previous work on bandpass loudspeakers in *Chapter 1, Section 1.320 History*). The Acoustic Lever is almost a cousin of the augmented passive radiator discussed in *Section 3.70*. If you look at the drawing in *Fig. 3.18*, the first thing you will notice is that the Acousic Lever uses a dual coupled passive radiator with dissimilar piston areas (S_d) for the two radiators, pretty much the same as Thomas Clarke's APR. However, here is where

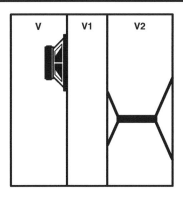

FIGURE 3.18: Diagram of Acoustic Lever (single lever).

the similarity ends. The APR uses the "lever" as an auxiliary device in parallel with the driver like a standard port and works primarily to extend the low end of the passband only, relying on the driver's direct radiation for the remainder of the bandwidth. Dr. Geddes Acoustic Lever uses the "lever" device in series with the driver in such a way that it acts to transform the radiating volume velocity, in some ways rather similar to a laminate core transformer as an electrical analogy, or a fulcrum and bar as a mechanical analogy. This means that when the outer radiator area is greater than the inner radiator area, then the output volume velocity of the Acoustic Lever will

be more than that of the transducer as a ratio of the radiator areas. From this it follows that an Acoustic Lever with radiating area ratios of 2:1 will have an approximately 6dB amplification factor over the direct output of the driver, which is indeed the case and is what makes the Acoustic Lever such an interesting device.

Figure 3.19 (note: *Figs. 3.19* and *3.21–3.24* were all supplied by Dr. Geddes and produced using Speak_32) shows a comparison of a standard passive radiator bandpass response to the same driver in an Acoustic Lever type enclosure. A standard bandpass, as seen in *Fig. 3.20*, will show a decreased bandwidth with increasing efficiency. As with all things, there is no free lunch, and indeed the AL has its price to pay, and that is somewhat greater volume. However, since one can use high density fiberglass to increase the effective volume of an enclosure, the AL chamber, V2 in *Fig. 3.18*, can be reduced considerably, making the price for an extra 6dB efficiency fairly reasonable.

There are a number of variables that control the response of the "Acoustic Lever," such as the volume and damping of the lever chamber, V2, the front chamber, V1, and the rear driver chamber, V. *Figure 3.21* shows that the volume of the lever chamber, V2, directly controls the low-frequency response of the AL. As V2 decreases in volume, the AL low-frequency f_3 increases in frequency.

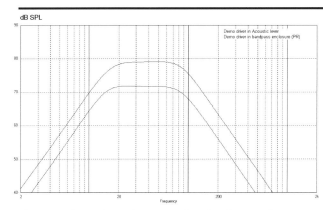

FIGURE 3.19: Comparison of passive radiator bandpass enclosure response to an Acoustic Lever with the same S_d woofer and passive radiator (exterior lever radiator for the AL).

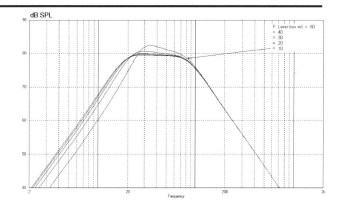

FIGURE 3.21: The effects of decreasing the lever chamber V2 on low-frequency response of the Acoustic Lever.

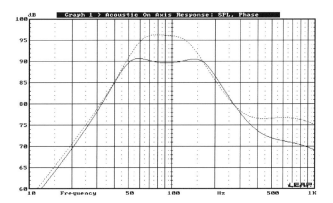

FIGURE 3.20: LEAP simulation showing the efficiency-bandwidth tradeoff for a standard bandpass enclosure.

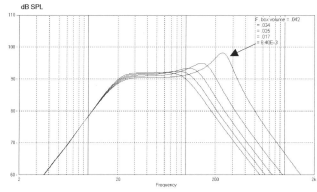

FIGURE 3.22: The effect of decreasing the volume of chamber V1 on the low-frequency response of the Acoustic Lever.

Somewhat like a standard bandpass, the front chamber, V1, controls the low-pass (upper frequency) roll-off of the system. However, looking at the graph in *Fig. 3.22*, you can see that something very interesting is going on with the AL. As the volume of the coupling chamber (front chamber in a standard BP) is decreased, the bandwidth increases, however, with only small changes in output. As Dr. Geddes put it in his AES article, a "designer's dream."

In terms of damping the individual chambers in an AL, these effects can be used in the final "tuning" of an AL system. Since the low-frequency extension is dominated by the lever chamber V2, increasing the damping in this chamber has similar effects on the high-pass roll-off of the AL system (see *Fig. 3.23*) as increased "stuffing" does in a standard sealed box system. Likewise since the coupling chamber V1 dominates the low-pass roll-off of the AL system, damping effects alter the knee of the upper frequency roll-off as illustrated in *Fig. 3.24*.

Since this is a relatively new design concept, there are no design tables to publish in order to create an AL. However, the criteria for building one are fairly straightforward. The formula, if you will, is simply to design a standard sealed rear chamber bandpass with a passive radiator. The two front chamber sizes for the AL, V and V1, will be the same as for the standard PR BP. For a 2:1 AL type design, the weight of the lever system will be ¼ the mass of the PR in a standard system. If the PR mass of the standard BP is 500 grams, then the lever system would be tuned with approximately 125 grams. The lever area of the larger piston is twice that of the smaller one

of the AL, the larger piston) that is larger than the transducer (i.e., for an 8″ driver, use a lever with approximately 6.5″ and 10″ diameter pistons). As for the volume of the lever chamber V2, the rule of thumb according to Dr. Geddes[10], is to make this chamber no smaller than the sum of the two other volumes V and V1 (V2 ≥ V + V1). Again, V2 could be decreased in volume by filling the box with fiberglass as long as the material does not interfere with the functioning of the lever. For example, you could line the interior of lever chamber V2 with Owens-Corning 703 fiberglass and partially fill the reamining area with R19 type fiberglass.

There are also dual lever incarnations of this device (cascaded levers are not viable) as seen in *Fig. 3.25*. In this case, both volumes V and V3 will effect the high-pass roll-off of the system.

Dr. Geddes asked me to remind readers that the AL is a patented device and not strictly legal to build, even for your own use, and absolutely illegal to manufacture without a license. You can simulate this unique type of enclosure with Earl Geddes' CAD design software Speak_32 (see *Chapter 9*). Purchasing this program also gives the user a "license" to build one AL for personal use.

REFERENCES

1. H. Olson, "Recent Developments in Direct Radiator High Fidelity Loudspeakers," *JAES*, October 1954.
2. Nomura and Kitamura, "An Analysis of Design Conditions for a Phase Inverter Speaker System with a Drone Cone," *IEEE Transactions Audio and Electroacoustics*, October 1973.
3. R. Small, "Passive-Radiator Loudspeaker Systems," *JAES*, Oct./Nov. 1974.
4. Douglas H. Hurlburt, "Complete Response Function and System Parameters for a Loudspeaker with Passive Radiator," *JAES* Volume 48, No. 2, March 2000.
5. D. B. Weems, *How to Design and Build and Test Complete Loudspeaker Systems*, Tab Books, no. 1064, 1978. (Out of print.)
6. G. R. Koonce, "Find f_p for Passive Radiator Speakers," *Speaker Builder*, 4/81, p. 25.
7. T. Clarke, "Augmented Passive-Radiator Loudspeaker Systems," *JAES*, June/July 1981.
8. E. R. Geddes, "The Acoustic Lever Loudspeaker Enclosure," *JAES*, Jan./Feb. 1999.
9. Acoustic Lever is a trademark of Gedlee Associates with exclusive license to Visteon Automotive products.
10. E-mail correspondence between Earl Geddes and Vance Dickason.

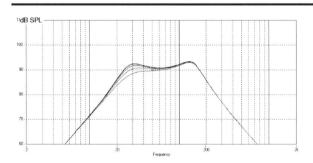

FIGURE 3.23: The effect of increasing internal damping (more fill material) in chamber V2 of the Acoustic Lever.

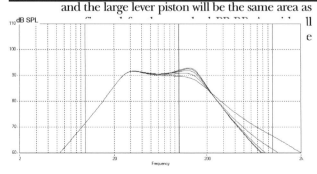

FIGURE 3.24: The effect of increasing internal damping (more fill material) in chamber V1 of the Acoustic Lever.

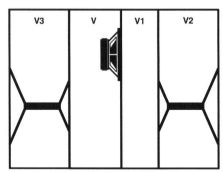

FIGURE 3.25: Diagram of Dual Acoustic Lever using levers on both front and rear radiating surfaces of the driver.

CHAPTER FOUR

TRANSMISSION LINE LOW-FREQUENCY SYSTEMS

4.10 DEFINITION.

Enclosed-box, vented-box, and PRs are all examples of enclosures which use various techniques to manipulate the resonant peak inherent in moving coil loudspeakers. Transmission lines (TLs) represent a class of supposedly non-resonant enclosures. TLs are a class of the device that perform as a phase inverter for the low-frequencies, allowing the energy of the rear of the woofer cone to be combined with the energy of the front of the cone (which is the goal of a vented or PR design), and simultaneously provide an approximately first-order low-pass transfer function to attenuate high frequencies[1]. However, at the outset, it is probably significant to note that transmission line loudspeakers are not particularly characteristic or analogous to an electrical transmission line. Electrical transmission lines generally are long in comparison to the wavelength and minimize standing waves caused by reflected energy. TL speakers are short compared to the wavelength being produced and have horrendous resonant modes that require significant damping (the inclusion of a seriously resistive element). A much better term for this class of loudspeaker would be "Damped Line," but at this point I don't think the term DL is going to "catch on." This being said, transmission line speakers can roughly be characterized by:

1. Low cabinet resonance
2. Relatively loud deep bass (below 50Hz)
3. Highly damped impedance peak
4. Decreased cone motion in the 40Hz region, accompanied by increased subsonic cone motion (as with vented, PR, and APR systems, this problem is easily overcome with an appropriate filter)
5. Low degree of mid-bass coloration
6. Low overall efficiency, somewhat like acoustic suspension sealed box speakers
7. High-pass roll-off and group delay similar to a well damped ($Q_{TC} = 0.6$–0.7) sealed box[2, 3]

The TL has gained a moderate degree of success in America, but is very popular in Britain, where its design was pioneered. A typical British TL features:

1. KEF drivers
2. tapered urethane or long fiber wool damped line
3. acoustical undercoating on interior walls
4. double wall cabinet construction
5. a speaker stand to isolate the speaker from the floor.

Not all audio professionals agree on this design. Martin Colloms, founder of Monitor Audio in England, believes the TL performance is no better than a properly-constructed vented enclosure, and that it is difficult to achieve uniform low frequency performance without exciting coloration-inducing line resonances in the mid-bass region.[4]

Despite an avid following among amateur speaker builders in the U.S., TLs have never been popular among U.S. speaker manufacturers, main stream or high-end. The last successful U.S. speaker company that specialized in TLs that I remember was Audionics which had a series of folded line towers in the mid '70s. Over the last ten years I have designed numerous well-reviewed home audio products for a number of speaker companies including Atlantic Technology, MB Quart, NEAR, Parasound, Signet, Snell (when my friend Kevin Voecks was running the engineering department) and a few other well known manufacturers I am not at liberty to mention. Not only have I never offered or recommended these companies incorporate TL designs into their product lineup, no one has ever even asked. It's not that TLs cannot provide accurate and musical bottom end sound quality, because they do, but that the device does not easily fit into modern manufacturing processes and is generally more expensive to build and significantly larger than other design methodologies. The typical volume of a full-size TL goes squarely against the current trend in compact subwoofers such as Bob Carver's Sunfire product. It appears from the lack of TLs used in commercial loudspeakers that the marginal sound quality difference between that and other well executed designs is mostly not worth the cost or effort. However, this often misunderstood device still provides a fruitful pursuit for those questing to have the very best loudspeaker ever produced, and I know there are still a lot of you out there. For me, I still have memories of TLs I was involved with in the '70s and the solid low-frequency sound one speaker in particular produced on the heart-beat in Pink Floyd's "Dark Side of the Moon" (and no I don't believe "Dark Side of the Moon" was written to be in-sync with the sound track of the *Wizard of Oz* movie). So while I do not incorporate this type of device into my consulting practice, I have attempted to collect in this chapter what I feel is the most up to date, accurate and useful information to guide you to the successful building of a TL.

4.20 HISTORY.

The antecedent to the present day transmission line design was the original patent on the Acoustic Labyrinth enclosure design authored by

FIGURE 4.1: Comparison of the measured SPL output of an undamped tapered, straight, and flared lines (reprinted from the Chalmers Report 74-35).

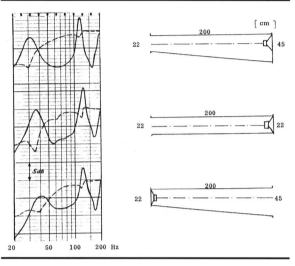

FIGURE 4.2: Comparison of the measured SPL output of an undamped straight line, single 90° bend line, and folded line (reprinted from the Chalmers Report 74-35).

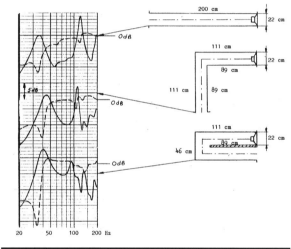

FIGURE 4.3: Comparison of measured SPL output for an undamped straight line, straight tapered line, folded tapered line, and Bailey type folded tapered line (reprinted from the Chalmers Report 74-35).

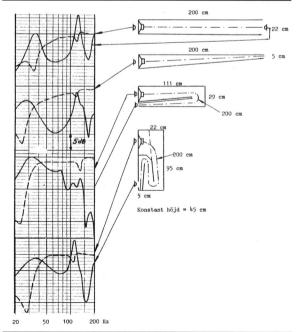

Benjamin Olney and described in his *JASA* article published in 1936[5]. This was incarnated in a commercial product by Stromberg-Carson from the '30s up until shortly after WWII. The original designs consisted of pipe length fixed to be about 25% of the wavelength of the driver free-air resonance and a cross sectional about the same as the driver S_d. This "pipe" was then folded to produce an acceptable shape that could fit into the "American" living room of the era and become an acceptable piece of furniture (oddly enough, an important design goal often time neglected in the pursuit of the ultimate speaker). The AL was lined with damping material, the idea being to filter out the upper frequency line resonance problems and then ideally combine the front and rear energy of the woofer at the low frequencies.

Working with the same basic concept in the early 1960s, A. R. Bailey experimented with different damping materials and techniques in folded labyrinth lines.[6, 7, 8] This work has since become the basic bible for most TL designs. Using Bailey's density criteria of 0.5 lb/ft[2], A. T. Bradbury published his 1976 paper[9] which described changes in the speed of sound for different types of damping material (fiberglass and long fiber wool).

Following Bailey's work and the success of several commercial transmission lines (not to mention the strong interest in TLs among *Speaker Builder* readers and other similar type publications worldwide) several attempts were made to produce a Novak/Thiele/Small type model that would allow the computer design of transmission lines in a fashion similar to sealed, vented, bandpass, and passive radiator loudspeakers. Two of the most notable attempts were by Robert Bullock[10] and Juha Backman[2]. Dr. Bullock and Peter Hillman met with some success in mathematically modeling a TL and while the math was turned into a software program (TL Box Model), the authors' own admission was that the model did not do well in predicting low-frequency behavior. Another significant step came with Juha Backman's attempt to describe TLs mathematically. This attempt was certainly notable in its effort to present a modeling that included group delay and cone excursion curves, but again, by the author's own admission, the work was never empirically validated, plus the model also suggested that tapered lines raise the cutoff frequency of the line, which is clearly not the case.

To this date, certainly the most impressive and empirically validated work on transmission lines comes from George Augspurger. George is a well known industry professional who began his loudspeaker career when he joined James B. Lansing Sound Inc. in 1958, becoming manager of JBL Pro prior to leaving in 1968 to form his own company, Perception Inc. If you read the *JAES*, George has written the patent review column for a number of years (this included a wonderful sense of humor that is more than appropriate when writing about audio patents). Most of the analysis and construction recommendations in this section are from Mr. Augspurger's work on TLs[1,3,11,12,13].

4.25 TRANSMISSION LINE BEHAVIOR and MODELING.

So exactly how do woofers perform when placed in an open-ended tube? One of the best studies done on transmission line behavior, and one which I have never seen referenced in literature on the subject, was written for an engineering masters thesis by Sven Tyrland and published in 1974 by the Chalmers University of Technology, Goteborg, Sweden[14]. The graphs and diagrams in *Fig. 4.1* show the empirical comparison of three undamped lines: a straight line, a tapered line, and a flared line (the solid curves are the nearfield response of the woofer and the dashed lines are the nearfield response of the line exhaust). Two things are immediately apparent. First, all of these straight-line variants suffer from resonant response anomalies above 100Hz. Second, and most important, is that line taper lowers the f_3 of the woofer in comparison to both a straight pipe and a flared pipe. The flared pipe would seem to be entirely undesirable as it tends to raise the f_3 in respect to the performance of a straight pipe. In terms of the upper frequency response anomalies caused by a line, folding the line seems to provide an immediate benefit aside from the obvious one of making the device more compact. This is illustrated in *Fig. 4.2* for two variants on a straight line. Note that folding the line also interferes with low-frequency performance and increases the f_3 frequency marginally.

Figure 4.3 compares three successful TL variants: a straight line, a tapered line, a simple folded tapered line, and a more complex folded line. None of these examples contain fill material, so the problem remains of how to eliminate the upper frequency modes and resonances that are transmitted by the line.

Attacking the line damping issue, Mr. Tyrland (see *Fig. 4.4*) experimented with a three tube folded line analyzing successive iterations of stuffing beginning with no stuffing to several complex stuffing methodologies. The solid line is the line exhaust output and the dashed line is the woofer cone output, both measured nearfield. Obviously, as the line gets more complex, the variety of ways to damp the line and achieve different results makes this a speaker tweaking dream.

What makes this data so interesting, besides the obvious one of explicating the effects of different line considerations with and without line damping, is that this study obtained very similar results to George Augspurger's work with TLs. *Figure 4.5* shows the measurements that were made in 1999 of an offset woofer in a tapered tube. Note how closely the shape of the two curves (the dark line is cone nearfield output and the light line is the line exhaust output also measured nearfield) compares to the

tapered line curves in *Fig. 4.3* for the Chalmers study. While the shapes are the same, it is obvi-

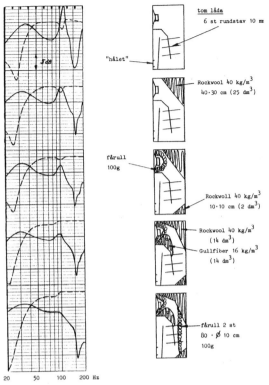

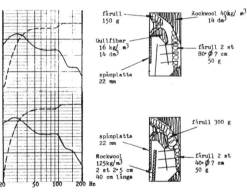

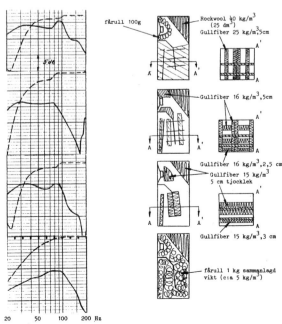

FIGURE 4.4
Comparison of measured SPL output for various stuffing methodologies incorporated into a Bailey style folded and tapered line (reprinted from the Chalmers Report 74-35).

ous that the resonant modes in the lines are occurring at very different frequencies. The Chalmers study used a KEF B139 woofer in a 78.75″ line, while Mr. Augspurger used a 3″ woofer in a 28″ line, which accounts for the frequency differential of the line harmonic output

that you are observing. Mr. Augspurger also notes that his results very closely mimicked the results achieved by another thesis study in 1975 at the University of Sydney Electrical Engineering department[15].

With a firm grasp of what a pipe does in terms of empirical results, Mr. Augspurger set out to come up with some type of computer simulation that would not only predict behavior, but would allow computer investigation of the methodology. George's tack on this departed significantly from previous failed attempts at computer simulation of TLs[2,10]. Instead of trying to work with models that focused upon air speed reductions due to stuffing material and effective line length changes, Mr. Augspurger started with the assumption that a damped pipe can be analyzed as a horn with losses. From here he adapted a basic model of an analog electrical transmission line developed by Bart Locanthi (Mr. Locanthi passed away several years ago, but remains one of the best speaker engineers of this era)[16]. The results are impressive as can be seen in *Fig. 4.6*. Comparing this response simulation of an undamped line to the measured line shown in *Fig. 4.5* shows a degree of validity never achieved by earlier attempts. To this was added a fairly complex method of simulating the various types of stuffing materials traditionally used in TL construction such as fiberglass, polyester fill, microfiber, and Acousta-Stuf®, a nylon polyamide available from Mahogany Sound. Combined analysis of damping losses

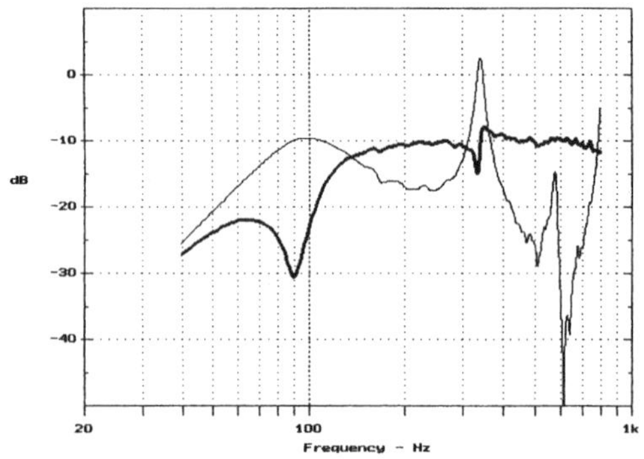

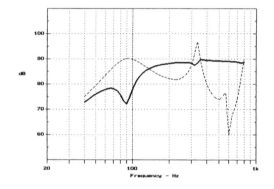

FIGURE 4.5: Measured response of an undamped offset line (reprinted from AES preprint 5011).

FIGURE 4.6 Computer simulation of the same undamped offset line in *Fig. 4.5* (reprinted from AES preprint 5011).

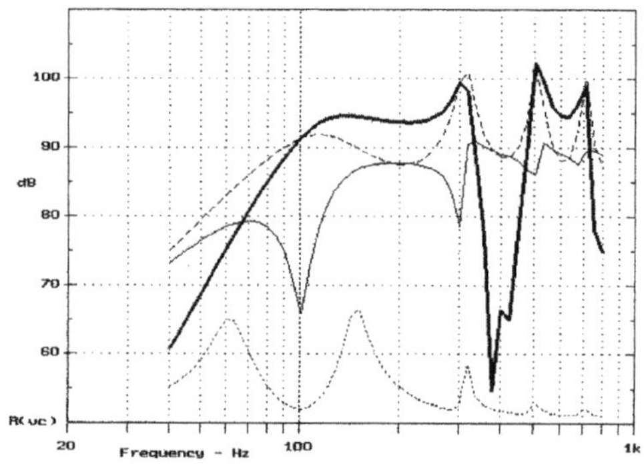

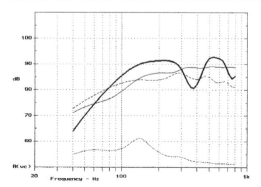

FIGURE 4.7: Computer simulation of an undamped straight line (reprinted from *SB* 2/00).

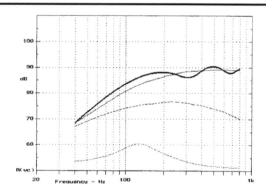

FIGURE 4.9: Computer simulation of a moderately damped straight line (reprinted from *SB* 2/00).

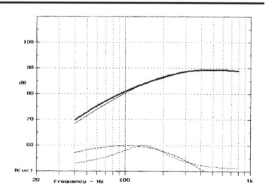

FIGURE 4.8: Computer simulation of a lightly damped straight line (reprinted from *SB* 2/00).

FIGURE 4.10: Computer simulation of a heavier damped straight line (reprinted from *SB* 2/00).

included four empirical parameters: fixed losses; the corner frequency of the variable losses (frequency dependent losses); the slope of the variable losses; and relative sound speed through the material. It is interesting to note that Mr. Augspurger discovered that speed of sound through the damping material, a real focus of most TL work done prior to this was not a relevant control parameter and that there is no direct correlation between pipe length and stuffing density. It seems that everybody read Bailey's paper, whose convincing assumptions made it a little hard to think "outside of the box," and all of us were pretty much guilty of "chasing the wrong rabbit" as George put it in his *SB* article[1]. Further, the primary control parameter of transmission line speakers is the Q_{TS} of the driver, with pipe shape coming in a close second.

Armed with a simulation that reasonably describes the functioning of a damped line, it is now possible to simulate the effects of different stuffing densities. *Figures 4.7–4.10* show computer simulations of a small woofer in a 3′ pipe with progressively greater stuffing densities. As stuffing density increases, several changes can be observed: the response gets smoother as the material damps upper harmonic anomalies; the exhaust port output decreases in magnitude as density increases; and the impedance becomes progressively damped. Analogous to a progressively over-stuffed sealed box, as the density of stuffing increases in the line, the knee (corner) of the high-pass roll-off becomes more shallow and f_3 increases (one can probably assume a decrease in group delay at this point as well).

The results shown in *Figs. 4.7–4.10* compare well with the empirical results of iterative types of stuffing in a more complex line shown in *Fig. 4.4* from the Chalmers study. One caveat to this simulated and empirically measured data has to do with actual system response when placed in a room. Both the nearfield empirical measured results in *Fig. 4.4* and the system response curves in the computer simulation in *Figs.4.7–4.10* assume phase correlation between the cone output and line exhaust. Phase correlation means that both cone and exhaust are emanating from pretty much the same point in space and time. Both Mr. Augspurger and Mr. Letts suggest that if the exhaust port is separated from the cone output for any substantial distance, the acoustic summation in a room will not have predictable results. From this, you can conclude that a folded line with fairly coinci-

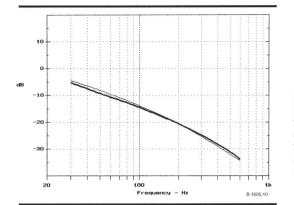

dent cone and exhaust port radiation may have a different subjective sound quality from a straight line whose cone and port are substantially non-correlated.

4.30 LINE DAMPING CONSIDERATIONS.

Previous empirical evidence suggested that shorter lines were possible with increased stuffing densities[17], which is another way of saying that shorter lines need increased stuffing density to match the performance of a longer line. Augspurger's study also confirms this. Although increased densities of line damping material in short lines can be made to produce the same transfer function (rate of frequency dependent attenuation) as lesser densities in longer line, this is not the primary control over cutoff frequency. *Figure 4.11* shows a computer simulation comparison of a 1 lb per ft^3 density in a 6′ line versus 3 lb. per ft^3 in a 3′ line. As can be seen, the results are nearly identical, however, path length is the dominant factor and can not be compensated for with speed changes due to increasing stuffing densities. The important conclusion made in this study is that the line cutoff, f_3, will occur between 0.7 and 1.4 times the line resonance, f_p, of the undamped line. Further, adjustments to response are better accomplished by changing driver parameters, not stuffing densities.

Since stuffing density relates primarily to damping the passband response anomalies (ripple), it would then be possible to provide an equivalent stuffing density criterion that establish the density requirements for various type materials that would result in approximately the same response. This is exactly what was done and a comparison of the different material densities for different line lengths to produce the same acceptable level of ripple (±1dB) is given in *Table 4.1*. This data was empirically confirmed for fiberglass, polyester

FIGURE 4.11: Computer simulation showing the transfer function difference between a 1 lb density Acousta-Stuf filled 6′ pipe and 3 lb density Acousta-Stuff filled 2′ pipe (reprinted from *SB* 3/00).

TABLE 4.1

LINE LENGTH	F_P	ACOUSTA-STUF	POLYESTER	FIBERGLASS	MICROFIBER
24″	140Hz	1.70	1.80	0.90	0.65
36″	94Hz	1.30	1.40	0.70	0.55
48″	71Hz	1.00	1.10	0.60	0.45
72″	48Hz	0.75	0.85	–	0.35
96″	36Hz	0.50	0.65	–	0.27

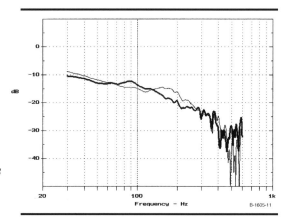

FIGURE 4.12: Measured transfer function difference between 1 lb Acousta-Stuf and 0.5 lb density R19 fiberglass in a 6′ pipe (reprinted from *SB* 3/00).

fill (Dacron), and Acousta-Stuf (to a lesser extent with Microfiber) and is valid for tapered, offset, and coupled chamber type lines. This data was not confirmed for simple straight lines.

While the line stuffing densities in *Table 4.1* are approximately equivalent in terms of suppressing upper harmonic levels, each of these materials will provide a different frequency dependent attenuation of the rear wave of the woofer. *Figure 4.12*[1] shows a measured response comparison depicted as a normalized transfer function done in the Augspurger study for fiberglass and Acousta-Stuf with obvious variances in the frequency response. In terms of anecdotal experience, long-fiber wool has always been the weapon of choice for experienced TL builders, with Acousta-Stuf coming in as a close second, while Dacron and fiberglass generally reported to provide inferior results in terms of subjective sound quality. Given the nature of "anecdotal" evidence and its typical inconsistencies, it's hard to make a valid statement about the subjective nature of

these materials other than to say that each of these materials will have its own subjective sound quality. In terms of pure application, long fiber wool is generally not easily available and difficult to work with, while fiberglass is also a somewhat itchy material to use. This leaves Dacron and Acousta-Stuf that are relatively easy to install.

As can be seen in *Fig. 4.4*, the variations in the ways to stuff a TL are many. Damping material can be suspended on dowels or nylon cord in the body of the line, or the surface of the line, such as the urethane foam damped lines from several manufacturers in the U.K. (foam was not mentioned as a line damping material in the Augspurger study). The other most common line stuffing variation has to do with changing the density along the length of the line. Typical formats include lighter densities behind the woofer progressing to full density at the exit (resulting in a constant resistive loading over a broad spectrum), or more commonly, to leave the last 20% of line length undamped. Final tuning then amounts to adjusting the stuffing density of the last few feet of the line, often seen in simple two tube folded tapered lines. This is usually done subjectively along with perhaps successive impedance curves to get some indication of damping (see *Fig. 4.13*).

4.35 LINE SHAPES.
The second part of Mr. Augspurger's study included looking at variations in line types. Although he didn't cover all the variations I have seen over the years, his data confirmed the utility of the most commonly used formats. These, shown in *Fig. 4.14*, are the tapered line, coupled chamber line, and offset driver line. The tapered line has found wide application in manufactured TLs and its easiest incarnation is a tower with an offset inner baffle as shown in the diagram in *Fig. 4.15* "c". Augspurger's work recommended a taper variation of 1:3 to 1:4. The taper used in the Chalmers study shown in *Fig. 4.3* has a taper ratio of about 1:4.4. Looking at the graph in *Fig. 4.16*[14], again from the Chalmers work, you can see the result progressing from a straight pipe to increasingly smaller exhaust ports from the resulting increase in taper ratio in an undamped line. The effect is pretty easy to see. As the taper ratio increases, the f_3 decreases in frequency while the efficiency goes down. In this comparison, f_3 went from 36Hz to 29Hz with the loss of about 2dB for a 1:4.4 taper ratio. Also apparent is a decrease in the height of the harmonic resonance at 125Hz, also beneficial.

Augspurger's study assumes a beginning line area within probably 25% of the S_d of the driver. Previous anecdotal work has varied the beginning area from 1.25 to 2.5 times driver S_d, with reports of different subjective sound quality for the changes[18]. Although Mr.

FIGURE 4.13: Typical impedance plot for tuning a transmission line.

FIGURE 4.14: Transmission line formats used in the Augspurger study (A = Tapered; B = Coupled Chamber; C = Offset).

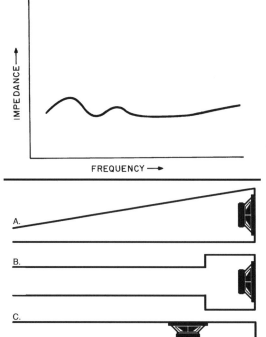

Augspurger has given a unified methodology for creating successful lines, other taper ratios and initial line areas are certainly worthy of consideration and experimentation.

A coupled chamber type line (type B in *Fig. 4.14*) has described a number of successful lines over the years and a typical incarnation would resemble the box lettered "e" in *Fig. 4.15* (this also combines a tapered line). Loading the driver into a small chamber provides an additional 6dB low-pass filter that further attenuates the upper harmonics generated by the line. Augspurger's suggestion is to make this volume about ⅓ of the total line volume.

The last enclosure variation investigated in this study was the offset type line. In this case the woofer drives the line from a distance between ⅓ to ⅕ of the line from the line beginning, similar to the box drawing "c" in *Fig. 4.14*. At a ⅕ offset, driver f_3 is set to be about 20% higher than the line f_p[12] for a nominally flat frequency response.

4.40 TL ALIGNMENT TABLES.

Using the software developed as a result of this project, Mr. Augspurger derived a set of alignment tables that are for specific line types. Two alignment groups are offered, a standard alignment resulting in more compact lines where f_3 is higher than f_p, and a group of

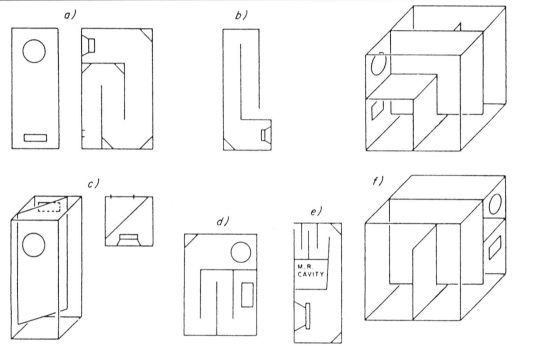

FIGURE 4.15: Examples of various transmission line formats.

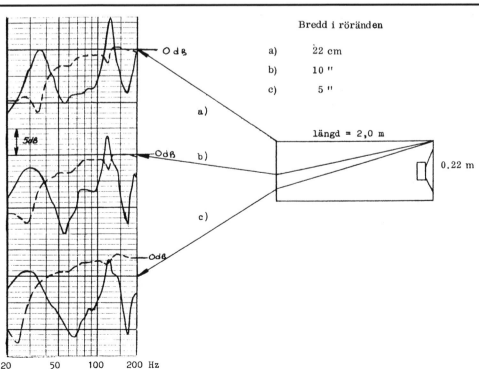

Bredd i röränden

a) 22 cm

b) 10 "

c) 5 "

längd = 2,0 m

0.22 m

FIGURE 4.16: Comparison of measured SPL output of a straight line, a 2.2:1 tapered line, and a 4.4:1 tapered line (reprinted from the Chalmers Reports 74-35).

larger volume lines with lower f_3s for a given driver Q_{TS}, about ⅓ of an octave for tapered lines and ⅙ of an octave for coupled chamber lines. Ratios that are used to determine cutoff frequency, line length, and line volume are given in *Table 4.2* for Augspurger's Standard alignments and *Table 4.3* for his Extended alignments. Each alignment set offers three different Q_{TS} choices. Obviously for points in-between, interpolation is appropriate.

For an example, an 8″ KEF B-200 has a Q_{TS} of 0.41, $f_S = 25.0$Hz, and a V_{AS} of 131 ltr. Using the Extended Alignment *Table 4.3*, $f_3/f_S = 1.3$, so $f_3 = 1.3 \times 25.0 = 32.5$Hz, which is not bad considering the shallow rolloff of a TL. The nominal ¼ wavelength resonance frequency of the line is also presented as a ratio to either f_3 or f_S. Using either ratio yields a f_P of about 41Hz. Since TLs are fairly forgiving devices, you could interpolate a line length from *Table 4.1* to be 84″ with an approximate stuffing density of 0.6 lbs per square foot of Acousta-Stuf, or just go with the next longest length and it will probably work just fine. Total line volume is determined by the ratio to V_{AS} of the driver. In this case the V_{AS} of the 8″ B-200 woofer was 131 ltr, and the ratio from *Table 4.3* is 0.6, so the line volume will be $131/0.6 = 218.3$ ltr, or 7.71ft³. The opening area of the pipe would be 16″ × 16″ with a 4.0″ × 16″ exhaust port, so this is a pretty large device, even if the line is folded. An f_3 of 38Hz can be had with the same driver (with the limited dynamic range characteristic of this woofer) in a 2.0ft³ sealed box ($Q_{TC} =$

0.73), so the TL version is a lot of work for that extra 5.5Hz. You really must like the subjective sound quality of a TL to make it worth the effort.

4.50 CONSTRUCTION ARTICLES.
This current chapter on transmission line construction relies heavily on George Augspurger's methodology, much in the same way the sealed and vented chapters rely on Neville Thiele and Richard Small. Transmission lines, however, have a wider variety of possible variations. Mr. Augspurger's work covers a lot of TL territory, but you may wish to investigate other "prior art," to look at some of the successful designs that have appeared over the years.

Speaker Builder

1. R. Sanders, "An Electrostatic Speaker System," 2, 3, 4/80.
2. G. Galo, "Transmission Line Loudspeakers," 1, 2/82.
3. C. Cushing, "A Compact Transmission Line Subwoofer," 1/85.
4. T. Cox, "An Experimental Transmission Line," 4/85.
5. C. Bauza, "The Modified Daline," 4/85.
6. D. Weems, "Experiments with Tapered Lines," 2/87, p. 18.
7. J. Cockroft, "The Octaline: A Small Transmission Line," 3/87, p. 9.
8. J. Cockroft, "The Shortline: A Hybrid Transmission Line," 1/88, p. 18.
9. J. Cockroft, "The Unline: Designing Shorter Transmission Lines," 4/88, p. 28.
10. P. Hillman, "Symmetrical Speaker System with

TABLE 4.2
STANDARD TRANSMISSION LINE ALIGNMENTS

LINE TYPE	Q_{TS}	F_3/F_S	F_3/F_P	F_S/F_P	V_{AS}/V_P
1:4 Taper	0.36	2.0	0.8	0.40	3.10
	0.46	1.6	0.8	0.50	2.00
	0.58	1.3	0.8	0.63	1.20
Coupled chamber	0.31	2.00	0.80	0.40	2.14
	0.39	1.60	0.80	0.50	1.35
	0.50	1.30	0.80	0.63	0.84
Offset	0.36	2.00	1.20	0.60	3.10
	0.46	1.60	1.20	0.74	2.00
	0.58	1.30	1.20	0.94	1.20

TABLE 4.3
EXTENDED TRANSMISSION LINE ALIGNMENTS

LINE TYPE	Q_{TS}	F_3/F_S	F_3/F_P	F_S/F_P	V_{AS}/V_P
1:4 taper	0.25	2.00	0.80	0.40	1.50
	0.33	1.60	0.80	0.50	1.00
	0.41	1.30	0.80	0.63	0.60
Coupled chamber	0.31	1.75	0.70	0.40	1.10
	0.39	1.40	0.70	0.50	0.68
	0.50	1.10	0.70	0.63	0.42
Offset	0.25	2.00	1.20	0.60	1.50
	0.33	1.60	1.20	0.74	1.00
	0.41	1.30	1.20	0.94	0.60

Dual Transmission Lines," 5/89, p. 10.

11. J. Cockroft, "The Microline," 5/89, p. 28.

12. G. DeMichele, "Cylindrical Symmetric Guitar TLs," 1/90, p. 22.

13. S. Ellis, "An Apartment TL," 1/91, p. 32.

14. C. Cushing, "The Pipes," 2/91, p. 18.

15. W. Wagaman, "Octaline Meets D'Appolito," 2/91, p. 38.

16. S. Wolf, "Pipe and Ribbon Odyssey," 3/91, p. 28.

17. R.J. Spear and A.F. Thornhill, "Fibrous Tangle Effects on Acoustical TLs," 5/91, p. 11.

18. Spear and Thornhill, "A Prize-Winning Three-Way TL," Part I, 4/92, p. 10; Part II, 5/92, p. 22; Part III, 1/93, p. 44.

19. J. Cockroft, "The Simpline," 1/93, p. 14.

20. D.K. Johns, "A 15″ Transmission Line Woofer," 7/94, p. 10.

21. K.W. Ketler, "The Achilles: A Two-Way Transmission Line," 1/95, p. 22.

22. J. Cockroft, "The Simpline Sidewinder Woofer," 4/95, p. 8.

23. J. Cockroft, "The Super Simpline," 1/96, p. 18.

24. R. Watson, "Design a Three-Way TL with PC Audiolab," 2/96, p. 12.

25. J Cockroft, "The Squatline, HDOLLP & ALL," 3/96, p. 16.

26. J. Viola, "Doing the Daline," 8/97, p. 16.

27. J. Cockroft, "The B-Line," 1/98, p. 8.

28. J. Mattern, "Another Look at TL Design," 3/99, p. 28.

Audio Amateur–
Loudspeaker Projects 1970–1979

1. T. Jastak, "A Transmission Line Speaker," p. 39.

2. T. Jastak, "A Jolly TL Giant," p. 42.

3. B.J. Webb, "A Proven TL Loudspeaker," p. 53.

4. D. Ruether, "The Big Bass Box," p. 126.

REFERENCES

1. G. L. Augspurger, "Part 2, Transmission Lines Updated—Stuffing Characteristics," *Speaker Builder* 3/00.

2. J. Backman, "A Computational Model of Transmission Line Loudspeakers," 92nd AES Convention, 1992, preprint no. 3326.

3. G. L. Augspurger, "Loudspeakers on a Damped Pipe," *JAES*, May 2000.

4. M. Colloms, *High Performance Loudspeakers*, Pentech Press, 1978, 1986, 1991.

5. B. Olney, "A Method of Eliminating Cavity Resonance, Extending Low-Frequency Response and Increasing Acoustic Damping in Cabinet Type Loudspeakers," *Journal* of the Acoustical Society of America, Oct., 1936.

6. A.R. Bailey, "A Non-Resonant Loudspeaker Enclosure Design," *Wireless World*, October 1965.

7. T. Jastak, "A Transmission Line Speaker," *Audio Amateur*, January 1973.

8. A.R. Bailey, "The Transmission Line Loudspeaker Enclosure," *Wireless World*, May 1972.

9. A. J. Bradbury, "The Use of Fibrous Materials in Loudspeaker Enclosures," *JAES*, April 1976.

10. R. M. Bullock and P. E. Hillman, "A Transmission-Line Woofer Model," 81st AES Convention, 1986, preprint no. 2384.

11. G. L. Augspurger, "Part 1, Transmission Lines Updated," *Speaker Builder* 2/00.

12. G. L. Augspurger, "Part 3, Transmission Lines Updated—Pipe Geometry and Optimized Alignments," *Speaker Builder* 4/00

13. G. L. Augspurger, "Loudspeakers on a Damped Pipe, Part One: Modeling and Testing; Part Two: Behavior," 107th AES Convention, 1999, preprint no. 5011.

14. Sven Tyrland, "KONSTRUCTION AV EN MONITORHOGTALARE," Rapport 74-35, Chalmers University of Technology, Goteborg, Sweden (www.chalmers.se) (this book was recommended to me by Mats Jarstrom, an engineering associate and friend).

15. G. Letts, "A Study of Transmission Line Loudspeaker Systems," Honors Thesis, University of Sydney, School of Electrical Engineering, Australia, 1975. Note this reference was included for completeness, but not viewed by this author.

16. B. N. Locanthi, "Application of Electric Circuit Analogies to Loudspeaker Design Problems," *JAES*, October 1971.

17. J. Cockroft, "The Unline: Designing Shorter Transmission Lines," *Speaker Builder* 4/88.

18. Vance Dickason, *Loudspeaker Design Cookbook*, 5th Edition, copyright 1997.

CABINET CONSTRUCTION: SHAPE AND DAMPING

5.10 ENCLOSURE SHAPE AND FREQUENCY RESPONSE.

The majority of low-frequency cabinets are rectangular in shape. This not only makes for reasonably aesthetic-looking loudspeakers/furniture, but is the easiest shape to construct for both amateur and manufacturer. The rectangular loudspeaker enclosure is, however, often judged less than optimal as a radiating surface because of edge diffraction issues and also less than optimal regarding internal standing wave modes.

A. The Olson Loudspeaker Enclosure Shape Study

Harry Olson's 1951 *JAES* article titled "Direct Radiator Loudspeaker Enclosures" is the classic work illustrating the effect of enclosure shape on cabinet diffraction. The article described a study done to determine the effect of different shapes on the frequency response of a speaker. Twelve shapes were used, which included a sphere, a hemisphere, a cylinder with the driver mounted in the end, a cylinder with the driver mounted on the curved surface, a cube, a rectangle, a cone (driver mounted in the tip), and double cone, a pyramid (the driver mounted in the tip), a double pyramid, a cube with beveled edges (the bevel equal to the width/height of the baffle), and a rectangle with beveled edges (the bevel equal to the width of the baffle). A $7/8''$ driver was mounted on each enclosure and measured in an anechoic chamber. The results showed the different shapes provided anywhere from a nearly flat response to a constantly undulating one with a ±5dB variation. The various shapes and their associated on-axis curves reprinted from Mr. Olson's AES articles are shown in *Fig. 5.1*. A summation of the SPL variations from the various cabinet shapes that might be practically used in loudspeaker design follows:

Shape	Variation
Sphere	±0.5dB
Cube	±5dB
Beveled Cube	±1.5dB
Rectangle	±3dB
Beveled Rectangle	±1.5dB
Cylinder	±2dB

From this, it is obvious that an enclosure in the shape of a sphere gives the least amount of "ripple" in the response. While this is good news, the sphere is a somewhat difficult shape to manufacture and there have never been many examples to reach the market, a few exceptions being the Gallo Acoustics Micro, the Morel Soundspot, and the satellite speakers from Orb Audio. While Dr. Olson's is still

the best study of its type, and indeed reveals much about enclosure shape and the resulting SPL, it has limitations in terms of both driver location vs. SPL, as well as not including some other enclosure shapes that have become popular over the years since 1951. This fact prompted me to undertake a second study in enclosure shape that takes up somewhat where Dr. Olson left off.

B. Olson's Enclosure Shape Study Extended

Since the publication of the 6th Edition of the *Loudspeaker Design Cookbook*, LinearX has released the Windows version of LEAP, LEAP 5. One of the many important new features of this software was the addition of a very powerful diffraction engine. The analysis mode for the box design part of LEAP 5, titled EnclosureShop, now includes what is literally an anechoic chamber in your computer. With the ability to accurately simulate up to 8th-order diffraction, LEAP 5 can quickly perform extensive diffraction analysis on a variety of shapes as well as be able to locate the transducer anywhere on the baffle surface, making this extended enclosure shape study much easier to undertake.

The shapes studied include some of the same ones done in Dr. Olson's original 1951 paper plus a few that weren't on Harry's list. Included in this 2005 study are a cube ($15'' \times 15'' \times 15''$), a beveled cube with 2" bevel, a beveled cube with a 4" bevel, a rectangle ($18'' \times 12'' \times 9''$), a beveled rectangle on four sides (Olson's was only beveled on three sides) with 2" bevel, a beveled rectangle on four sides with 4" bevel, a pyramid with the driver located on a facet (Olson's were located on the apex) (18" height with 4" width at top and 10" width at the bottom), a cylinder (18" height, 16" diameter), a sphere (16" diameter) and an egg-shaped enclosure (18" height, 14" diameter).

If you look at the shapes in *Fig. 5.1*, you will notice that the driver was located in the center of square-shaped types, the sphere, and the cylinder, and at different locations between the center and top of the enclosure on rectangles. I have also chosen to ignore the shapes in which the driver was located at the apex in this section because they are either not likely to be used as a commercial speaker enclosure or never have been to my knowledge. Because location of the driver on the baffle has such a strong effect on the response smoothness (this is investigated extensively in Chapter 6), I decided to include more than one driver location for each shape so as to better investigate the use of this shape for the different driver formats being used today.

Since Dr. Olson was trying to define the SPL response across the relevant frequency range, he designed a very special driver that had a response from below 100Hz to above 4kHz, but that had a power (combined off-axis) response that would not affect the results by "beaming" at the higher frequencies. The 7⁄8″ cone driver he used was essentially a miniature woofer that could produce energy at 100Hz. Although Mr. Olson never published the absolute

SPL of the unique device he designed for the study, you can assume that the SPL was very low so as not to cause the tweeter-sized woofer to overexcurse and distort at low frequencies. Since I did not have data available on that particular unique transducer, I instead substituted two drivers, a 2″ wide range cone driver and a 1″ dome tweeter. Between these two drivers, you can get a very good idea about what is happening with each of these enclosure shapes.

Each shape was used to produce four SPL curves, one curve for each of the two drivers at the two different baffle locations. The two locations were defined as mounting the simulated driver mid point in the center of the baffle for the one reference point, and the other location at the top of the baffle, centered between the right and left sides of the enclosure. The center position was used because many of the current woofer-tweeter-woofer (often referred to as the D'Appolito configuration for Dr. Joseph D'Appolito, who first published this design concept in *Speaker Builder* magazine) designs usually have a driver located at the center of the baffle. The second location at the top of the baffle was used because it is a typical location for tweeters. The exact location at the baffle top was roughly far enough from the baffle edge to allow for a grille frame to be installed.

Pictures from LEAP 5 for each shape and baffle location can be seen in *Figs. 5.2* through *5.19* (these are for the tweeters only; however, the 2″ full-range was placed in the exact same baffle locations). Because there is only a center position for a sphere- or egg-shaped box, only one baffle location was used for each of these two shapes. You will also notice that some of the shapes appear to be faceted; however, this is a necessary part of the technology employed in LEAP 5 that enables the software to analyze these shapes.

Before considering what each of these different-shaped cabinets does to alter the SPL of a driver, we need to establish some kind of reference with which to judge all the changes that can be observed. The choices would be to look at the driver mounted with no baffle in open air or mounted on an infinitely large baffle. Whatever any baffle does to a transducer's overall SPL, it will fall somewhere between these two extremes.

Figure 5.20 shows the 1″ tweeter (solid curve) and 2″ cone full-range (dashed curve) suspended in open air (anechoic) with no baffle, and *Fig. 5.21* depicts the response of the same two devices mounted on an infinitely large baffle, which is the same as saying they are being measured in half space (note, all data simulated at 2.83V/1m on-axis). As you can see, when mounted on an infinitely large baffle, the response is substantially flattened out and the anomalies become washed out by the full-range reflective nature of the baffle. As you go through the various examples, it will be helpful to keep these two extremes in mind.

Each of the different shapes except for the sphere and egg-shaped enclosure have four curves placed on two graphs, one graph for the 1″ tweeter with the SPL generated at 2.83V/1m on-axis at both the top (dashed curve) center baffle location and the center middle baffle location (solid

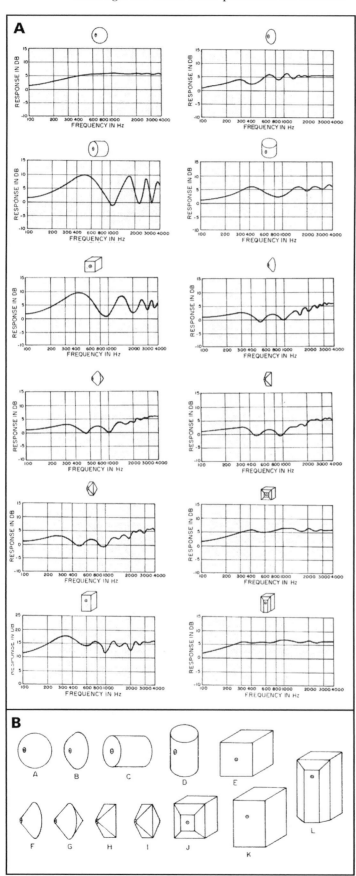

FIGURE 5.1A (right) and B (below). After Direct Radiator Loudspeaker Enclosures (Harry Olson, JAES, November 1951).

curve), and the same presentation for the 2″ cone full-range driver in the second graph. The sphere and egg-shaped box have only one graph with both the 1″ dome (solid curve) and 2″ cone (dashed curve) on the same scale. This graphic series is depicted in *Figs. 5.22–5.39*. The data shown in *Tables 1* and *2* summarizes the ±SPL range for each of these. Because the tweeter begins to rolloff below 1.25kHz, its data was calculated in two ranges, 1kHz–10kHz and 2kHz–10kHz, and given in *Table 1*. The 2″ full-range driver was examined from 500Hz to 10kHz, with the data displayed in *Table 2*.

There are some general conclusions to be drawn from this. First, a word of caution, the SPL data for this type of study can vary substantially by the choice of dimensions, so no matter what, the best you can hope for is to observe some general trends. That said, the following can be concluded from these graphs:

1. Cubes have the most SPL variation, followed by the standard rectangle, pyramid, egg shape, cylinder, and finally the sphere.

2. Beveled edges do decrease SPL variation, but it takes a substantial bevel to be really effective.

3. While a sphere may be the best performer in terms of minimal SPL variation, egg-shaped and cylinder-shaped enclosures are also quite good in this respect. You will notice a drawing in Chapter 6 of a cylindrical-shaped dual enclosure loudspeaker.

Table 1. 1″ Dome SPL dB Variations for Different Enclosure Shapes.

	1″ Soft Dome 1kHz-10kHz		1″ Soft Dome 2kHz-10kHz	
	Center	Top	Center	Top
Cube	4.73	3.29	1.99	1.15
Cube 2″ Bev	3.33	2.77	3.33	1.32
Cube 4″ Bev	2.54	2.65	2.54	0.92
Rectangle	3.09	2.22	1.64	1.84
Rect. 2″ Bev	2.33	2.72	1.90	1.83
Rect. 4″ Bev	3.05	3.89	1.71	1.38
Pyramid	1.76	2.78	1.49	1.18
Cylinder	2.82	2.60	0.81	0.78
Sphere	2.72	NA	1.11	NA
Egg	2.18	NA	0.55	NA

Table 2. 2″ Full-Range SPL dB Variations for Different Enclosure Shapes.

	2″ Full-Range 500Hz-10kHz	
	Center	Top
Cube	4.35	2.22
Cube 2″ Bev	3.22	1.92
Cube 4″ Bev	2.82	1.20
Rectangle	3.03	2.26
Rect. 2″ Bev	2.05	2.28
Rect. 4″ Bev	1.13	1.60
Pyramid	2.71	2.30
Cylinder	1.29	1.15
Sphere	1.08	NA
Egg	1.51	NA

This drawing is a representation of a loudspeaker I introduced with my first company, SRA (Speaker Research Associates) at CES (Consumer Electronics Show) in Las Vegas, 1978.

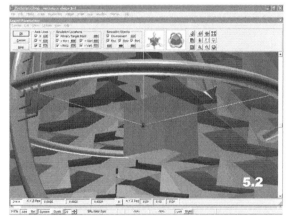

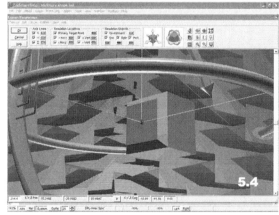

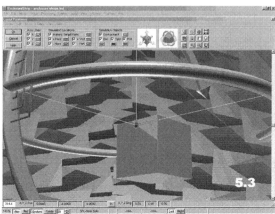

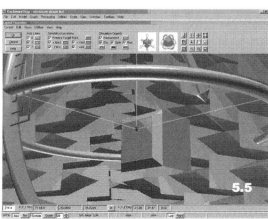

FIGURE 5.2: Cube Enclosure (driver center).

FIGURE 5.3: Cube Enclosure (driver top).

FIGURE 5.4: 2″ Beveled Cube Enclosure (driver center).

FIGURE 5.5: 2″ Beveled Cube Enclosure (driver top).

4. Pyramid-shaped enclosures are not appreciably better than rectangle box-type enclosures.

5. Drivers mounted near the top or bottom of an enclosure have substantially less SPL variation than drivers mounted in the center. This will be investigated in much greater detail in Chapter 6, in-

cluding a subjective study of this type of diffraction phenomenon.

Besides the simulated data shown in this enclosure shape study, I also published an empirical look and the difference between standard enclosures

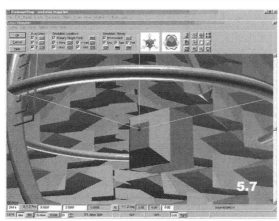
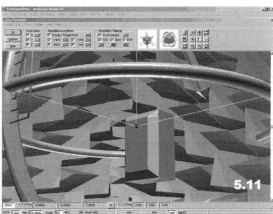
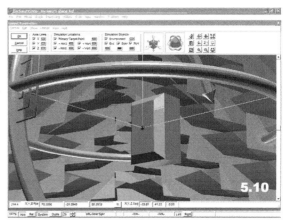
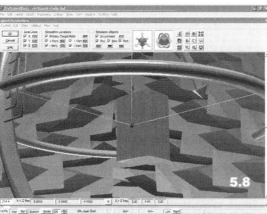
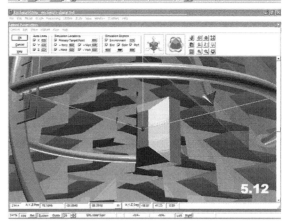
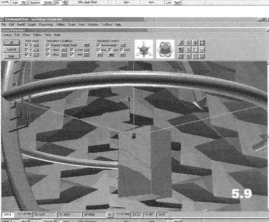
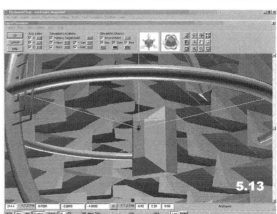

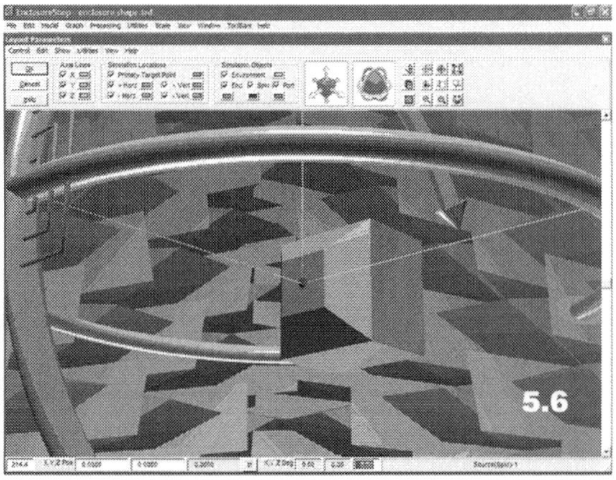

FIGURE 5.6: 4″ Beveled Cube Enclosure (driver center).

FIGURE 5.7: 4″ Beveled Cube Enclosure (driver top).

FIGURE 5.8: Rectangle Enclosure (driver center).

FIGURE 5.9: Rectangle Enclosure (driver top).

FIGURE 5.10: 2″ Beveled Rectangle Enclosure (driver center).

FIGURE 5.11: 2″ Beveled Rectangle Enclosure (driver top).

FIGURE 5.12: 4″ Beveled Rectangle Enclosure (driver center).

FIGURE 5.13: 4″ Beveled Rectangle Enclosure (driver top).

and a more exotic enclosure in *Voice Coil* magazine. The October 1990 issue featured a comparison between a rectangular-shaped enclosure and a flat baffled cylindrical shape, shown in *Figs. 5.40* and *5.41* (drivers were not inset on either enclosure). The flat-sided cylinder is manufactured by Cubicon, who makes geometric cardboard shapes for the furniture, display, and the speaker industry. The response differences shown in *Figs. 5.40* and *5.41* are not very dramatic for either the tweeter response or the woofer response (the woofer test was made without enclosure fill material, so part of the deviation is due to unsuppressed internal standing wave modes), but some deviation is apparent. The measurement was done with the MLSSA FFT analyzer and is windowed at 10mS, making the measurement essentially anechoic in nature. The data was moved from MLSSA into LEAP 4.0 to facilitate PostScript printout.

Since the anechoic measurement of these two different-shaped enclosures is so close using identical drivers, a subjective judgment of which shape "sounded" best in a room would be difficult. The final subjective response to any enclosure is influenced by the location of the speakers on the baffle and the way in which the various response variations in the driver combine with the variations caused by whatever enclosure diffraction effects are subjectively apparent in the listening environment. Although exotic enclosure shapes would seem intuitively to offer some of the best alternatives, the reality is that it isn't critically as important as has been claimed by some manufacturers. It remains true that some of the best reviewed and successfully marketed loudspeakers used simple rectangular shapes.

One thing that many loudspeaker companies have ignored over the years when designing off-wall loudspeaker enclosures is that loudspeakers are more than just sonic reproducers, but also a piece of furniture that will ultimately have to reside in someone's home. A good example of this is the Spica Angelus from the 1980s, a loudspeaker design optimized for minimal diffraction, but undoubtedly also with a very low WAF (Wife Acceptance Factor)

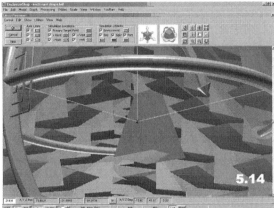

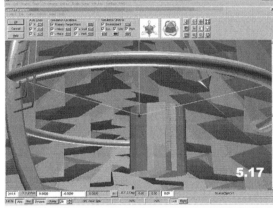

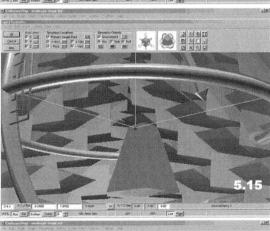

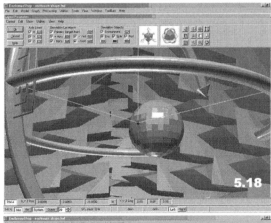

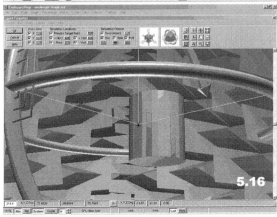

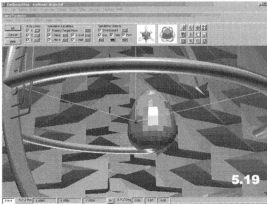

FIGURE 5.14: Pyramid Enclosure (driver center).

FIGURE 5.15: Pyramid Enclosure (driver top).

FIGURE 5.16: Cylinder Enclosure (driver center).

FIGURE 5.17: Cylinder Enclosure (driver top).

FIGURE 5.18: Sphere Enclosure (driver center).

FIGURE 5.19: Egg Enclosure (driver top).

because of its unusual shape. However, there is no question that diffraction is certainly measurable using a single point microphone measurement, but the implication is generally that this has a negative effect on sound quality.

All the data presented here and presented in Harry Olson's 1951 study of SPL variation and enclosure shape is done on-axis. If you think of baffles as being analogous to the reflector on a flashlight, then indeed, baffle diffraction is primarily an on-axis phenomenon that is diminished off-axis[1] and certainly somewhat swamped by the ambient field produced by placing the loudspeaker in a room. Although often thought of as strictly an on-axis event, the effect on the horizontal and vertical polar response is also relevant and will be further

FIGURE 5.20: Frequency response of 2″ cone woofer (A) and 1″ dome tweeter (B) simulated with no baffle.

FIGURE 5.21: Frequency response of 2″ cone woofer (B) and 1″ dome tweeter (A) simulated with infinite baffle.

FIGURE 5.22: Frequency response for cube enclosure with 1″ dome tweeter (A = tweeter mounted center; B = tweeter mounted top).

FIGURE 5.23: Frequency response for cube enclosure with 2″ woofer (A = woofer mounted center; B = woofer mounted top).

FIGURE 5.24: Frequency response for 2″ beveled cube enclosure with 1″ dome tweeter (A = tweeter mounted center; B = tweeter mounted top).

FIGURE 5.25: Frequency response for 2″ beveled cube enclosure with 2″ woofer (A = woofer mounted center; B = woofer mounted top).

FIGURE 5.26: Frequency response for 4″ beveled cube enclosure with 1″ dome tweeter (A = tweeter mounted center; B = tweeter mounted top).

FIGURE 5.27: Frequency response for 4″ beveled cube enclosure with 2″ woofer (A = woofer mounted center; B = woofer mounted top).

FIGURE 5.28: Frequency response for rectangle enclosure with 1″ dome tweeter (A = tweeter mounted center; B = tweeter mounted top).

FIGURE 5.29: Frequency response for rectangle enclosure with 2″ woofer (A = woofer mounted center; B = woofer mounted top).

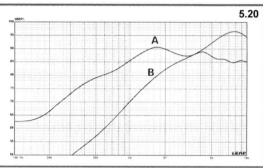

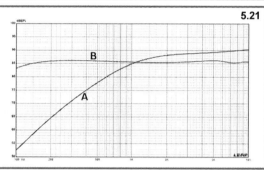

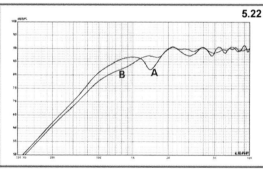

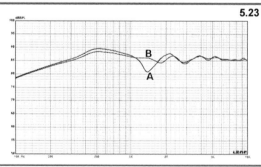

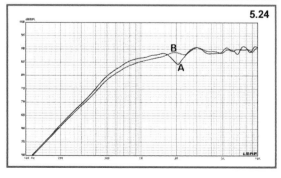

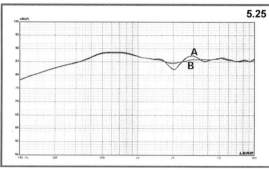

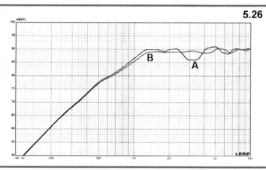

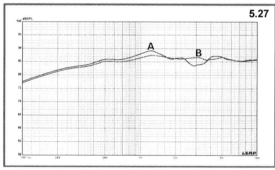

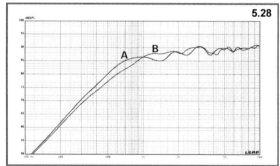

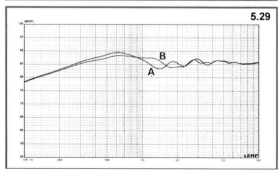

considered as this diffraction study continues in Chapter 6.

5.20 MIDRANGE ENCLOSURES.

You will face two major considerations when configuring a midrange enclosure: the type of enclosure you will use, and how to minimize internal reflections. Your enclosure will be determined by the crossover frequency and network slope you have chosen. If the crosspoint is above 300Hz, if you have kept the driver resonance one to two octaves below the crosspoint (which would require a minimum midrange cavity resonance of 75–150Hz), and if you use a second-order or higher low-pass filter, the driver will not be operating significantly in its piston range and a simple sealed enclosure will be

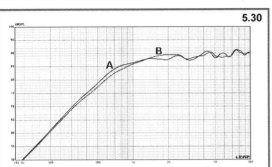

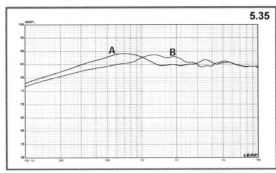

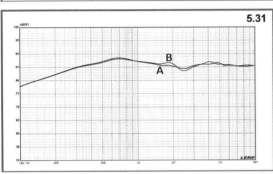

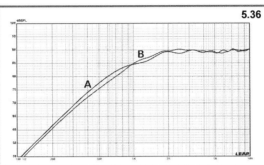

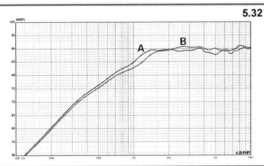

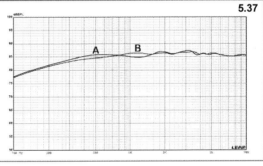

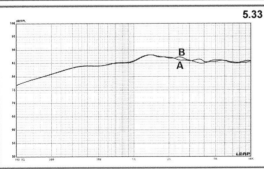

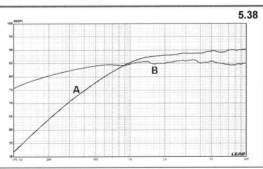

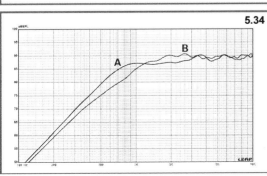

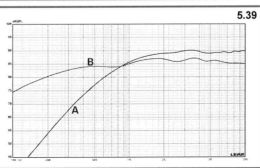

FIGURE 5.30: Frequency response for 2″ beveled rectangle enclosure with 1″ dome tweeter (A = tweeter mounted center; B = tweeter mounted top).

FIGURE 5.31: Frequency response for 2″ beveled rectangle enclosure with 2″ woofer (A = woofer mounted center; B = woofer mounted top).

FIGURE 5.32: Frequency response for 4″ beveled rectangle enclosure with 1″ dome tweeter (A = tweeter mounted center; B = tweeter mounted top).

FIGURE 5.33: Frequency response for 4″ beveled rectangle enclosure with 2″ woofer (A = woofer mounted center; B = woofer mounted top).

FIGURE 5.34: Frequency response for pyramid enclosure with 1″ dome tweeter (A = tweeter mounted center; B = tweeter mounted top).

FIGURE 5.35: Frequency response for pyramid enclosure with 2″ woofer (A = woofer mounted center; B = woofer mounted top).

FIGURE 5.36: Frequency response for cylinder enclosure with 1″ dome tweeter (A = tweeter mounted center; B = tweeter mounted top).

FIGURE 5.37: Frequency response for cylinder enclosure with 2″ woofer (A = woofer mounted center; B = woofer mounted top).

FIGURE 5.38: Frequency response for sphere enclosure (A = tweeter; B = 2″ woofer).

FIGURE 5.39: Frequency response for egg enclosure (A = tweeter; B = 2″ woofer).

adequate. If the crossover frequency is 100–300Hz (or lower than 450Hz using a first-order low-pass filter), the driver will be at least partially operating in its piston range, and will benefit from a properly optimized low-frequency enclosure (*Fig. 5.42*).

You can use a vented enclosure or transmission-line configuration if your driver resonance is at least two octaves below the crossover frequency. If your driver enclosure resonance can be only an octave or less from the crosspoint, a sealed enclosure, with its shallow rolloff, will cause less phase disturbance in the low-pass filter stopband (the region where filter attenuation takes place). In other words, if you cannot get an enclosure resonance from a TL or vented-type configuration at least two octaves below the crossover frequency, use a sealed enclosure. The benefits of using the TL and vented enclosure

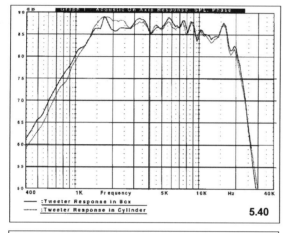

5.40

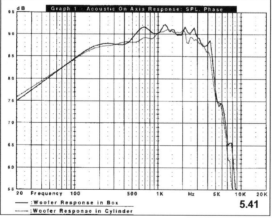

5.41

include less rear reflection (in the case of the TL), and less midrange cone excursion, with less Doppler distortion (vented).

All types of midrange enclosures, except the TL, will benefit from an enclosure with non-parallel walls, because it will minimize reflections in the critical range of driver operation. In addition, the proper use of fibrous damping material, such as fiberglass, Dacron, or long fiber wool, will go a long way to make for optimal midrange driver operation. The wall damping techniques described in Section 5.40 will be appropriate for midrange enclosures that are located separate from the woofer enclosure.

Last, and an often-overlooked option for midrange enclosures, is no enclosure at all, otherwise referred to as unbaffled mounting[2,3]. Probably the most popular example of an unbaffled midrange was the Dahlquist DQ-10 loudspeaker. The benefits include complete freedom from internal box reflections (critical to midrange drivers) and bipolar radiation in the mid-frequency range. You may find it somewhat surprising how low in frequency an unbaffled driver is able to perform.

With a fairly large baffle of one square meter, you can make a mid-woofer type driver operate down to 100Hz, with a 6dB octave rolloff from 100Hz to driver resonance[4]. Since power-handling capacity for this type of configuration is lowered if the driver is operating below 300Hz, you should provide minimal acoustic loading by affixing a small acoustic "blanket" over the rear of the driver. This can be the usual fibrous material or the felt type of acoustic material used in automobiles.

5.30 ENCLOSURE SHAPE AND STANDING WAVES.

Standing-wave modes within a rectangular enclosure can cause amplitude variations in driver response. The problem of standing-wave reflections into the driver cone is substantially eliminated by the use of damping materials such as those described in Chapter 1, Section 1.82. This is illustrated by the comparison of a rectangular enclosure with 100% fill and without enclosure fill shown in *Fig. 5.43*. The response with the enclosure filled with sound-absorbing material has substantially less amplitude deviation than the empty enclosure. Because the

FIGURE 5.42:
A variety of midrange enclosures.

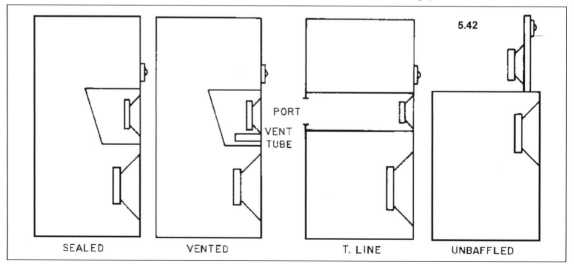

5.42

inclusion of this type of material is so effective in damping standing-wave modes, any other considerations such as box shape and dimension ratios tend to be secondary. This applies especially to closed box designs which often use 100% fill with damping material. Vented boxes seldom have any greater than 50% fill so are somewhat more affected by box modes.

Standing waves in rectangular speaker enclosures are supposedly minimized by choosing appropriate ratios for box dimensions. These ratios usually coincide with ratios chosen to eliminate standing-wave modes in room environments. The most commonly quoted box dimension ratio is one suggested by Thiele, and also happens to be an artifact from the golden rule of architectural design dating back to the pyramids of Egypt[5]. The ratio of height/width/depth is given as 2.6/1.6/1. Other ratios have been suggested such as 2/1.44/1[6] and 1.59/1.26/1[7], but any improvement attributed to box dimension ratios will be a secondary effect as long as the enclosure is appropriately damped with absorbent material.

These ratios are still a good guideline (given the limitations of driver dimensions and layout) since it precludes making excessively long and narrow enclosures which can be prone to pipe resonances (which can, if necessary, be "broken" up by using internal reflecting baffle panels). Other types of enclosure shapes which have nonparallel sides, such as pentangle-shaped enclosures and enclosures with slanted front baffles, will have different and probably less pronounced standing-wave modes, but the

attraction is often more cosmetic than pragmatic.

Location of a low-frequency driver on the baffle also affects the severity of standing waves in an untreated enclosure. Locating the driver in the exact center and somewhere just below this point will minimize standing waves across the height and width (but not the depth) of the enclosure, according to one study[8]. The semi-cylindrical and cylindrical enclosure shapes will reduce standing waves across the depth of the enclosure and reduce the pressure response up to about 800Hz. This effect was determined by finite element analysis of unfilled enclosures, but will become less important when absorbent material is included.

In the example in *Fig. 5.41*, the difference between rectangular and cylindrical shapes is minimal even though the enclosures included absolutely no filling material. Rather than diffraction and standing-wave suppression, the real benefit is more cosmetic than anything, although the Cubicon cardboard tube does have good vibration damping qualities when compared to the same thickness of MDF (medium density fiberboard).

5.40 BOX DAMPING.

It is a well-established fact that typical veneered MDF and particleboard loudspeaker boxes resonate in conjunction with the woofer and radiate nearly as much sound pressure as the driver itself at certain frequencies[9]. The primary reason for the success of the Celestion SL-600 speaker (no longer in production) is the honeycomb aircraft aluminum enclosure which eliminates much of the coloration re-transmitted by most wood enclosures. A number of materials and techniques can be used to minimize enclosure vibration. This includes the choice of wall material, wall resonance damping material, bracing technique, driver mounting technique, and enclosure floor-coupling techniques.

A. Wall Materials

There are two basic schools of thought when it comes to choosing wall materials. One is the brute-force technique, which dictates the use of thick-walled high-density materials, such as 1″ MDF in conjunction with extensive bracing and sometimes additional wall damping compounds. Speakers like those from Thiel Audio, Wilson Audio, and Aeriel

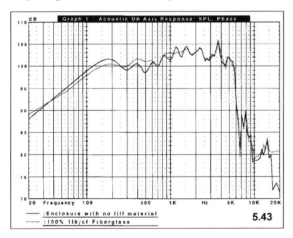

5.43

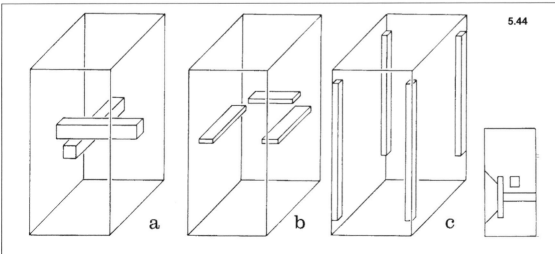

FIGURE 5.44:
Cross brace (a), horizontal brace (b), and corner brace (c) for rectangular enclosure.

Acoustics use this type of construction.

Another school suggests the use of moderately stiff and lighter thin-walled material such as ½″–¾″ marine plywood with the application of heavy damping materials to achieve low-level coloration in the 100-500Hz region. The Leak Sandwich speaker built in the late 1960s used this type of construction. It was made from ½″ plywood and damped with thick layers of roofing felt. Both formats seem to work and examples of both are found in the industry.

Constrained layer materials are another technique. Constrained layer construction board is made of two layers of MDF or similar material with a layer of wall resonance damping material sandwiched in between. This product is highly specialized and not generally available for amateur construction. An interesting alternative was suggested in a 3/89 *Speaker Builder* article[10] consisting of two layers of ¼″ veneered plywood with two layers of ½″ sheetrock sandwiched between, with each layer bonded with construction adhesive. Another constrained layer example came from a 4/82 *Speaker Builder* article[11], which suggested the use of sand-filled panels for wall material (originally proposed by G.A. Briggs, founder of the British Wharfedale Company).

The relatively poor damping of untreated MDF and the superior characteristics of constrained layer materials was quantitatively analyzed by Nokia engineer Juha Backman in a paper presented at the 101st AES Convention[12]. This study included both accelerometer measurements and nearfield cabinet measurements and clearly showed the superiority of constrained layer damping over extentional (external) damping.

B. Wall Resonance Damping Materials

If panel resonances are raised to a higher frequency, either by the choice of moderately stiff thin-walled material or by bracing, the higher frequency resonances can be damped by means of extensional damping compounds. Examples of these types of materials were discussed in previous issues of *Voice Coil*[13,14] and include two extremely effective products, Antiphon Type A-13 and EAR type CN-12. Antiphon Type A-13 is a bituminous felt/clay composite damping product. It is sold primarily to the automobile industry to damp resonance vibration in the roof panels of cars. It comes in ⅟16″ thick self-adhesive sheets. Two layers of this material applied to 50% or more of a cabinet's wall area is quite effective.

The EAR product is a graphite-filled vinyl product developed by EAR for the US Navy to damp hull vibration in nuclear submarines. It comes in ⅟16″–¼″-thicknesses and is applied the same way as Antiphon and is likewise very effective. These materials are not as yet available to amateurs, but will likely be in the future. The cost of these products in quantity is from $1.60–$5/ft².

Less expensive alternatives are the use of multiple layers (4–6) of 30 lb. roofing felt stapled to the enclosure wall. The walls and inside of the front baffle should be 50–75% covered with the material and stapled in four corners and the center of each panel.

Liquid materials made for car undercoating have been applied in speakers, but the solvent-based products are likely to be hazardous to driver adhesives, surrounds, and cone materials. Another alternative is a 50/50 mix of sand and roofing cement, but the application is tedious and time consuming.

C. Bracing Techniques

Bracing effectively divides the wall into two quasi-independent panels, each having its own resonant frequency. The three basic bracing types are shown in *Fig. 5.44*. They are the horizontal, corner, and the cross-brace. The horizontal brace can be used to break up the enclosure resonance around the girth of the box.

Typical material used is ¾″ × 2″ lumber, although angle iron has been used in the same application. A variation, used in commercial manufacturing, is the shelf brace[15], which is a combination horizontal and cross-brace. The shelf brace is basically a solid panel which is attached to three or four sides of the enclosure and with large cutouts to allow for air flow within the box (*Fig. 5.45*).

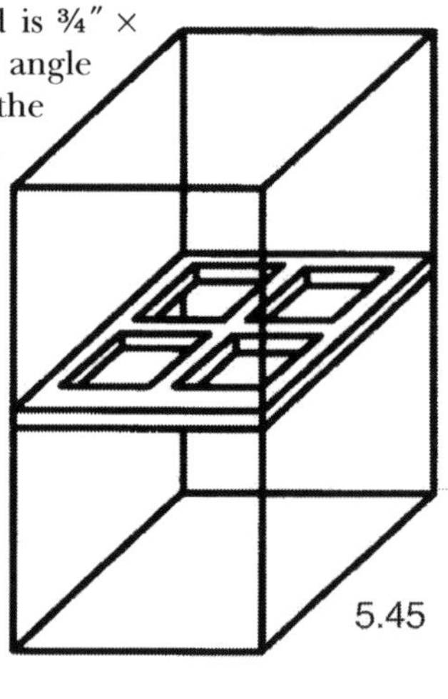

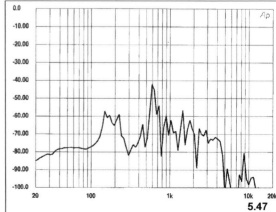

The corner brace increases the mutual coupling of adjacent walls and helps dissipate energy. The

FIGURE 5.45: Example of shelf brace.

FIGURE 5.46: Accelerometer measurement of untreated 0.75″ PB box.

FIGURE 5.47: Accelerometer measurement of damped and braced 1″ MDF box.

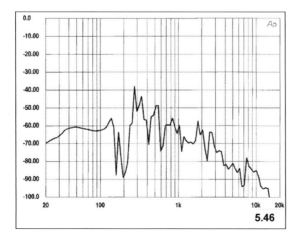

cross-brace is used to connect opposite walls, left-to-right and front-to-back. Braces can be $2'' \times 2''$ lumber or large diameter dowels ($1''$–$1.5''$), which are stiff for the amount of enclosure space they occupy and can be used effectively. Cross-bracing placed in the center of a panel divides the resonance by two; however, staggering the brace so the two panel resonances are at different frequencies prevents their acoustic summation to some extent and spreads out the "noise" at a somewhat lower level[16].

D. Driver Mounting Techniques

Isolating the speaker frame's vibration has also been shown to be useful in lowering cabinet noise. A commercial product, Well-Nut Fasteners (USM Corp., Molly fastener division), works quite well in this technique. Well-Nuts are a rubber insert with a brass nut embedded in the base. This free-floating fastener is often used to damp vibration in electric motors and has the same effect with drivers. Well-Nuts also makes it easy to remove drivers without wearing out screw holes. Small rubber grommets located on the mounting screws or bolts have also been used[17] with some success. This plus an air-tight damping rubber, foam, or putty type gasket will help isolate driver vibration.

Another simple technique is to mount drivers with silicone adhesive. A ¼" bead of silicone placed on the driver mounting flange will provide an air-tight seal as well as a degree of vibration damping. The downside is the difficulty in removing the driver if it has been inset into the cabinet baffle.

E. Enclosure Floor Coupling

Floor-standing enclosures can transmit substantial vibration into the floor which in turn couples to the air. The fad for the last several years has been to use some type of metal spike to physically stabilize the speaker (usually three) to isolate it from the floor. While the spikes may provide some degree of isolation by limiting physical contact, they can be made even more effective by applying additional mass at the base. A new technique seen in the marketplace consists of providing some type of energy "sink" for the enclosure to rest upon. This takes the form of a heavy stone or marble platform which simply does not vibrate in any fashion and cannot transmit vibration to the floor.

A combination of all of these techniques can be quite effective in lowering the coloration caused by a vibrating enclosure. *Figures 5.46* and *5.47* show the results of an accelerometer measurement done on two enclosures, one ¾" particleboard enclosure (*Fig. 5.46*) and one 1" MDF enclosure with dowel cross-bracing and Antiphon Type A-13 extensional damping material (*Fig. 5.47*). The test was done with an Audio Precision System 1 sine wave sweep analyzer using a PVDF (polyvinylidene) accelerometer[18]. The PVDF accelerometer is not a calibrated unit, but the relative difference is apparent. Although the higher frequencies still persist, the level below 150Hz has been substantially attenuated. It is also apparent that some of the resonances have been shifted slightly higher, but not attenuated.

REFERENCES

1. Robert C. Kral, "Diffraction: The True Story," *Speaker Builder* 1/80, p. 28.

2. H. Olson, "Gradient Loudspeakers," *JAES*, March 1972.

3. J. Backman, "A Model of Open-Baffle Loudspeakers," 107th AES Convention, September 1999, preprint no. 5025.

4. R.J. Newman, "Dipole Radiator Systems," *JAES*, Jan./Feb. 1980.

5. David Weems, *How to Build & Test Complete Speaker Systems*, Tab Books No. 1064 (out of print).

6. From the enclosure library menu in LEAP 4.0 computer software, manufactured by LinearX.

7. Lubos Palounek, "Enclosure Shapes and Volumes," *Speaker Builder* 3/88, p. 22.

8. Shinichi Sakai, "Acoustic Field in an Enclosure and Its Effect on Sound-Pressure Responses of a Loudspeaker," *JAES*, April 1984.

9. J.K. Iverson, "The Theory of Loudspeaker Cabinet Resonances," *JAES*, April 1973.

10. Allan Millikan, "Dynaudio Drivers and Sheetrock," *Speaker Builder* 3/89, p. 15.

11. M. Lampton, "A Three-Way Corner Loudspeaker System," *Speaker Builder* 4/82, p. 7.

12. J. Backman, "Effect of Panel Damping on Loudspeaker Enclosure Vibration," 101st AES Convention, November 1996, preprint no. 4395.

13. *Voice Coil*, October 1989.

14. *Voice Coil*, January 1991.

15. Mike Chin, "Cabinet Bracing," *Speaker Builder* 2/91, p. 71.

16. Peter Muxlow, "Loudspeaker Cabinets," *Speaker Builder* 2/88, p. 24.

17. S. Linkwitz, "A Three-Enclosure Loudspeaker System," *Speaker Builder* 2/80, p.12.

18. *Voice Coil*, February 1991.

CHAPTER SIX

LOUDSPEAKER BAFFLES: DRIVER LOCATION, SEPARATION, AND OTHER CONSIDERATIONS

6.10 SPL VARIATION DUE TO DRIVER BAFFLE LOCATION.

Loudspeaker front baffles provide a "launch" area for the wave front developed by a transducer. As was discussed in Chapter 5, enclosure (baffle) shape—be it rectangular, cylindrical, spherical, square, egg-shaped, etc.—can cause substantial variation in the measured SPL of the driver. In Section 5.10B, the effect of different locations was also discussed, but only from a limited perspective of the difference from the center of a baffle to the top of a baffle. The reality is that as woofers, midranges, and tweeters are mounted in different physical locations on a baffle, and as the distance from each baffle edge changes, the on- and off-axis SPL will be very different for each discrete location.

Where to locate a set of drivers on a new design has always been an unanswered question for speaker designers. Most manufacturers opt for the drivers centered on the baffle between the left and right sides of the speaker and usually build the array starting with the tweeter at the top of the baffle (or a woofer at the top if it's a woofer-tweeter-woofer design). Variations have been numerous and include such concepts as mirror-imaged drivers in which the tweeters are located close to the left for one channel of a stereo pair and on the right side of the cabinet for the other channel. Because this is such an open question to be answered when designing a loudspeaker, what is needed is a comprehensive set of examples that will give you some idea of what to expect when selecting a mounting location for a woofer, midrange, or tweeter. Using the virtual anechoic chamber provided in the LEAP 5 EnclosureShop software, I conducted an extensive study describing the consequence of the various possible mounting locations on different-size baffles. This is intended to serve as a guideline for this part of the loudspeaker design process.

This study of SPL variation vs. driver location is broken down into two parts: two-way loudspeaker SPL variations due to different driver mounting locations, and three-way midrange SPL variations due to different driver locations. Because the majority of loudspeakers built both by manufacturers and by amateur builders are standard rectangular types, and because trying to do baffle location variations for all the different possible enclosure shapes would be too exhaustive for the scope of this book, I consider only rectangular baffles.

A. Two-Way Baffle Woofer and Tweeter Location SPL Variation.

Because baffle size and driver diameter both affect the diffraction that causes SPL changes, this study includes four different baffle sizes and four different woofer diameters. The enclosures were modeled after enclosure volumes and dimensions of loudspeakers that were in production in 2005:

Woofer Diameter	Enclosure Dimensions (H × W × D)	Simulated Woofer
4.5″	8.75″ × 5.25″ × 5.5″	Peerless 830516
5.25″	10.75″ × 7.25″ × 8.5″	Vifa C13WG-19-08
6.5″	13.5″ × 8.75″ × 11.25″	Vifa P17WJ-00-08
8″	15.75″ × 10″ × 10.5″	Vifa P21WO-10-08

While the LEAP 5 EnclosureShop simulations are very good, they could not simulate the SPL anomalies that often occur in a tweeter's response and are caused by the reflections in a woofer or midrange cone. While there was no way to simulate this response affectation, if you care to know whether or not a particular response anomaly in a tweeter measurement is being caused by a reflection out of a woofer or midrange cone, cover the cone with a thin flat piece of cardboard and repeat the measurement. The question will be answered in the comparison.

Because there are an infinite number of discrete baffle locations for any transducer, for the purposes of this study, I used just three different locations for the woofer and eight for the tweeter. These are pictured for the 8″ enclosure woofer SPL simulations in *Figs. 6.1–6.3* and for the 8″ enclosure tweeter SPL simulations in *Figs. 6.4–6.11*. These driver locations were relatively identical for all four different woofer and enclosure sizes.

I determined the exact locations with both the woofer and tweeter mounted as close together as possible on the baffle and with the measurement axis placed between the two mounting positions. The three positions for the woofer are with the two drivers located such that the tweeter was at the top of the enclosure with enough clearance for a grille frame, with the two drivers located at the midpoint between the top and the bottom of the enclosure; the last position was with the woofer located at the bottom of the enclosure, again with enough practical clearance for a grille frame. Because contemporary rectangular cabinets tend to be just wide enough for the woofer and grille frame, there is no excess baffle space to study the left or right offset of a woofer, so this consideration is not part of this presentation. However, because tweeters can easily be offset to the left or right side of the enclosure, more variations can be considered.

If you look at the tweeter baffle locations in *Figs. 6.4–6.11*, you can see that this includes the same three locations as the woofer (the tweeter mounted just above the woofer with minimal spacing between the two drivers) down the center of the baffle, plus the same three locations with the tweeter offset to the far right, as far as possible while still maintaining some clearance for a grille. Also included is a tweeter mounting location for a WTW (woofer-tweeter-woofer) format with the tweeter located in the exact middle of the baffle plus the same center/middle position offset to the right. In this case of the WTW tweeter location, there is not sufficient room on the baffle for dual woofers, but you could easily use this baffle size for a dual woofer design by using the next smaller size woofers (dual 6.5″ on the 8″ baffle, dual 5.25″ on the 6.5″ baffle,

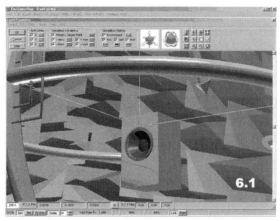

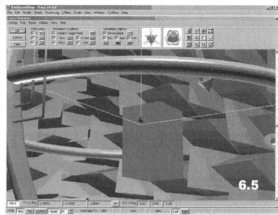

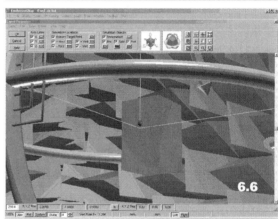

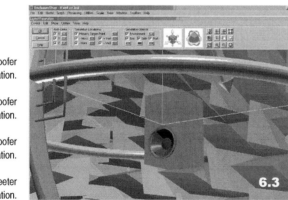

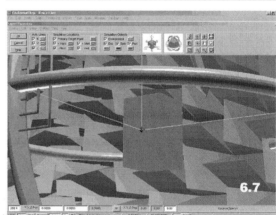

FIGURE 6.1: 8″ woofer top baffle location.

FIGURE 6.2: 8″ woofer mid baffle location.

FIGURE 6.3: 8″ woofer bottom baffle location.

FIGURE 6.4: 1″ tweeter top center baffle location.

FIGURE 6.5: 1″ tweeter mid center baffle location.

FIGURE 6.6: 1″ tweeter bottom center baffle location.

FIGURE 6.7: 1″ tweeter WTW center baffle location.

FIGURE 6.8: 1″ tweeter top offset baffle location.

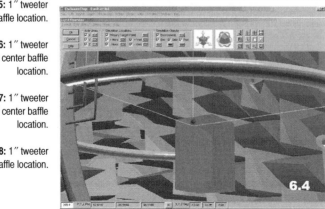

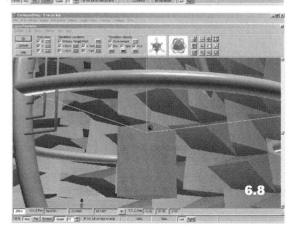

dual 4.5″ on the 5.25″ baffle, and dual 3″ on the 4.5″ baffle). For this reason, I included the middle tweeter position to provide tweeter data for possible dual woofer formats, albeit with a somewhat wider than normal baffle for that format.

In order to fully understand what edge diffraction does to the driver response when measured with a microphone, you need to look at both the on-axis response as well as the horizontal and vertical polar responses. That said, the basic format for this diffraction study includes the following graphic information for each combination of baffle size and driver locations:

1. as a reference, the on-axis half-space response graph plus the half-space horizontal and vertical polar plots

2. a composite graph that compares the different anechoic on-axis response curves—one for the three woofer locations, one graph for four tweeter locations placed on the baffle centerline, and one graph for the four tweeter locations placed on the right side baffle location

3. individual on-axis and horizontal and vertical polar plots for each baffle location.

Given the number of baffle locations and enclosure examples, the number of graphs and plots required totals 168. Because the space required for this on the printed page is somewhat excessive (we are trying to keep this volume substantially smaller than Tolstoy's *War and Peace*!), you will find a mail-in coupon for receiving the data on CD-ROM that contains the complete graphic set for this entire diffraction study in full-color PDF format, along with some other useful information. While there are sim-

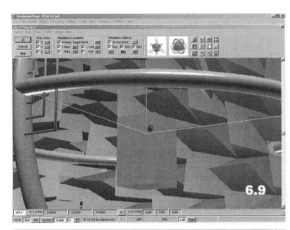

6.9

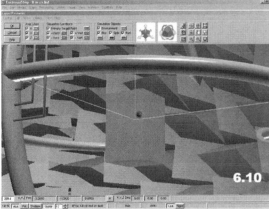

6.10

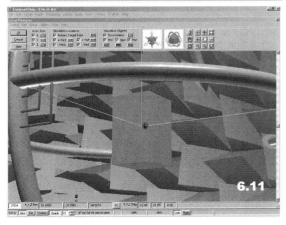

6.11

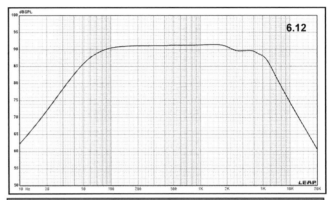

6.12

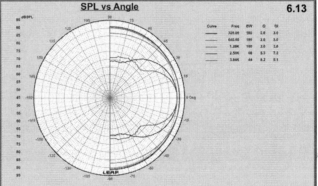

SPL vs Angle 6.13

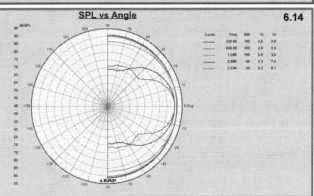

SPL vs Angle 6.14

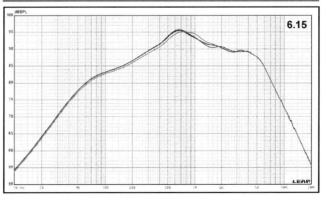

6.15

LOUDSPEAKER BAFFLES

FIGURE 6.9:
1″ tweeter mid offset baffle location.

FIGURE 6.10:
1″ tweeter bottom offset baffle location.

FIGURE 6.11:
1″ tweeter WTW offset baffle location.

FIGURE 6.12:
Frequency response of 8″ woofer mounted on an infinite baffle.

FIGURE 6.13:
Horizontal polar plot for *Fig. 6.12* (320Hz = solid, 640Hz = dash, 1.25kHz = dot, 2.56kHz = solid, 3.84kHz = dash).

FIGURE 6.14: Vertical polar plot for *Fig. 6.12* (320Hz = solid, 640Hz = dash, 1.28kHz = dot, 2.56kHz = solid, 3.84kHz = dash).

FIGURE 6.15:
Comparison of 8″ woofer frequency response for all three center baffle locations (top = solid, mid = dash, bottom = dot).

119

ilarities in the effects of baffle location between the different enclosure sizes, the differences are much greater than you might expect, so as an exercise in overall understanding, you will find it very worthwhile to examine all the graphs and conclusions for each size enclosure. You will also find the polar plots much easier to read in color. That aside, data on the 8″ enclosure is provided in *Figs. 6.12–6.24* for the woofer locations and *Figs. 6.25-6.53* for the tweeter locations.

For the 8″ woofer example, start by examining *Figs. 6.12–6.14*. This is the half-space response of the woofer (mounted on an infinitely large baffle)

on-axis and horizontal and vertical polar plots. As you can see, the response is mostly flat with a 2dB decrease in SPL in the octave from 2–4kHz, and the polar plots are identical and perfectly symmetrical. These plots are the result of having no edge diffraction or frequency-dependent-shaped baffle reflection to affect the single point microphone measurement.

The on-axis comparison graph of all three 8″ woofer baffle locations on the 15.75″ × 10″ baffle in *Fig. 6.15* shows the absolute SPL differences on-axis to be within a 1–1.5dB range. In terms of affecting the crossover from 2kHz and higher, the differenc-

FIGURE 6.16: Frequency response for 8″ woofer at bottom baffle location.

FIGURE 6.17: Horizontal polar plot for *Fig. 6.16* (320Hz = solid, 640Hz = dash, 1.28kHz = dot, 2.56kHz = solid, 3.84kHz = dash).

FIGURE 6.18: Vertical polar plot for *Fig. 6.16* (320Hz = solid, 640Hz = dash, 1.28kHz = dot, 2.56kHz = solid, 3.84kHz = dash).

FIGURE 6.19: Frequency response for 8″ woofer at mid baffle location.

FIGURE 6.20: Horizontal polar plot for *Fig. 6.19* (320Hz = solid, 640Hz = dash, 1.28kHz = dot, 2.56kHz = solid, 3.84kHz = dash).

FIGURE 6.21: Vertical polar plot for *Fig. 6.19* (320Hz = solid, 640Hz = dash, 1.25kHz = dot, 2.56kHz = solid, 3.84kHz = dash).

FIGURE 6.22: Frequency response for 8″ woofer at top baffle location.

FIGURE 6.23: Horizontal polar plot for *Fig. 6.22* (320Hz = solid, 640Hz = dash, 1.28kHz = dot, 2.56kHz = solid, 3.84kHz = dash).

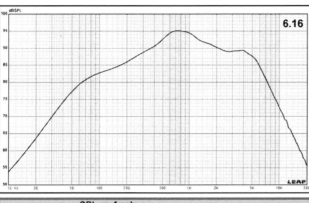

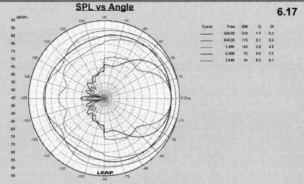

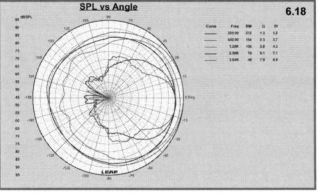

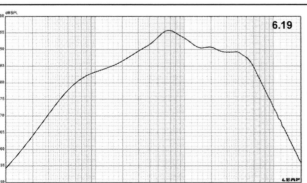

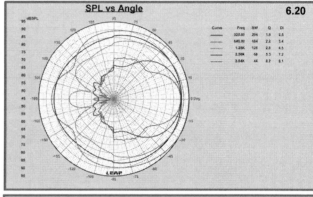

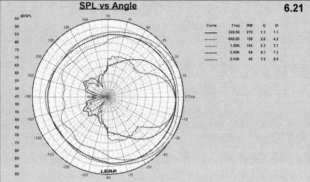

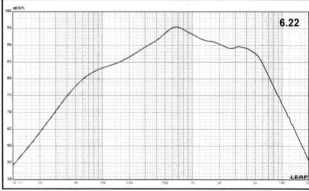

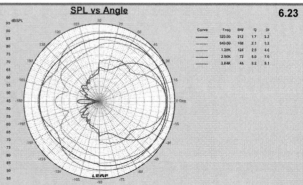

es are not particularly significant. In terms of the horizontal polar plots (*Figs. 6.17, 6.20,* and *6.23*), any position on the baffle provides a symmetrical pattern, which means no "lobing" to one side or the other in the horizontal plane, so this is not an issue. Incidentally, the 8″ enclosure data is repeated on the CD-ROM and is much easier to read in color on a computer screen than the black and white graphic rendering on these pages. Baffle position does affect, however, the vertical polar response (*Figs. 6.18, 6.21,* and *6.24*) and causes some degree of frequency dependent lobing for each location.

As a generalization, frequencies above 3kHz, which ideally will be in the stopband of the low-pass filter section of a two-way crossover, are affected about the same for all three locations, which happens to be a downward tilt of about 6°. For the frequencies below 3kHz, the bottom position causes a 15° upward tilt, the middle position has some unevenness—but primarily does not cause any significant tilt (lobing)—and the top position causes a 15° downward tilt.

If you consider that non-coincident driver mounting of the tweeter above the woofer (both woofer and tweeter mounted on the baffle surface and the baffle oriented perpendicular to the floor and

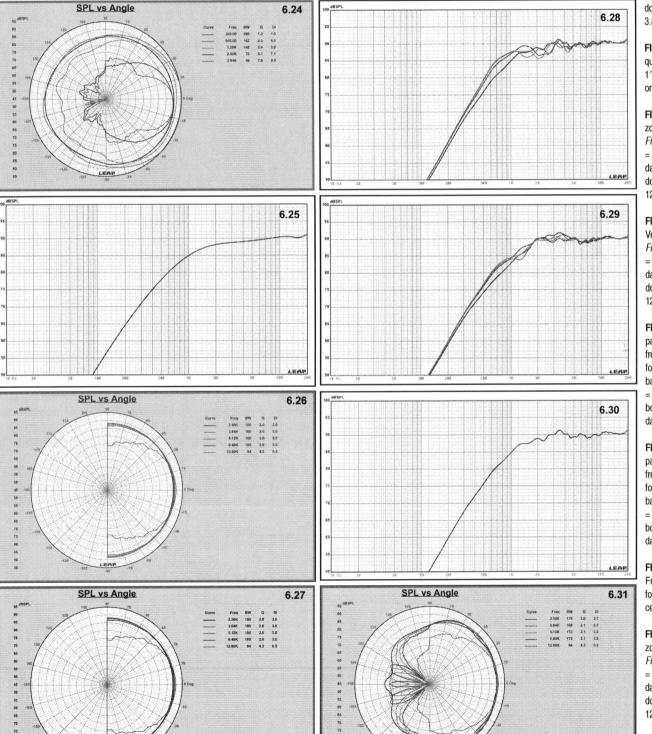

FIGURE 6.24: Vertical polar plot for *Fig. 6.22* (320Hz = solid, 640Hz = dash, 1.28kHz = dot, 2.56kHz = solid, 3.84kHz = dash).

FIGURE 6.25: Frequency response of 1″ tweeter mounted on an infinite baffle.

FIGURE 6.26: Horizontal polar plot for *Fig. 6.25* (2.56kHz = solid, 3.84kHz = dash, 5.12kHz = dot, 6.40kHz = solid, 12.8kHz = dash).

FIGURE 6.27: Vertical polar plot for *Fig. 6.25* (2.56kHz = solid, 3.84kHz = dash, 5.12kHz = dot, 6.40kHz = solid, 12.8kHz = dash).

FIGURE 6.28: Comparison of 1″ tweeter frequency response for all four center baffle locations (top = solid, mid = dash, bottom = dot, WTW = dash/dot).

FIGURE 6.29: Comparison of 1″ tweeter frequency response for all four offset baffle locations (top = solid, mid = dash, bottom = dot, WTW = dash/dot).

FIGURE 6.30: Frequency response for 1″ tweeter at top center baffle location.

FIGURE 6.31: Horizontal polar plot for *Fig. 6.30* (2.56kHz = solid, 3.84kHz = dash, 5.12kHz = dot, 6.40kHz = solid, 12.8kHz = dash).

not tilted at an angle) causes a downward lobing (discussed in Chapter 7), then the top mounting position for the woofer will minimize the overall lobing and is the preferred mounting position that provides an overall more consistent vertical polar response for the system. If you mount the tweeter below the woofer, as the PSB Mini Stratus two-way (now out of production), the bottom mounting position for the woofer would make more sense, because the driver orientation would cause an upward vertical polar response tilt in the crossover region, and the baffle orientation of the woofer would also cause an upward tilt, again giving a more consistent

vertical polar response for the system. This is important, as different types of lobing affect the perceived sound quality. The subjective perceived effect of lobing is discussed more in Section 6.20.

For the tweeter, you should again start by looking at the half-space on-axis response and the half-space horizontal and vertical polar plots in *Figs. 6.25–6.27*. As you can see, the response of this tweeter is relatively flat and the polar plots in either plane are totally symmetrical. There are two comparison on-axis plots given—one for the center baffle locations in *Fig. 6.28* and one for the offset positions in *Fig. 6.29*. This ±SPL data is quantified in *Table 6.1*.

FIGURE 6.32:
Vertical polar plot for
Fig. 6.30 (2.56kHz
= solid, 3.84kHz =
dash, 5.12kHz =
dot, 6.40kHz = solid,
12.8kHz = dash).

FIGURE 6.33:
Frequency response
for 1″ tweeter at mid
center baffle location.

FIGURE 6.34: Horizontal polar plot for
Fig. 6.33 (2.56kHz
= solid, 3.84kHz =
dash, 5.12kHz =
dot, 6.40kHz = solid,
12.8kHz = dash).

FIGURE 6.35:
Vertical polar plot for
Fig. 6.33 (2.56kHz
= solid, 3.84kHz =
dash, 5.12kHz =
dot, 6.40kHz = solid,
12.8kHz = dash).

FIGURE 6.36:
Frequency response
for 1″ tweeter at
bottom center baffle
location.

FIGURE 6.37: Horizontal polar plot for
Fig. 6.36 (2.56kHz
= solid, 3.84kHz =
dash, 5.12kHz =
dot, 6.40kHz = solid,
12.8kHz = dash).

FIGURE 6.38:
Vertical polar plot for
Fig. 6.36 (2.56kHz
= solid, 3.84kHz =
dash, 5.12kHz =
dot, 6.40kHz = solid,
12.8kHz = dash).

FIGURE 6.39:
Frequency response
of 1″ tweeter at the
WTW center baffle
location.

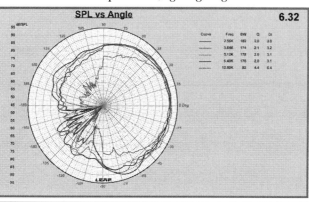

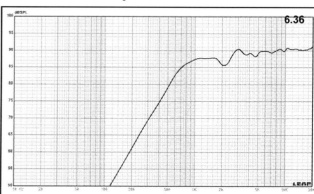

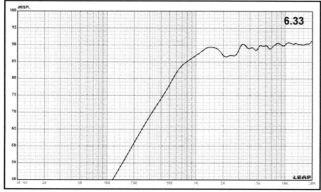

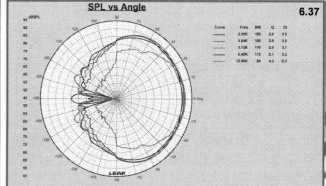

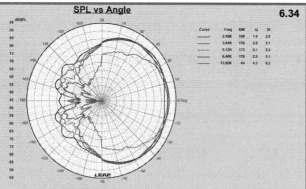

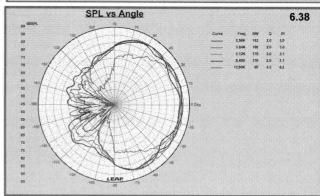

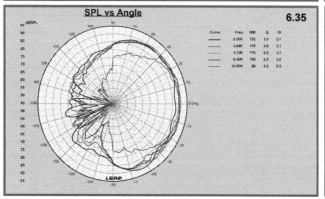

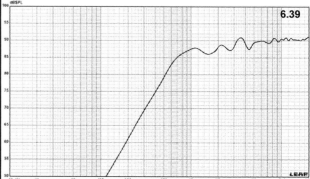

TABLE 6.1 Tweeter SPL Variations from 2kHz-10kHz for Different Baffle Locations (dB).

	Baffle Location			
	TOP	MID	BOTTOM	CENTER (WTW)
Center	1.57	1.89	2.41	1.87
Offset	1.51	1.04	1.17	1.11

From an on-axis single point microphone measurement standpoint, offsetting the tweeter to the far side of a baffle unquestionably results in a smoother response with less SPL variation on-axis, although it's not as significant as is often assumed, and the difference varies somewhat depending on the vertical placement (top, mid, bottom, or WTW). In the vertical polar plots (*Figs. 6.32, 6.35, 6.38, 6.41, 6.44, 6.47, 6.50,* and *6.53*), both center and offset, there is a small upward tilt for the top, mid, and bottom positions of 5, 4, and 3°, respectively, for the frequencies from 3.8kHz to 12.8kHz, while the WTW is centered on the 0 axis with no tilt. While the upward tilt for the top, mid, and bottom positions is not great, the symmetry of the tilt is somewhat better for the mid and bottom positions. Given that the top position was judged optimal for

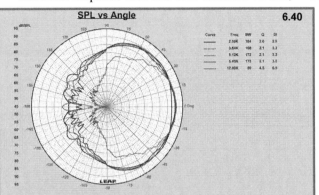

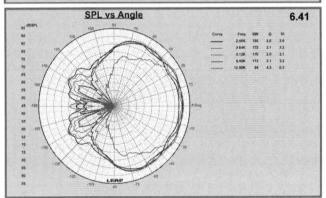

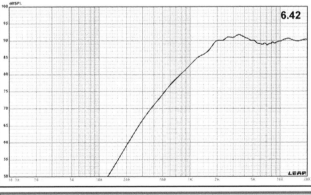

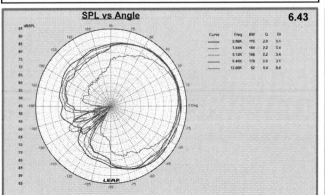

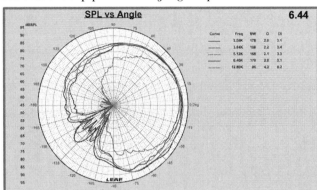

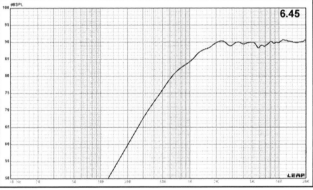

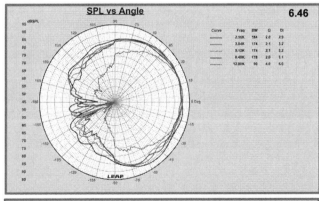

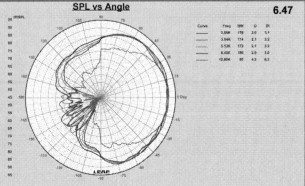

FIGURE 6.40: Horizontal polar plot for *Fig. 6.39* (2.56kHz = solid, 3.84kHz = dash, 5.12kHz = dot, 6.40kHz = solid, 12.8kHz = dash).

FIGURE 6.41: Vertical polar plot for *Fig. 6.39* (2.56kHz = solid, 3.84kHz = dash, 5.12kHz = dot, 6.40kHz = solid, 12.8kHz = dash).

FIGURE 6.42: Frequency response for 1″ tweeter at top offset baffle location.

FIGURE 6.43: Horizontal polar plot for *Fig. 6.42* (2.56kHz = solid, 3.84kHz = dash, 5.12kHz = dot, 6.40kHz = solid, 12.8kHz = dash).

FIGURE 6.44: Vertical polar plot for *Fig. 6.42* (2.56kHz = solid, 3.84kHz = dash, 5.12kHz = dot, 6.40kHz = solid, 12.8kHz = dash).

FIGURE 6.45: Frequency response for 1″ tweeter at mid offset baffle location.

FIGURE 6.46: Horizontal polar plot for *Fig. 6.45* (2.56kHz = solid, 3.84kHz = dash, 5.12kHz = dot, 6.40kHz = solid, 12.8kHz = dash).

FIGURE 6.47: Vertical polar plot for *Fig. 6.45* (2.56kHz = solid, 3.84kHz = dash, 5.12kHz = dot, 6.40kHz = solid, 12.8kHz = dash).

the woofer, somewhere between what has been shown as the top and mid positions would likely be optimal from a system viewpoint, but I really don't believe that small changes such as this are extremely critical subjectively, so that the top position for both woofer and tweeter would still be a good choice. Another important observation regarding the WTW tweeter position is that the lack of lobing, coupled with the very symmetrical woofer and midrange lobing that you will see in sections 6.10B and 6.20, is undoubtedly at least one of the reasons the WTW D'Appolito format speaker has been so successful and tends to be subjectively well liked.

As far as the horizontal polar response of these different tweeter positions is concerned, like the woofer, all the center locations (*Figs. 6.31, 6.34, 6.37,* and *6.40*), whether mounted at the top, mid, bottom, or center (WTW) of the enclosure, have totally symmetrical profiles. However, if you consider all four offset tweeter positions (*Figs. 6.43, 6.46, 6.49,* and *6.52*), then the horizontal polar plots resemble the tweeter top vertical polar plots in terms of lobing. In this case the lobing is about a 15° tilt toward the side the tweeter is mounted on. Also important is that the amplitude spread at the various frequencies is more even on this side of the horizontal response.

In the '70s, mirror-imaged speakers were popular. "Mirror Image" meant that the tweeters were offset to the far side of the baffle, but on opposite sides for each stereo channel. The left channel speaker had the tweeter mounted on the right side of the baffle, and the right channel speaker had its tweeter mounted on the left side of the baffle. This at least was the preferred orientation, because by switching the location of the stereo pair, the tweeters would be on the outside instead of the inside.

From the various horizontal polar plots for the offset tweeters it becomes obvious why the inside orientation was the correct subjective choice. This way both horizontal polar responses tilt to the inside toward the center "sweet spot" listening position, with the added benefit of more consistent SPL and reportedly improved sonic imagery. This would also account for why non-mirror-image stereo and home theater LRs (Left/Right channels) are frequency canted to the inside listening area, because pointing the speaker at the listener is a similar function to the offset lobing pointing at the listener with mirror-image speakers, except there is an enhanced measured SPL consistency with the offset.

Tweeter offset to the baffle edge probably makes

FIGURE 6.48: Frequency response for 1″ tweeter at bottom offset baffle location.

FIGURE 6.49: Horizontal polar plot for *Fig. 6.48* (2.56kHz = solid, 3.84kHz = dash, 5.12kHz = dot, 6.40kHz = solid, 12.8kHz = dash).

FIGURE 6.50: Vertical polar plot for *Fig. 6.48* (2.56kHz = solid, 3.84kHz = dash, 5.12kHz = dot, 6.40kHz = solid, 12.8kHz = dash).

FIGURE 6.51: Frequency response of 1″ tweeter at the WTW offset baffle location.

FIGURE 6.52: Horizontal polar plot for *Fig. 6.51* (2.56kHz = solid, 3.84kHz = dash, 5.12kHz = dot, 6.40kHz = solid, 12.8kHz = dash).

FIGURE 6.53: Vertical polar plot for *Fig. 6.51* (2.56kHz = solid, 3.84kHz = dash, 5.12kHz = dot, 6.40kHz = solid, 12.8kHz = dash).

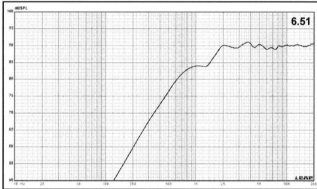
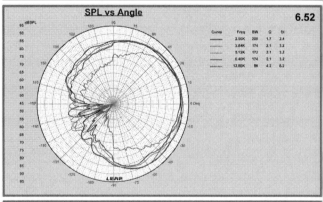
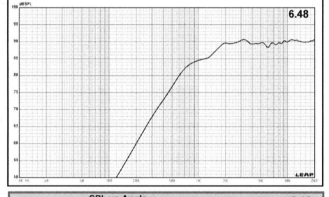
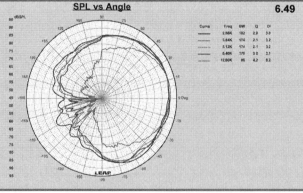
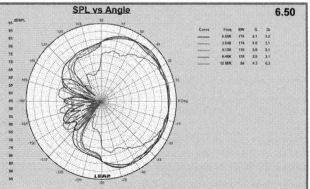
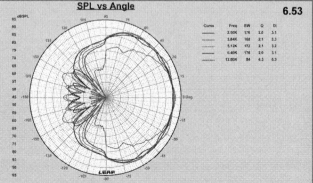

more sense with close-in listening positions in stereo than further back listening positions with home theater, especially with larger screens. The higher degree of ambient content at the more distant listening positions makes a symmetrical horizontal polar response more desirable. Of course, there is also the manufacturing issue of matching up left and right mirror offset channels, which makes such a practice more trouble than most manufacturers are willing to subject themselves to.

As a "guiding" principle, I think that keeping both vertical and horizontal polar responses as sym-

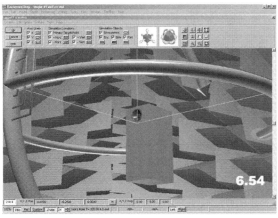

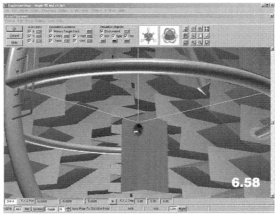

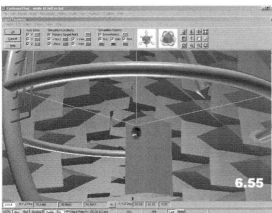

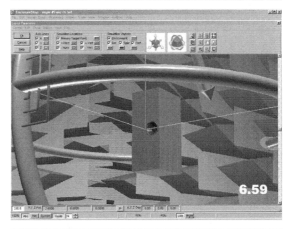

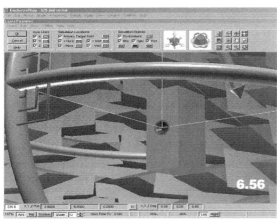

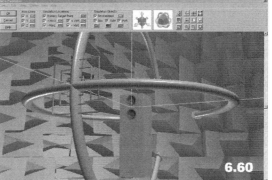

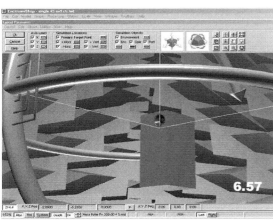

FIGURE 6.54: Single 4.5″ midrange top center baffle location.

FIGURE 6.55: Single 4.5″ midrange mid center baffle location.

FIGURE 6.56: Single 4.5″ midrange WTW center baffle location.

FIGURE 6.57: Single 4.5″ midrange top offset baffle location.

FIGURE 6.58: Single 4.5″ midrange mid offset baffle location.

FIGURE 6.59: Single 4.5″ midrange WTW offset baffle location.

FIGURE 6.60: Dual 5.25″ midrange top center baffle location.

FIGURE 6.61: Dual 5.25″ midrange WTW center baffle location.

metrical as possible results in a better subjective experience, which is at least one of the reasons I undertook this diffraction study. This, of course, runs contrary to some engineering personalities who maintain that home speakers should have some type of directivity. Controlled directivity is a standard practice for loudspeakers designed for use in large venues and is done to provide specific coverage patterns. Enhanced directivity loudspeakers are, however, the exception rather than the rule in speakers intended for small rooms.

B. Three-Way Midrange Location SPL Variation.
Section 6.10A discussed SPL differences for woofers and tweeters in two-way systems located at different positions on a front baffle. This section will discuss the same information for midranges in three-way systems. I simulated two examples—a single 4.5″ midrange, and dual 5.25″ midranges—and chose the cabinet volumes and dimensions for this exercise from current production 2005 loudspeakers.

Midrange Diameter	Enclosure Dimensions (H × W × D)	Simulated Woofer
4.5″	20″ × 8.25″ × 10.5″	Peerless 830516
5.25″ (2)	32″ × 7.10″ × 10.5″	Vifa C13WG-19-08

I outlined six different positions for the 4.5″ midrange: three mounted on the baffle centerline and three offset to the right from those positions. These six baffle locations are depicted in *Figs. 6.54–6.59*. As you can see, the midrange was located in the center of the baffle (this is the same configuration as the production speaker with 6.5″ woofers mounted above and below this position), between the center and the top of the baffle (both 6.5″ woofers would be mounted below the midrange and a tweeter mounted above), and at the top of the baffle (both woofers at the bottom of the baffle and the tweeter mounted just below the midrange).

For the dual 5.25″ midrange format, only four positions are examined and are illustrated in *Figs. 6.60-6.63*. Only two basic positions are available for this format with this size enclosure. One configuration is with the tweeter mounted in the center of the baffle with one midrange mounted above and below and an 8″ woofer mounted at the top and bottom of the baffle (plus the offset variation). The other is with the midrange tweet-

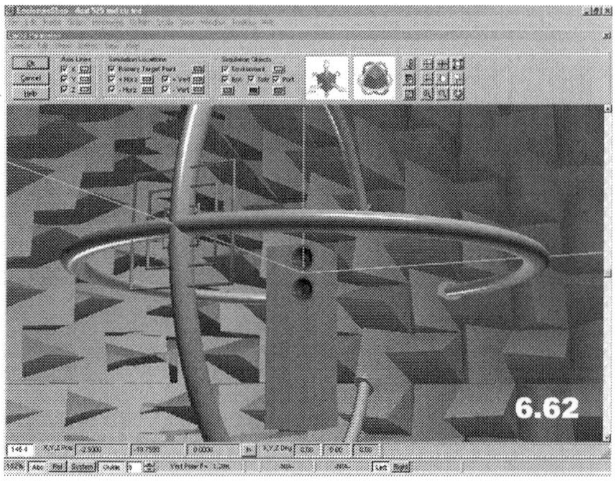

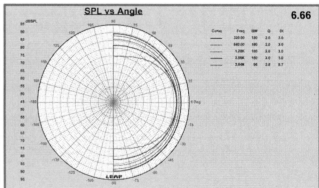

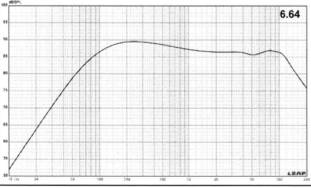

FIGURE 6.62: Dual 5.25″ midrange top offset baffle location.

FIGURE 6.63: Dual 5.25″ midrange WTW offset baffle location.

FIGURE 6.64: Frequency response of single 4.5″ midrange mounted on an infinite baffle.

FIGURE 6.65: Horizontal polar plot for *Fig. 6.64* (320Hz = solid, 640Hz = dash, 1.25kHz = dot, 2.56kHz = solid, 3.84kHz = dash).

FIGURE 6.66: Vertical polar plot for *Fig. 6.64* (320Hz = solid, 640Hz = dash, 1.25kHz = dot, 2.56kHz = solid, 3.84kHz = dash).

FIGURE 6.67: Comparison of single 4.5″ midrange frequency response for all three center baffle locations (top = solid, mid = dash, center = dot).

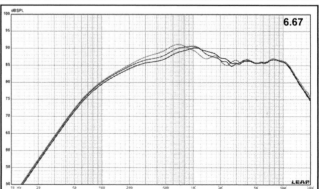

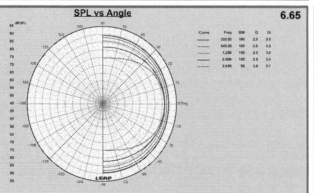

er MTM array located at the top of the baffle with two 8″ woofers located together at the bottom of the baffle. The graphic presentation for both three-way examples is similar to section 6.10A, as follows:

1. a set of reference graphs, on-axis half-space plus horizontal and vertical polar plots in half space

2. a composite graph that compares the different anechoic on-axis response curves, one for the three center 4.5″ single midrange locations, one for the three offset 4.5″ single midrange locations, plus one for the two dual 5.25″ midrange locations and one for the two 5.25″ midrange offset locations

3. individual on-axis and horizontal and vertical polar plots for each baffle location

As before, you should again start by looking at the half-space on-axis response and the half-space horizontal and vertical polar plots in *Figs. 6.64-6.66* for the single 4.5″ mid and in *Figs. 6.87-6.89* for the dual 5.25″ midrange format. Both midrange drivers have a smooth on-axis response, relatively flat for the dual 5.25″ set and with a somewhat declining SPL with increasing frequency for the single 4.5″. As with any half-space measurement, the horizontal and vertical polar plots are totally symmetrical, although you can see the cancellation effects and

FIGURE 6.68: Comparison of single 4.5″ midrange frequency response for all three offset baffle locations (top = solid, mid = dash, center = dot).

FIGURE 6.69: Frequency response for single 4.5″ midrange at top center baffle location.

FIGURE 6.70: Horizontal polar plot for *Fig. 6.69* (320Hz = solid, 640Hz = dash, 1.25kHz = dot, 2.56kHz = solid, 3.84kHz = dash).

FIGURE 6.71: Vertical polar plot for *Fig. 6.69* (320Hz = solid, 640Hz = dash, 1.25kHz = dot, 2.56kHz = solid, 3.84kHz = dash).

FIGURE 6.72: Frequency response for single 4.5″ midrange at mid center baffle location.

FIGURE 6.73: Horizontal polar plot for *Fig. 6.72* (320Hz = solid, 640Hz = dash, 1.25kHz = dot, 2.56kHz = solid, 3.84kHz = dash).

FIGURE 6.74: Vertical polar plot for *Fig. 6.73* (320Hz = solid, 640Hz = dash, 1.25kHz = dot, 2.56kHz = solid, 3.84kHz = dash).

FIGURE 6.75: Frequency response for single 4.5″ midrange at WTW center baffle location.

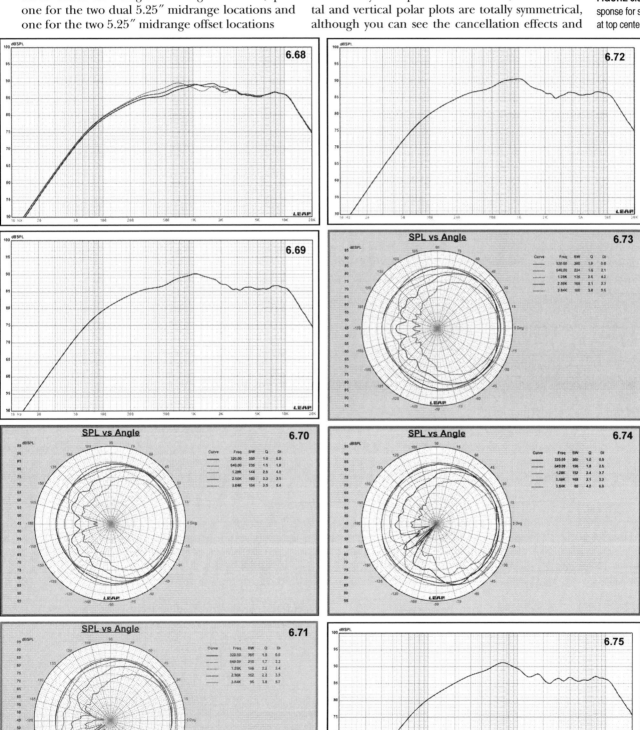

resulting lobing of the dual 5.25″ midrange drivers in the vertical polar plot.

The on-axis comparison graph for the different centerline locations and the different offset locations are given in *Figs. 6.67–6.68* for the single 4.5″ midrange and in *Figs. 6.90–6.91* for the dual 5.25″ scenario. The ±SPL data for both the 4.5″ and 5.25″ drivers is quantified in *Tables 6.2* and *6.3*, respectively.

Table 6.2 Single 4.5″ SPL Variation from 500Hz–3kHz for Different Baffle Locations (dB).

| | Baffle Location | | |
	TOP	MID	CENTER (WTW)
Center	2.30	2.95	3.07
Offset	1.69	1.07	1.45

Table 6.3 Dual 5.25″ SPL Variation from 500Hz–3kHz for Different Baffle Locations (dB).

| | Baffle Location | |
	TOP	CENTER (WTW)
Center	1.56	1.13
Offset	0.97	0.76

Results for the single 4.5″ midrange were very similar to the tweeter conclusions in Section 6.10A. Offsetting this midrange to the far side of a baffle unquestionably results in a smoother response with less SPL variation (see *Table 6.2* and *Figs. 6.67–6.68*), although, as before, not as significant as is often assumed. In the vertical polar plots (*Figs. 6.71, 6.74, 6.77, 6.80, 6.83,* and *6.86*), the baffle also tilts the response 5° upward for the placement at the top, about 3° upward for the mid location between the top and center, and no tilt at all with the midrange mounted in the center position. With no offset and the midrange located on the centerline of the baf-

FIGURE 6.76: Horizontal polar plot for *Fig. 6.75* (320Hz = solid, 640Hz = dash, 1.25kHz = dot, 2.56kHz = solid, 3.84kHz = dash).

FIGURE 6.77: Vertical polar plot for *Fig. 6.75* (320Hz = solid, 640Hz = dash, 1.25kHz = dot, 2.56kHz = solid, 3.84kHz = dash).

FIGURE 6.78: Frequency response for single 4.5″ midrange at top offset baffle location.

FIGURE 6.79: Horizontal polar plot for *Fig. 6.78* (320Hz = solid, 640Hz = dash, 1.25kHz = dot, 2.56kHz = solid, 3.84kHz = dash).

FIGURE 6.80: Vertical polar plot for *Fig. 6.78* (320Hz = solid, 640Hz = dash, 1.25kHz = dot, 2.56kHz = solid, 3.84kHz = dash).

FIGURE 6.81: Frequency response for single 4.5″ midrange at mid offset baffle location.

FIGURE 6.82: Horizontal polar plot for *Fig. 6.81* (320Hz = solid, 640Hz = dash, 1.25kHz = dot, 2.56kHz = solid, 3.84kHz = dash).

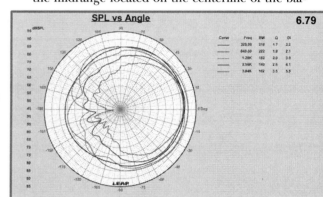

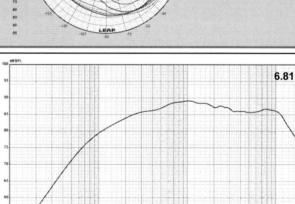

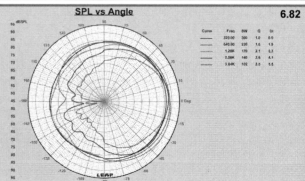

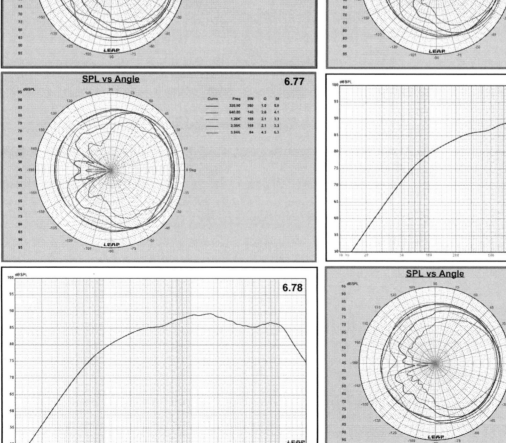

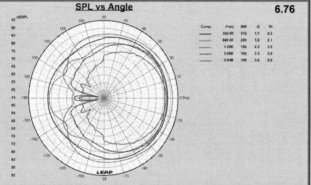

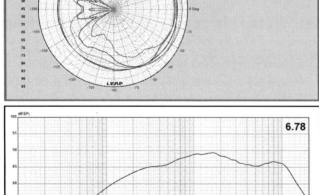

fle, all three locations have symmetrical horizontal polar plots (*Figs. 6.70, 6.73, 6.76*), again the same as the tweeter analysis in the previous section. The response change in the horizontal plane due to offset to the right side of the enclosure (*Figs. 6.79, 6.82, and 6.85*) results in a 10–15° tilt toward the right side of the enclosure plus a tighter SPL spread on that side as well, again, very similar to the tweeter example.

Results for the 5.25″ dual midrange example were somewhat different, although the offset response given in *Table 6.3* (also see *Figs. 6.90–6.91*) also showed some small improvement over the cen-

ter baffle location. In the vertical polar plots (*Figs. 6.94, 6.97, 6.100, and 6.103*), the drivers located in the center of the baffle have a totally symmetrical response. When the two midranges are relocated to the top of the baffle, the response change is not extreme, but it also is not as symmetrical as the center location. Obviously, both locations exhibit a degree of cancellation due to the use of two sources operating in the same frequency range.

The offset location for the dual 5.25″ midranges resulted in less "tilt" toward the same side of the baffle they were located on—about 5° compared to 15° for the single 4.5″ midrange—but also ex-

FIGURE 6.83: Vertical polar plot for *Fig. 6.81* (320Hz = solid, 640Hz = dash, 1.25kHz = dot, 2.56kHz = solid, 3.84kHz = dash).

FIGURE 6.84: Frequency response for single 4.5″ midrange at WTW offset baffle location.

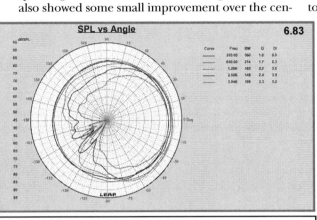

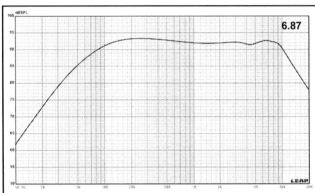

FIGURE 6.85: Horizontal polar plot for *Fig. 6.84* (320Hz = solid, 640Hz = dash, 1.25kHz = dot, 2.56kHz = solid, 3.84kHz = dash).

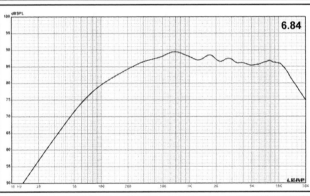

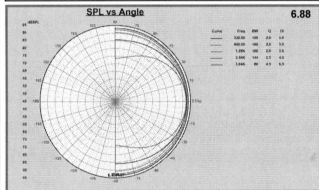

FIGURE 6.86: Vertical polar plot for *Fig. 6.84* (320Hz = solid, 640Hz = dash, 1.25kHz = dot, 2.56kHz = solid, 3.84kHz = dash).

FIGURE 6.87: Frequency response of dual 5.25″ midranges mounted on an infinite baffle.

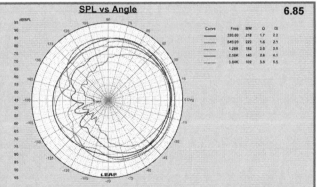

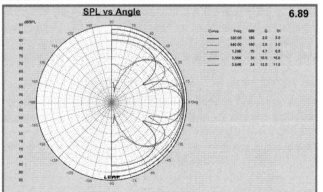

FIGURE 6.88: Horizontal polar plot for *Fig. 6.87* (320Hz = solid, 640Hz = dash, 1.25kHz = dot, 2.56kHz = solid, 3.84kHz = dash).

FIGURE 6.89: Vertical polar plot for *Fig. 6.87* (320Hz = solid, 640Hz = dash, 1.25kHz = dot, 2.56kHz = solid, 3.84kHz = dash).

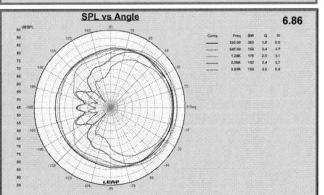

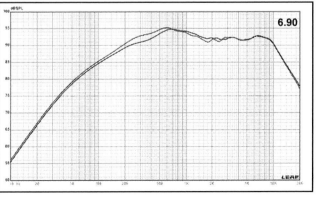

FIGURE 6.90: Comparison of dual 5.25″ midranges frequency response for both center baffle locations (top = solid, center = dash).

hibited the same tight SPL grouping for the various frequencies in the plot.

6.20 RESPONSE VARIATION DUE TO DRIVER SEPARATION.

Any time there are more than two radiating sources operating in the same frequency range on a single baffle, the combined output will produce complex interference patterns like those seen in the vertical polar plots of the dual 5.25″ example in section 6.10B. This produces areas of reinforcement ("hot spots") and cancellation ("dead spots") in the sound field. The further apart you place the two radiat-

ing points (woofers, midranges, or tweeters), the more complex the pattern will become. *Figure 6.104* compares the number of nodes in the cancellation pattern when the drivers are separated by one wavelength at a given frequency and at four times that same wavelength (for the physical distance, wavelength, of a given frequency, consult *Table 6.4*).

TABLE 6.4   Wavelength Chart

Frequency (Hz)	Distance
5000	2.7
3000	4.5
1500	9.0
750	18.1
500	27.1
300	45.2
200	67.8
100	135.6

In terms of subjective listening, interference patterns, also called comb filtering, are very hearable and can be distracting, especially if you happen to be moving around a room. This is particularly true of multiple tweeters on a single baffle. This was probably one of the biggest criticisms of the early THX

FIGURE 6.91: Comparison of dual 5.25″ midranges frequency response for both offset baffle locations (top = solid, center = dash).

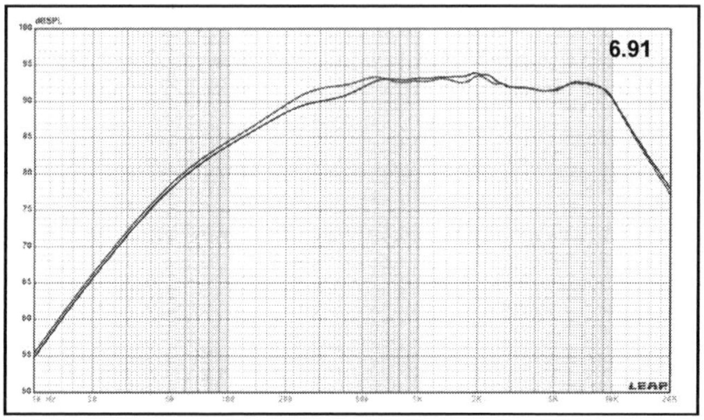

FIGURE 6.92: Frequency response for dual 5.25″ midranges at top center baffle location.

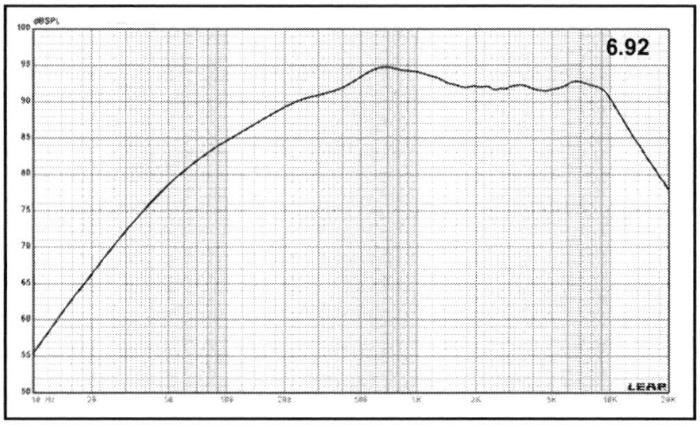

FIGURE 6.93: Horizontal polar plot for *Fig. 6.92* (320Hz = solid, 640Hz = dash, 1.25kHz = dot, 2.56kHz = solid, 3.84kHz = dash).

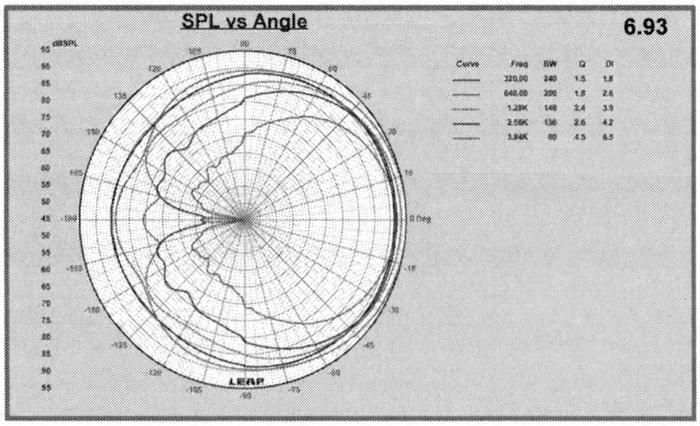

FIGURE 6.94: Vertical polar plot for *Fig. 6.92* (320Hz = solid, 640Hz = dash, 1.25kHz = dot, 2.56kHz = solid, 3.84kHz = dash).

FIGURE 6.95: Frequency response for dual 5.25″ mid-ranges at WTW center baffle location.

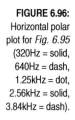

FIGURE 6.96: Horizontal polar plot for *Fig. 6.95* (320Hz = solid, 640Hz = dash, 1.25kHz = dot, 2.56kHz = solid, 3.84kHz = dash).

FIGURE 6.97: Vertical polar plot for *Fig. 6.95* (320Hz = solid, 640Hz = dash, 1.25kHz = dot, 2.56kHz = solid, 3.84kHz = dash).

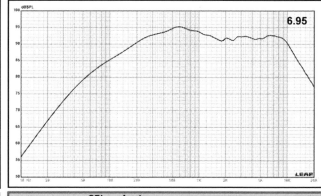

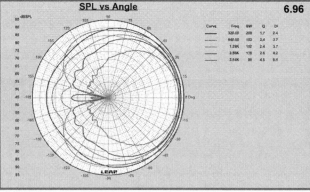

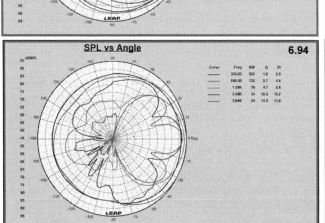

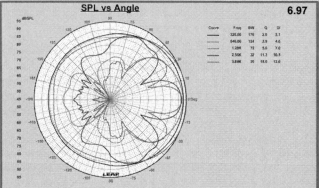

speakers that used a vertical array of two or three tweeters to obtain enhanced vertical directivity.

However, this format was abandoned following the introduction of the first 3-way THX speaker that used a vertical MTM array, which incidentally was the Atlantic Technology 350, another of my designs (the concept of getting the same vertical directivity from a MTM array rather than a multiple tweeter array came from Tommy Freedman, then chief engineer of Altec). The guiding principles, however, are to keep multiple woofers and midrange transducers grouped as closely as possible, and to generally avoid using multiple tweeters. If you can't avoid using multiple tweeters, or if you are doing so to obtain increased power handling and lower distortion at a high SPL, at least put a low-pass filter (8kHz would be a good choice of frequency for this) on the outboard members of the tweeter array, which will substantially reduce the comb filtering at high frequencies.

There are two specific 2-way design formats that deserve attention regarding the subject of multiple driver separation, and both have to do with home theater loudspeakers, although the principles to be discussed would apply to any circumstance using a similar driver format. The first concerns the viabil-ity of full-range vertical WTW (D'Appolito) speaker arrays used as LRs (left/right) in home theater sys-

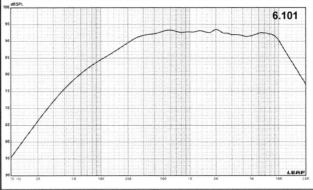

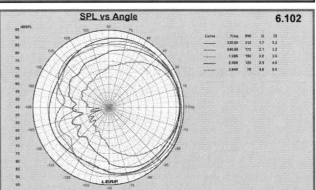

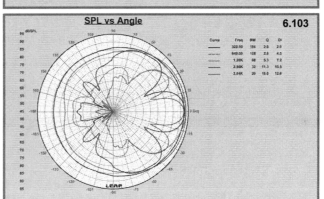

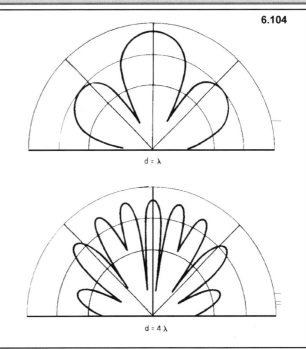

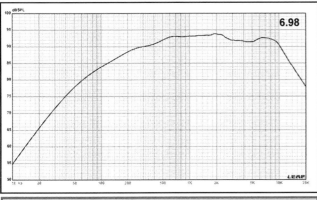

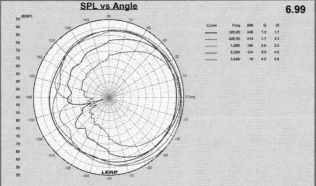

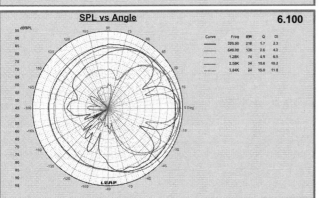

FIGURE 6.98: Frequency response for dual 5.25″ midranges at top offset baffle location.

FIGURE 6.99: Horizontal polar plot for *Fig. 6.98* (320Hz = solid, 640Hz = dash, 1.25kHz = dot, 2.56kHz = solid, 3.84kHz = dash).

FIGURE 6.100: Vertical polar plot for *Fig. 6.98* (320Hz = solid, 640Hz = dash, 1.25kHz = dot, 2.56kHz = solid, 3.84kHz = dash).

FIGURE 6.101: Frequency response for dual 5.25″ midranges at WTW center baffle location.

FIGURE 6.102: Horizontal polar plot for *Fig. 6.101* (320Hz = solid, 640Hz = dash, 1.25kHz = dot, 2.56kHz = solid, 3.84kHz = dash).

FIGURE 6.103: Vertical polar plot for *Fig. 6.102* (320Hz = solid, 640Hz = dash, 1.25kHz = dot, 2.56kHz = solid, 3.84kHz = dash).

FIGURE 6.104: Driver interference patterns.

tems, and the second involves the design of horizontal WTW arrays used for center channel speakers in home theater systems.

A. 2.5-Way WTW vs. Full-Range 2-Way WTW.

Over the last several years I have heard criticism leveled at the use of full-range two-way WTW speaker formats in home theater with the suggestion that 2.5-way formats are superior. A 2.5-way speaker has two woofers like a regular WTW format, but instead of crossing both of them over at the same frequency to

blend with a tweeter, one woofer uses a low-pass filter set over an octave or so lower in frequency, while the other is crossed normally with the tweeter. The concept being promoted is that the 2.5-way format will reduce the "undesirable" interference (lobing) due to the separation between the two woofers that are both operating together with the tweeter. The claim is that by reducing the lobing, the resulting 2.5-way format will produce a subjectively superior-sounding loudspeaker.

Fortunately, the occasion came up in the last year (sometime in 2004) for me to compare the exact same loudspeaker (cabinet and driver set) opti-

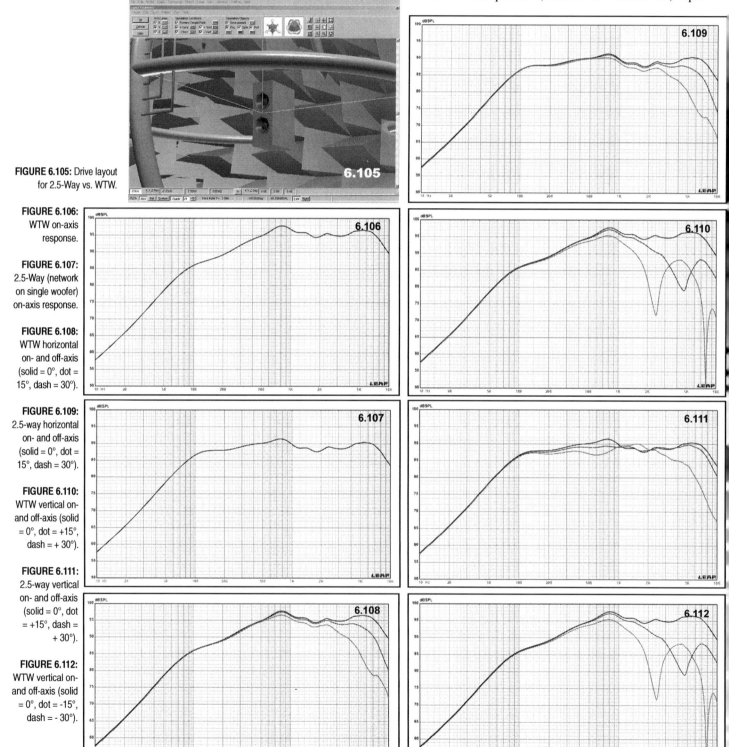

FIGURE 6.105: Drive layout for 2.5-Way vs. WTW.

FIGURE 6.106: WTW on-axis response.

FIGURE 6.107: 2.5-Way (network on single woofer) on-axis response.

FIGURE 6.108: WTW horizontal on- and off-axis (solid = 0°, dot = 15°, dash = 30°).

FIGURE 6.109: 2.5-way horizontal on- and off-axis (solid = 0°, dot = 15°, dash = 30°).

FIGURE 6.110: WTW vertical on- and off-axis (solid = 0°, dot = +15°, dash = + 30°).

FIGURE 6.111: 2.5-way vertical on- and off-axis (solid = 0°, dot = +15°, dash = + 30°).

FIGURE 6.112: WTW vertical on- and off-axis (solid = 0°, dot = -15°, dash = - 30°).

mized for a full-range WTW and the same speaker optimized as a 2.5-way. This gave me the opportunity to very critically compare the 2.5-way incarnation side by side with the same drivers in a full-range D'Appolito format. My observation was that although the timbre of both formats was very similar, the 2.5-way lacked the perceived image depth of the standard WTW format (this was a mono A/B test—for more on this, see Chapter 7.90, Loudspeaker Voicing). What follows explains the subjective difference using LEAP 5 simulations and offers some more guidance in terms of the optimal polar response in a loudspeaker.

Figure 6.105 gives the cabinet and driver setup that was configured in LEAP 5 EnclosureShop. What you see in the illustration is a dual 6.5″ speaker with the woofers spaced at the distance required to fit a small faceplate neodymium tweeter between them. Because of the methodology employed in LEAP 5 that allows you to add passive filter sections for diffraction analysis, the only network employed in the simulations was the low-pass on the single woofer in the 2.5-way example, so both the full-range WTW and the 2.5-way do not have a crossover at the tweeter crossover frequency.

The full-range WTW woofer example (without

any crossover) and the 2.5-way (with a bottom woofer 1.5kHz LP filter) analysis resulted in the production of an on-axis curve, a 30° off-axis curve in both the horizontal and vertical planes, and both horizontal and vertical polar plots for each example. Because no crossover was used in conjunction with the WTW example, you see the woofer "step" response (this will be discussed in Section 6.30), but the overall response above the "step" is identical to the 2.5-way. Curves for the full-range WTW woofers and the 2.5-way examples are as follows:

	WTW	2.5-way
On-axis	6.106	6.107
On-axis, 15° H, 30° H	6.108	6.109
On-axis, +15° V, +30° V	6.110	6.111
On-axis, –15° V, –30° V	6.112	6.113
Vertical polar plot	6.114	6.115

When examining these two curve sets, you see that the horizontal on- and off-axis (*Figs. 6.108* and *6.109*) curves are very symmetrical for both formats. However, if you compare the vertical off-axis curves both up (+) and down (–) from the measurement axis, the pair of graphs for the WTW woofers is identical (*Figs. 6.110* and *6.112*), but for the 2.5-way (*Figs. 6.111* and *6.113*), definitely not at all symmetrical. This is to be expected when one woofer is below the other and playing in a different frequency range. This is confirmed by comparing the vertical polar plots in *Figs. 6.114* and *6.115*.

While the 2.5-way speaker does not have the lobing that is typical of the full-range WTW, its vertical response is very asymmetrical. My conclusion is that having a radiating field that is symmetrical, lobing or not, sounds better than an asymmetrical radiat-

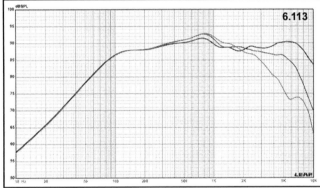

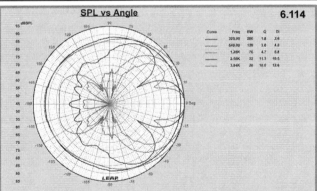

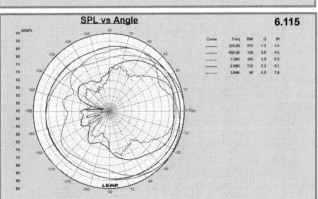

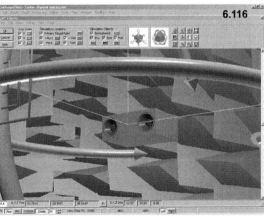

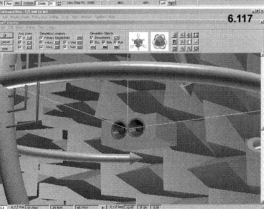

FIGURE 6.113: 2.5-way vertical on- and off-axis (solid = 0°, dot = -15°, dash = - 30°).

FIGURE 6.114: Vertical polar plot for WTW (320Hz = solid, 640Hz = dash, 1.25kHz = dot, 2.56kHz = solid, 3.84kHz = dash).

FIGURE 6.115: Vertical polar plot for 2.5-way (320Hz = solid, 640Hz = dash, 1.25kHz = dot, 2.56kHz = solid, 3.84kHz = dash).

FIGURE 6.116: Center channel dual woofer layout with wide spacing.

FIGURE 6.117: Center channel dual woofer layout with close spacing.

ing field, and this is the reason the full-range WTW gave a better sense of ambience in a room than the 2.5-way speaker did.

B. Woofer Spacing for 2-Way WTW Center Channel Speakers.

While three-way center channel speakers that have vertical MTM arrays with the same acoustic polarity as their accompanying LR speakers are by far one of the best solutions for home theater, the majority of center channel loudspeakers are 2-way horizontal aspect ratio dual woofer WTW arrays. If you survey the variety of the horizontal

FIGURE 6.118: Frequency response comparison of on-axis close and wide spaced center channels (solid = close spaced woofers, dash = wide spaced woofers).

FIGURE 6.119: On- and off-axis horizontal frequency response for wide spaced woofers (solid = 0°, dot = 15°, dash = 30°).

FIGURE 6.120: On- and off-axis horizontal frequency response for close spaced woofers (solid = 0°, dot = 15°, dash = 30°).

FIGURE 6.121: Vertical polar plot for wide spaced woofers (320Hz = solid, 640Hz = dash, 1.25kHz = dot, 2.56kHz = solid, 3.84kHz = dash).

FIGURE 6.122: Vertical polar plot for close spaced woofers (320Hz = solid, 640Hz = dash, 1.25kHz = dot, 2.56kHz = solid, 3.84kHz = dash).

FIGURE 6.123: Horizontal polar plot for wide spaced woofers (320Hz = solid, 640Hz =dash, 1.25kHz = dot, 2.56kHz = solid, 3.84kHz = dash).

FIGURE 6.124: Horizontal polar plot for close spaced woofers (320Hz = solid, 640Hz = dash, 1.25kHz = dot, 2.56kHz = solid, 3.84kHz = dash).

FIGURE 6.125: Horizontal polar plot for wide spaced woofers (640Hz = dash, 1.25kHz = dot, 2.56kHz = solid).

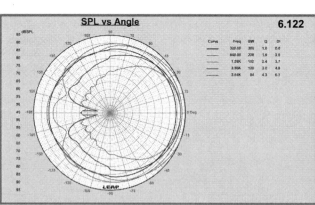

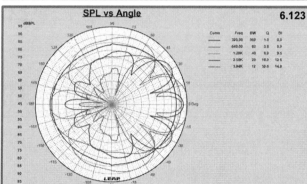

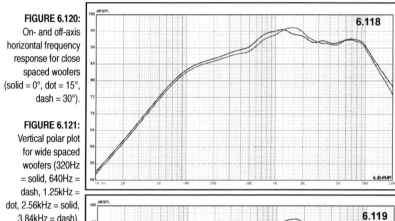

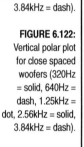

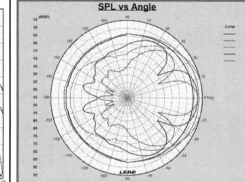

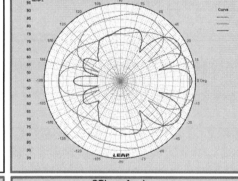

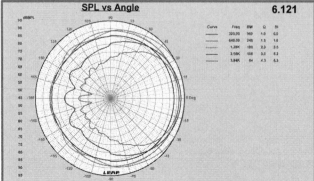

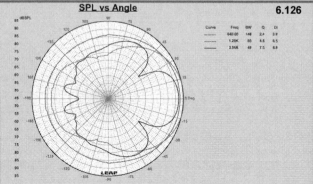

cabinet dual woofer center channel speakers being offered by the loudspeaker industry, you will notice that spacing between the woofers varies considerably from very close and nearly touching to spread apart at a considerable distance that can be as much as 5-8″ from cabinet center for each woofer. The consequence of wider spacing between the dual woofers is more complex lobing in the horizontal response of the speaker and can be avoided.

Figures 6.116 and *6.117* depict two horizontal center channel scenarios for a speaker with 5.25″ woofers, one with wide spaced woofers each mounted 5″ from the cabinet center, and the other with the two woofers nearly touching. For the speaker in *Fig. 6.117*, the tweeter would be mounted either at the top or bottom of the baffle, on the centerline where the two woofers are mounted. Generally, this requires the use of a small footprint neodymium type dome tweeter. Graphic data for the comparative analysis of these two center channel formats is as follows:

	Wide Spaced	Close Spaced
On-axis for both cabinets	6.118	
On-axis, Horizontal 15, 30°	6.119	6.120
Vertical Polar Plots	6.121	6.122
Horizontal Polar Plots	6.123	6.124
3 Freq. Horizontal Polar Plots	6.125	6.126

As you can see in *Fig. 6.118*, the on-axis response is somewhat different due to the spacing, but nothing that would indicate any kind of SPL problem. Also, in the two vertical polar plots in *Figs. 6.121* and *6.122*, there is no indication of a problem, as these are nearly identical. However, when you look at the horizontal on- and off-axis curves in Figs. *6.119* and *6.120*, it is obvious that the complexity of the off-axis cancellation nulls is much greater for the wide-spaced dual woofer example. However, with a 3kHz low-pass network, it doesn't really look like all that much of an issue.

If you now look at the two horizontal polar plots in *Figs. 6.123* and *6.124*, you can get a better feel for what is going on. *Figures 6.125* and *6.126* are the same polar plots as *Figs. 6.123* and *6.124*, but are somewhat easier to read and only display the 640Hz, 1.28kHz, and 2.56kHz frequency bands. The idea is that the closer-spaced woofers will give a more even coverage pattern across your listening audience, especially if they are fairly close to the screen and the speakers.

6.30 RESPONSE VARIATION DUE TO BAFFLE AREA (STEP RESPONSE).

Chapter 5, Cabinet Design: Shape and Damping, discussed the effect that different baffle shapes have on the SPL of a woofer, midrange, or tweeter. However, while exotic shapes are interesting, the fact remains that the majority of loudspeakers both currently and historically are built from simple rectangular boxes. The analogy of a baffle in anechoic space is similar to a flashlight reflector, except that the wavelength of light is at just one frequency (well, actually it's a grouping of wavelengths between 400–800nm), while the bandwidth of a loudspeaker relevant to typical baffle areas is actually quite wide. As the area of a loudspeaker baffle increases, it will offer more and more reinforcement to the very lowest frequencies right up to the point where the baffle becomes infinitely large and reinforces all frequencies from 1Hz to the upper limit of the audio spectrum. Step response is often used to describe this phenomenon, the step being the upper frequency at which the baffle supplies even reinforcement for all frequencies at that frequency and higher.

The example that was simulated in LEAP 5 EnclosureShop to demonstrate the overall SPL changes that occur with increasing total baffle area incorporates a 6.5″ driver. The extremes for any baffle-related response change, as discussed in Chapter 5, are from the speaker being mounted with no baffle in open air to the speaker being mounted on an infinitely large baffle, or half-space. For the 6.5″ example, the simulation curves for these two extremes are given in *Fig. 6.127*. Any realizable baffle response will fall somewhere between these two curves.

The simulation started with the 6.5″ woofer loaded into a small enclosure with a baffle that measured 9.25″ high by 6″ wide, just wide enough for the example driver to fit (*Fig. 6.128*). Keeping the

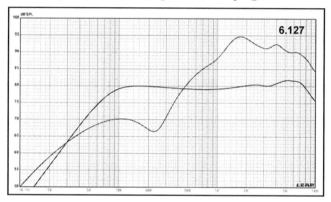

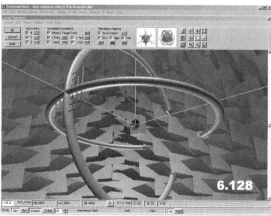

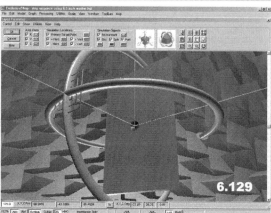

FIGURE 6.126: (*see previous page*) Horizontal polar plot for close spaced woofers (640Hz = dash, 1.25kHz = dot, 2.56kHz = solid).

FIGURE 6.127: Frequency response comparison of 6.5″ woofer with no baffle and with infinite baffle (solid = infinite baffle response, dash = no baffle response).

FIGURE 6.128: Smallest baffle layout for step response.

FIGURE 6.129: Largest baffle layout for step response.

aspect ratio (the ratio of height to width) the same, I increased the baffle width in 1″ increments from 6″ to 12″, and then from 12″ to 20″ in 2″ increments, plus added a final monster baffle that measured 60.6″ × 40″ (*Fig. 6.129*).

I programmed all the various baffle sizes into LEAP 5 and performed on-axis 2.83V/1m simulations for all 13 baffle sizes. The results are shown in *Figs. 6.130–6.133*. The series of graph curves begins with all of the SPL curves displayed simultaneously along with the reference half-space infinitely large baffle curve (*Fig. 6.130*). While this many curves on one graph are difficult to read, you can definitely see the emerging pattern. As the baffle mutates from a 6″ width to a 40″ width, two major features are apparent.

First, the peak at 1.1kHz in the 6″ wide baffle curve decreases in frequency as the baffle area increases. Next, the amplitude of the 100Hz corner frequency of the high-pass rolloff of this driver gradually increases from 70dB for the smallest baffle area to 83dB for the largest baffle area. Looking at *Fig. 6.131*, which represents the baffle widths from 6″ to 12″ in 1″ increments, the 1.1kHz peaking in the 6″ wide baffle not only decreases in frequency as the baffle area increases, but the peak also de-

creases somewhat in amplitude in this series.

The group of curves in *Fig. 6.132* shows the progression from 12″ wide to 20″ wide in 2″ increments, again showing an inverse relationship with the peaking in the response decreasing in frequency as area increases, but this time a small increase in amplitude occurs as the area increases. This SPL pattern is somewhat easier to see in *Fig. 6.133*, where the graph has three curves starting at a 10″ baffle width, doubling to 20″ and then 40″. *Figure 6.134* compares the largest 40″ wide baffle with a half-space measurement of the driver, showing that this process is definitely mutating to half-space.

This raises an interesting issue about which design format is subjectively superior: a loudspeaker in an off-wall cabinet with some kind of defined baffle area and shape, or an in-wall speaker with comparatively large baffle area the size of a house wall. Over the years I have designed a number of in-wall products for various companies including M&K Sound, Parasound, Posh Audio, coNEXTion, and a THX Ultra in-wall for Atlantic Technology, as well as a rather large number of off-wall speakers.

In a 2005 interview by Brent Butterworth in *Robb Report Home Entertainment* magazine[1], I was asked the question of which worked best, in-wall or off-

FIGURE 6.130: On-axis frequency response comparison of all 12 baffle sizes with infinite baffle base curve (thick solid = infinite baffle response).

FIGURE 6.131: On-axis frequency response comparison of seven baffle sizes in 1″ increments from 6″ wide to 12″ wide.

FIGURE 6.132: On-axis frequency response comparison of five baffle sizes in 2″ increments from 12″ wide to 20″ wide.

FIGURE 6.133: On-axis frequency response comparison of three baffle sizes, 10″ wide, 20″ wide, and 40″ wide (solid = 10″ wide baffle, dash = 20″ wide baffle, dot = 40″ wide baffle).

FIGURE 6.134: On-axis frequency response comparison of the largest baffle size with the infinite baffle response (solid = infinite baffle, dash = 40″ wide baffle).

FIGURE 6.135: On-axis frequency response comparison of an undamped and a foam damped baffle (solid = no surface damping on baffle, dash = foam damped baffle surface).

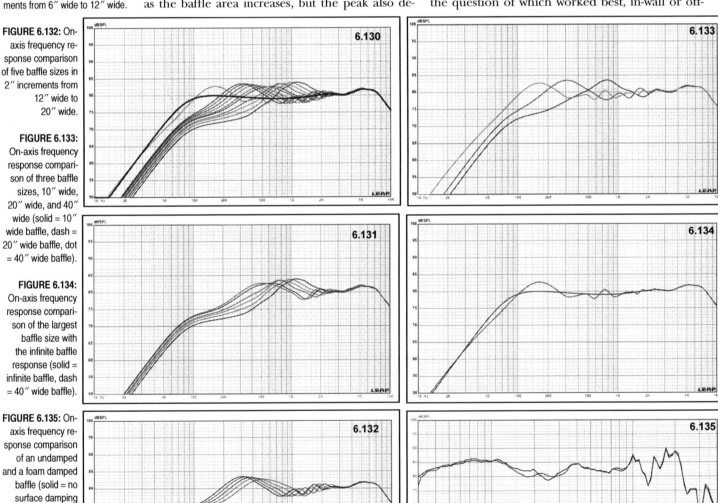

wall speakers. Because Mr. Butterworth has been measuring and reviewing home theater loudspeakers for a number of years, first in *Home Theater* magazine, and then for *Robb Report*, he has had the opportunity to observe the loudspeaker industry trend of home theater speakers being placed physically out of sight in a home theater, as they are in a real commercial theater. Rather than have the LCR (Left, Center, Right), surrounds, subwoofer, and rear channels all sitting on the floor (or placed on speaker stands) away from the walls or even mounted on the walls, many installations have speakers that are hidden behind curtains, behind grilles in home entertainment centers, mounted in the wall, or mounted in the ceiling. Chapters 5 and 6 have spent much effort defining the SPL modification that a discrete baffle has on the response of a speaker, but also made it apparent that the larger the baffle, the more even and smooth a response the drivers will be able to produce.

While it might be tempting to either conclude that in-wall baffles are superior for this reason, or, to the contrary, adapt an elitist attitude that in-wall speakers are inherently inferior (it's only been since perhaps the year 2000 that really high-end in-wall speakers have appeared on the market, and prior to that they were mostly distributed audio speakers intended for a muzak scenario, totally repugnant to any self-respecting audiophile), the truth is that neither of these would be correct. My answer to Brent's insightful question is that you can take any given set of drivers and make them sound musical in either design format, in-wall or off-wall; it's just a matter of design criteria. Each platform provides a launch vehicle for a wave front, and the fact remains that you have several choices on how to get sound into

a room. In my experience as a loudspeaker design consultant, you can easily make all of these formats (on-wall, off-wall, in-wall, or in-ceiling) work extremely well.

6.40 BAFFLE DAMPING.

All baffles reflect sound. As the initial wave-front propagates from the driver diaphragm, it travels over the surface of the baffle, and part of the energy is reflected and some diffracted off edges and protrusions. All of these incidental aspects of the composite wave front unavoidably involve some small time delay (up to $0.5mS^2$) compared to the initial wave front emanating from the driver diaphragm. This acoustic "clutter" tends to smear the sonic detail and muddy your subjective musical perception.

As a consequence, much effort over the years has gone into limiting this problematic consequence. Various ideas have been patented and incorporated for this purpose, but the two basic approaches are to either scatter the reflective energy in all directions, or to damp it as much as possible. B&W has used plastic baffles with a 3-D surface composed of hundreds of tiny pyramids that were supposed to scatter reflective energy, and numerous speaker designers, including models from Cizek and SRA (my first company) in the mid to late 1970s, have used die cut sheets of acoustic foam to cover all or part of a baffle.

The measured differences between an undamped and damped baffle can range from fairly impressive to not particularly significant, depending on the type and amount of material[3]. The on-axis curves shown in *Fig. 6.135* are for a $15.75'' \times 10'' \times 8''$ enclosure with a $3''$ full-range (the driver mounted about $3''$ down from the top of the baffle and centered) both with and without a ¼" thick foam-damping material covering the entire surface of the baffle up to the edges of the driver. The material was a specialized acoustic foam-damping product that came from Soundcoat, an OEM noise-control manufacturer.

As you can see, the primary damping effect occurred between 1kHz and 3.5kHz for this particular material; however, if you look at the on-axis CSD (Cumulative Spectral Decay) plots done with a CLIO MLS analyzer in *Figs. 6.136–6.137*, you can also see decay differences that occur at different parts of the bandwidth (see *Fig. 6.136* for the waterfall plot without foam material on the baffle and *Fig. 6.137* for the waterfall plot with the Soundcoat foam attached). This changes significantly off-axis, as seen in the 30° off-axis curve comparison in *Fig.*

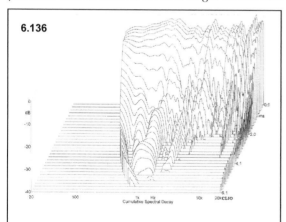

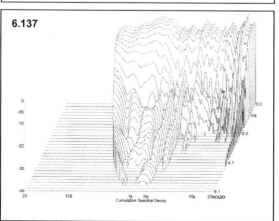

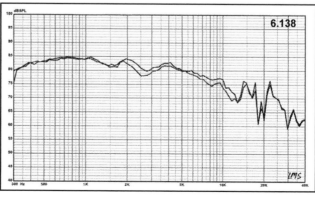

FIGURE 6.136: CLIO CSD plot of undamped baffle.

FIGURE 6.137: CLIO CSD plot of foam damped baffle.

FIGURE 6.138: 30° off-axis frequency response comparison of an undamped and a foam damped baffle (solid = no surface damping on baffle, dash = foam damped baffle surface).

6.138, where the attenuation from the foam material extends from 1.8kHz to above 12kHz. The actual subjective differences are discussed in Section 6.50, Subjective Evaluation of Diffraction.

6.50 SUBJECTIVE EVALUATION OF DIFFRACTION.

Virtually all published loudspeaker diffraction information discusses either the mathematics of simulating diffraction[2, 4–9] or measurements of various diffraction scenarios[10]. However, the really important aspect of diffraction is not how it measures with a microphone or simulates in a computer, but how it subjectively affects what you hear. To my knowledge, no one has ever published any kind of controlled listening test to determine how hearable different aspects of cabinet diffraction can be perceived, despite the fact that both amateur and professional loudspeaker designers still spend considerable effort trying to eliminate the deleterious effects of measured diffraction.

Peter Kates concluded at the end of his diffraction paper titled "Loudspeaker Cabinet Diffraction effects"[2] that "reflections can cause frequency response irregularities of up to 4dB, accompanied by group delays of up to 0.5mS. These irregularities contribute to spectral coloration, confuse localization, and increase the apparent source width of the loudspeaker system." To what extent this is true apparently has never been determined in any kind of published subjective study.

Before undertaking this project, I contacted Sean Olive, the manager of the Subjective Evaluation R&D Group at Harman International (JBL, Infinity, Revel, and so on). Mr. Olive works for and with Dr. Floyd Toole, Vice President of the Acoustic Engineering Group at Harman International, and both of them have been working on the science of listening since their work at the NRC (National Research Council) in Canada. Certainly this group, which includes Floyd Toole, Sean Olive, and Alan Devantier, has contributed more to the science of subjective listening than any group I know of in the industry. When I asked Mr. Olive whether he was indeed aware of any published works on the subjective evaluation of various diffraction phenomena, he replied, after consulting with Dr. Toole, that neither he nor Floyd was aware of any available published information on the subject, so if there is, it doesn't seem to be showing up on anybody's radar. My apologies ahead of time if we missed someone's work.

As a result of this communication, I decided to design my own informal subjective diffraction study with the goal of either confirming or denying the existence of some of the conventional wisdom and wives' tales regarding the sonic effects of diffraction. Before beginning, I would first like to emphasize the informal nature of the following undertaking.

This was not a double-blinded ABX study using a large group of trained and untrained listeners that was followed up with some kind of statistical analysis to reinforce conclusions. Rather, this was just two very experienced loudspeaker industry professionals doing what we have successfully done for a living for a number of years: listen to loudspeakers and describe differences. The two professionals were myself and my business associate and voicing partner, Nancy Weiner, Vice President of Marketing for coNEXTion Systems Inc. (www.conextionsystems.com). Nancy and I together have voiced over 30 products for Atlantic Technology, coNEXTion Systems, and several other well-known loudspeaker manufacturers, all well reviewed by the major industry publications such as *Robb Report Home Entertainment, Home Theater Magazine, Stereophile Home Theater*, and *Sound and Vision*.

Comparative analysis of complete systems is a difficult task and has been well documented in the industry[11-19]. Just placing speakers in a room to compare them can be a daunting task[16,18]. I reported on a unique device for rapid A/B comparison of loudspeakers that was created by Dr. Toole's group at Harman called the "speaker shuffler" in an August 1999 issue of *Voice Coil*[20]. This device, described in detail in AES Preprint 4842[21], was built by a high-tech aerospace company for Harman and effectively could switch a pair of speakers for a listening test in 2–3 seconds while keeping the speakers in the exact acoustic space. This is *very* important, because placing even two speakers next to each other in a test can cause timbre differences due to room modes.

Unfortunately, I really could not justify the $150,000 price tag of having my own "speaker shuffler" built for my office, so I came up with an extremely cost-effective alternative that only cost about $29! *Figure 6.139* shows a picture of my rapid A/B comparison fixture that will keep the speakers you are comparing in the exact same acoustic space and perform this task with an A/B switch time of less than 1 second. All you need are a couple of 24″ diameter MDF (Medium Density Fiberboard) platters from Home Depot and an 11″ lazy Susan bearing from Ace Hardware, two speaker stands, and a partner willing to rotate the platter while you are listening and switching the amplifier channels.

A total of five separate tests were performed to determine the subjective nature of the various aspects of diffraction.

Test #1—*Tweeter Inset*—the loudspeaker industry has spent probably millions of dollars recessing tweeters and other drivers over the years. The practice undoubtedly began because of measured differences in surface-mounted drivers and inset drivers, but also has probably continued as a cosmetic affectation that goes along with the ever-increasing industrial design aspect of loudspeaker manufacturing.

The test was simple: A/B compare two identical tweeters (Vifa DX25TG05-04 1″ soft domes)—one inset flush with the baffle and the other surface mounted. I mounted both tweeters on the front baffle of a 15.75″ × 10″ × 8″ enclosure and centered them 3″ down from the top of the baffle. I used the LinearX LMS analyzer to take 2.83V/1m measurements of both examples with the on-axis comparison depicted in *Fig. 6.140* and the 30° off-axis curves shown in *Fig. 6.141* (both curves were of the exact same driver). The on-axis difference is rather substantial, but looking at the 30° off-axis curve comparison in *Fig. 6.141*, it's obvious that this is very much an on-axis phenomenon.

Test #2—*Baffle Size*—it's a generally accepted fact that smaller baffles sound different than larger baffles, but exactly what subjective characteristics each has should be revealing. For this test, and all the remaining tests for this study, I used a pair of closely matched 3″ full-range woofers (Tang Band model

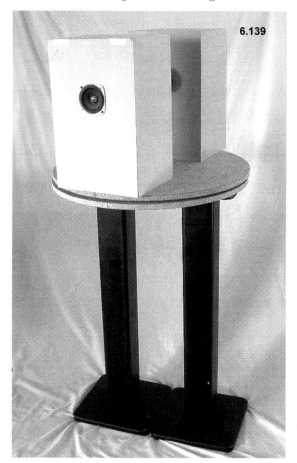

W3-594S). These have a frequency response from about 100Hz to beyond 10kHz, plus the off-axis performance of a relatively small diameter cone. I mounted one of the W3s in the 15.75″ × 10″ × 8″ enclosure baffle, 6″ from the top of the baffle and centered (vertically off-center from the middle of the baffle).

Inside the enclosure was another smaller sealed enclosure, the same volume as the second W3 enclosure. This was done to keep the bottom end response of the two speakers as close as possible. The second and smaller enclosure measured 7″ × 4″ × 4″ and had the W3 mounted 3″ down from the top of the baffle and centered (see *Fig. 6.142* for a photograph

FIGURE 6.139: Picture of the rapid A/B comparison fixture.

FIGURE 6.140: On-axis frequency response comparison of an inset tweeter and a surface mounted tweeter (solid = inset, dash = surface mounted).

FIGURE 6.141: 30° off-axis frequency response comparison of an inset tweeter and a surface mounted tweeter (solid = inset, dash = surface mounted).

FIGURE 6.142: Relative size comparison of baffles used for diffraction subjective Test #2.

FIGURE 6.143: Test #2 on-axis frequency response comparison of driver mounted on small baffle and driver mounted on larger baffle (solid = larger baffle, dash = small).

FIGURE 6.144: Test #2 30° off-axis frequency response comparison of driver mounted on small baffle and driver mounted on larger baffle (solid = larger baffle, dash = small).

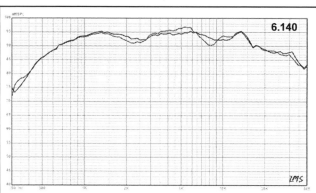

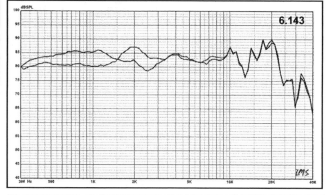

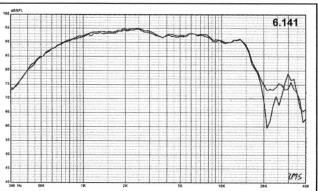

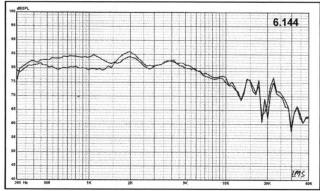

of both enclosures placed side by side with the drivers mounted).

Both small enclosures had 100% fill material, which in this case happened to be Acousta-Stuf, which is a good wide-range damping material for enclosure volumes. The W3s were A/B compared with the same driver height above the platter so that the perceived image location would be identical. Objective 2.83V/1m measurements of the 3″ driver on the different size baffle are shown in *Fig. 6.143* for the on-axis response and *6.144* for the 30° off-axis response (both curves were of the exact same driver). The differences in this case were strong both on- and off-axis.

Test #3—Baffle Shape—Over the years manufac-

turers and amateur builders alike have produced cabinets designed to defeat edge diffraction with bevels anywhere from ¾″ roundovers to large 3–6″ straight, compound, and curved bevel shapes. Obviously, such exotic additions to plain rectangular enclosures are both time-consuming and expensive, although sometimes from an industrial design aspect, very attractive cosmetically. This test compared the standard 15.75″ × 10″ × 8″ sharp-edged rectangular enclosure using the W3 driver in the same mounting position as Test #2 with the same enclosure, driver, and mounting position but with the addition of a 3″ compound bevel (2″ at a 45° angle and 1″ at a 60° angle).

The bevel-modified enclosure is depicted in *Fig. 6.145*. Objective 2.83V/1m measurements on-axis

FIGURE 6.145: Picture of the compound beveled edge baffle for diffraction subjective Test #3.

FIGURE 6.146: Test #3 on-axis frequency response comparison of a compound beveled baffle and plain sharp edged baffle (solid = plain baffle, dash = compound beveled baffle).

FIGURE 6.147: Test #3 30° off-axis frequency response comparison of a compound beveled baffle and plain sharp edged baffle (solid = plain baffle, dash = compound beveled baffle).

FIGURE 6.148: Picture of foamed damped baffle for diffraction subjective Test #4.

FIGURE 6.149: Test #4 on-axis frequency response comparison of an undamped baffle and foam damped baffle (solid = plain baffle, dash = foam damped baffle).

FIGURE 6.150: Test #4 30° off-axis frequency response comparison of an undamped baffle and foam damped baffle (solid = plain baffle, dash = foam damped baffle).

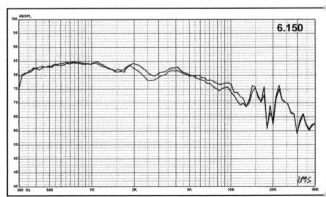

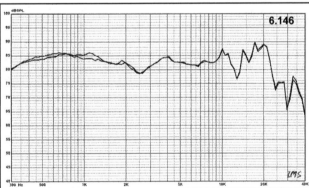

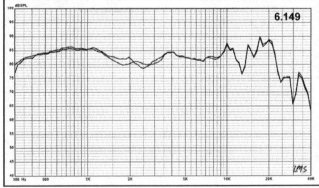

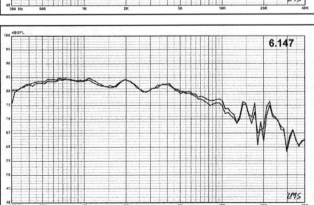

comparing the straight rectangular baffle with the compound beveled baffle are illustrated in *Fig. 6.146* with the 30° off-axis comparison shown in *Fig. 6.147* (both curves were of the exact same driver). Differences on-axis are primarily below 2.5kHz on-axis and extend to above 10kHz at 30° off-axis.

Test #4—Damped Baffle—the measurable effect of a damped baffle was discussed in Section 6.40, so this test was to confirm the subjective consequence of a foam-damped baffle. The test involved the same two enclosures, W3 drivers, and mounting positions as in Test # 3, but one baffle was 100% covered with the ¼" Soundcoat acoustic damping foam (*Fig. 6.148*). Objective 2.83V/1m measurements of the baffle with the foam and without the foam blanket are given in *Fig. 6.149* for the on-axis response, and *Fig. 6.150* for the 30° off-axis response (all curves produced with the exact same driver). Differences on-axis again are mostly below about 3kHz and extend to above 10kHz at 30° off-axis.

Test #5—Driver Baffle Location—The discussion and simulations in Section 6.10A and B were aimed at revealing measured SPL differences that occur when the same driver is located in different areas on a standard rectangular baffle. This listening test was designed to reveal the subjective differences that can be perceived from moving a driver to different locations on a baffle. Four locations were used—the middle and top of the baffle along the vertical centerline, and the same locations moved to the far right side of the baffle (*Fig. 6.151*). Objective 2.83V/1m measurements were made of the various baffle locations with the W3 full-range and are shown in *Fig. 6.152* for the on-axis response, and *6.153* for the 30° off-axis response. The SPL variations ranged up to 4dB, and were apparent on-axis out to about 3kHz and out to above 10kHz off-axis.

Testing required that the rapid A/B fixture be located 6′ from the nearest walls in a large 20 × 30 carpeted room with a vaulted ceiling, and oriented diagonally rather than parallel with the wall structures. Nancy and I took turns listening to each comparative test and used a simple 1–3 scale to evaluate the differences (note: this test was aimed only at establishing a level of perceptual difference between the two choices and not a preference). A score of 1 indicated that there was no discernible difference between the two choices. A score of 2 meant that the change was detectable, but not significant enough to matter. The highest score, 3, meant that the difference was both discernible and significant.

At the end of each test using the rapid A/B device, we then placed the two test speakers side by side and A/B-compared the two speakers a few times in this orientation and then reversed positions and repeated the procedure. It is interesting to note that with the really large amount of comparative listening and voicing we have done together using two samples in mono placed side by side in the same location as just described, we both found the high-speed A/B device to be very useful, but almost too slow. Acoustic memory is so brief that instantaneous comparison is almost a requirement to differenti-

ate between two sonic choices. The 0.5–1 second delay that it took to rotate the speakers into place was barely fast enough for either of us to feel totally comfortable, which is the reason we did the side-by-side comparison at the end of each test. However, at the end of each separate test, we also took the time to discuss what we each had heard and summarized these details.

The results were very interesting, but not unexpected.

Test #1—neither of us was able to distinguish any difference between the inset or the surface-mounted tweeter dome. Given the highly directional nature of the objective measurements, it is not surprising that this was the result. It is possible that the result could be different with a multi-way speaker with the woofer frames surface-mounted or recessed, but I tend to

FIGURE 6.151: Baffle locations for subjective diffraction Test #5.

FIGURE 6.152: Test #5 on-axis frequency response comparison of different baffle locations (solid = center location, dot = top offset location, dash = top center location, dash/dot/dot = center offset location).

FIGURE 6.153: Test #5 30° off-axis frequency response comparison of different baffle locations (solid = center location, dot = top offset location, dash = top center location, dash/dot/dot = center offset location).

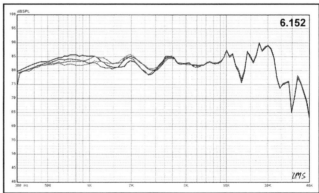

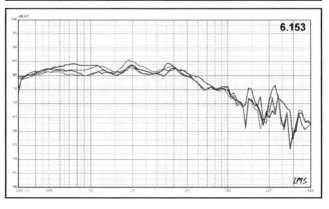

doubt it. The primary reason for recessing drivers would seem to be more cosmetic than acoustic.

Test #2—the large vs. small baffle test received a 3 from both of us: definitely discernible and significant. We both believe the larger baffle had more warmth, but definitely less detail and tending toward "muddy." The small baffle speaker had less "warmth," probably because there is less low-frequency reinforcement and low-frequency (approximately 50–200Hz) emphasis. We both believe that there was a significant impression of increased detail with the small baffle. It's no secret that small baffles have much more pinpoint imaging in stereo than larger baffles, but the increased detail is certainly a function of less reflection and delay, which are consequences of a comparatively small baffle area.

Test #3—the results of comparing a sharp edge baffle to a large beveled edge baffle was somewhat unexpected. We both rated it a 3, definitely detectable and significant, but neither of us believe it was so much an improvement as just a difference. The sharp edge, often criticized for all the diffraction it produces, actually seemed more "live." Nancy described it as having more "room tone" from the recording.

We also noticed that the large beveled edge made the image (listening in mono) seem larger and more spacious, but again somewhat dulled by comparison. Obviously, there have been many extremely well-reviewed and popular loudspeakers built from the lowly rectangular cabinet, and, frankly, neither of us thought that this was a serious handicap. As far as the large bevel goes, it definitely changes things, but for better or for worse I think is a matter of opinion.

Test #4—since I have used damped baffle configurations on numerous occasions over these years in my design work, I pretty much expected the results of this test. We both gave this a resounding score of 3, definitely discernible and very significant. The foamed damped baffle really made the driver sound smoother, less edgy, and increased the sense of detail in the music. Nancy noted it seemed to bring out the midrange more. Her perception was likely due to less high-frequency delayed reflection, and the decreased high-frequency "hash" would have the effect of making the midrange seem more pronounced.

Test #5—four A/B comparisons in terms of baffle placement were done for this test as follows:
   a. top center compared to middle center
   b. middle center compared to top right
   c. middle center compared to middle right
   d. top center compared to top right

Mounting location comparison (a) rated a score of 3 from both of us, definitely detectable and significant. The center baffle position had more perceived "warmth," but the top position had a more "open and airy" quality, undoubtedly caused by the asymmetrical vertical polar response and the slight upward tilt of the polar pattern. Mounting location comparison (b) rated a 3 as well, but seemed less prominent an effect than (a). Comparisons (c) and (d) both rated a 2, were discernible, but did not impress either Nancy or me as being very significant.

Diffraction has always seemed to me as being touted as more of a "boogie man" than reality would indicate. I have frequently commented when asked about the importance of diffraction that "the diffraction caused by cabinet edges and baffle protrusions is probably at least as hearable as the diffraction caused by the vase your wife or girlfriend put on top of your speaker, which is to say, not at all." While this may not be far from true, the benefit from damping a front baffle is still a very real and important tool for increasing the quality of the subjective listening experience, but at the same time does not mean the undamped baffles are so objectionable as to be unusable. Ultimately, it is an eclectic combination of driver timbre, driver placement, sharp or beveled edges, different baffle areas, crossover and enclosure low-frequency design, the degree of baffle damping and the room interface that describe the subjective experience.

REFERENCES

1. B. Butterworth, "The Speaker Sage Speaks," *Robb Report Home Entertainment*, March/April 2005.

2. J. Kates, "Loudspeaker Cabinet Reflection Effects," *JAES*, May 1979.

3. D. Ralph, "Diffraction Doesn't Have to Be a Problem," *audioXpress*, June 2005.

4. R. M. Bews, M. J. Hawksford, "Application of the Geometric Theory of Diffraction (GTD) to Diffraction at the Edges of Loudspeaker Baffles," *JAES*, October 1986.

5. J. Porter, E. Geddes, "Loudspeaker Cabinet Edge Diffraction," *JAES*, November 1989.

6. J. Backman, "Computation of Diffraction for Loudspeaker Enclosures," *JAES*, May 1989.

7. J. Vanderkooy, "A Simple Theory of Cabinet Edge Diffraction," *JAES*, December 1991.

8. J. Vanderkooy, "On Loudspeaker Cabinet Diffraction," *JAES*, March 1994.

9. J. R. Wright, "Fundamentals of Diffraction," *JAES*, May 1997.

10. J. Moriyasu, "Acoustic Diffraction: Does It Matter?," *Voice Coil*, February 2005 (reprinted from *audioXpress* February 2003).

11. F. E. Toole, "Listening Tests - Identifying and Controlling the Variables," Proceedings of the 8th International Conference, Audio Eng. Soc., 1990 May.

12. F. E. Toole and S.E. Olive, "Hearing is Believing vs. Believing is Hearing: Blind vs. Sighted Listening Tests and Other Interesting Things," 97th Convention, Audio Eng. Soc., Preprint No. 3894, Nov. 1994.

13. F. E. Toole, "Listening Tests, Turning Opinion Into Fact," *JAES*, June 1982.

14. F. E. Toole, "Subjective Measurements of Loudspeaker Sound Quality and Listener Performance," *JAES*, January/February 1985.

15. S. Bech, "Perception of Timbre of Reproduced Sound in Small Rooms: Influence of Room and Loudspeaker Position," *JAES*, December 1994.

16. S.E. Olive, P. Schuck, J. Ryan, S. Sally, M. Bonne-ville,

"The Variability of Loudspeaker Sound Quality Among Four Domestic-Sized Rooms," presented at the 99th AES Convention, preprint 4092, October 1995.

17. F. E. Toole, "Loudspeakers and Rooms for Stereophonic Sound Reproduction," Proceedings of the 8th International Conference, Audio Eng., Soc., May 1990.

18. S. E. Olive, P. Schuck, S. Sally, M. Bonneville, "The Effects of Loudspeaker Placement on Listener Preference Ratings," *J*AES, September 1994.

19. Antti Jarvinen, Lauri Savioja, Henrik Moiler, Veijo Ikonen, Anssi Ruusuvuori, "Design of a Reference Listening Room - A Case Study," AES 103rd Convention, New York, Preprint 4559, September 1997.

20. V. Dickason, "Harman's Moving Speakers," *Voice Coil*, August 1999.

21. S. Olive, B. Castro, and F. Toole, "A New Laboratory for Evaluating Multichannel Audio Components and Systems," presented at the 105th AES Convention, preprint 4842, September 1998.

CHAPTER SEVEN

PASSIVE AND ACTIVE CROSSOVER NETWORKS

7.10 PASSIVE NETWORKS.

Passive crossover network design is a rather complex subject involving a vast number of variables. In fact, it would be quite possible to write an entire book about it. In keeping with the "cookbook" approach, however, this section will deal mostly with examples of accepted methods used by the loudspeaker industry. You should remember that the final choice of what type crossover configuration you use with a particular set of drivers will depend not only on the application of these methods, but also on a certain amount of trial and error experimentation, and much subjective listening.

7.11 HISTORY.

With the inception of multi-way loudspeakers in the 1930s, the crossover network designs were based on the constant-K and M-derived filter theory of both G.A. Campbell and O.J. Zobel, Bell Telephone engineers. Some of the first published work on the subject was done by John K. Hilliard and H.R. Kimball for the sound department at Metro-Goldwyn-Mayer studios. This paper, published March 3, 1936, in the Academy Research Council Technical Bulletin, was titled "Dividing Networks for Loudspeaker Systems" (reprinted in the *JAES*, Nov. 1978). Hilliard published another definitive explication of network design in *Electronics* magazine, January 1941, titled "Loudspeaker Dividing Networks," which listed all the formulas for creating first- and third-order Butterworth parallel and series networks. By the 1950s, Butterworth networks were established as the preferred filter shape for loudspeaker crossovers.

C. P. Boegli, writing for *Audio* magazine in November 1956,[1] described the effects of driver offset for coaxial and coplanar systems using Butterworth first- and second-order filters. The article discussed both the second-order Butterworth in-phase null (out of phase peak), as well as the deteriorating frequency response caused by vertical driver separation.

Then, in the 1960s, Ashley and Small described the attributes of series-type filters with constant-voltage transfer properties. This introduced the so-called quasi second-order network. Unfortunately, though offering somewhat more attenuation than first-order filters and having the same "phase coherent" characteristics, the amount of attenuation was still inadequate to prevent modulation distortion in most drivers. This was the subject of Small's 1971 *JAES* paper,

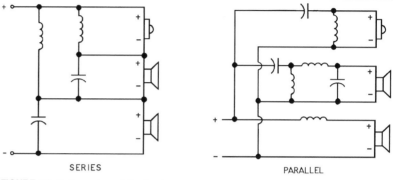

FIGURE 7.1: Series and parallel networks.

"Crossover Networks and Modulation Distortion," which suggested that 12dB/octave filters were the minimum slope attenuation needed to prevent excessive distortion due to high output levels at the driver's resonance frequency. At the same time, Ashley and Henne were extolling the flat amplitude response and all-pass phase characteristics of third-order Butterworth networks.[2]

In 1976, Siegfried Linkwitz described the vertical polar response of various types of filters, concluding that the second- and fourth-order all-pass filters (dubbed Linkwitz-Riley or Squared Butterworth filters) yielded a symmetrical on-axis response.[3] Somewhat after that, Peter Garde pulled all of the previous work together with his description of the entire family of all-pass networks and their derivation.[4] Using some of the same ideas, Dennis Fink, working with Ed Long on the Urei Time-Align studio monitor, developed further the method of correcting driver horizontal offset with delay networks.[5]

The next major contributions came when Marshall Leach[6] and Robert Bullock[7] established the need for speaker builders to consider driver resonance and relative inter-driver separation (both vertical and horizontal) when determining what type and order of filter will produce optimum results. Continuing with his excellent work with loudspeaker networks, Bullock[8] described the proper derivation of symmetrical three-way networks, showing that three-way networks are not just combined two-way networks (as so many of us had assumed).

Other recent contributions of interest include the series of papers by Stanley Lipshitz and John Vanderkooy which present a number of possible, although not necessarily practical, alternatives for achieving minimum phase network responses. The most important advances in the last few years, however, have been the availability of com-

FIGURE 7.2

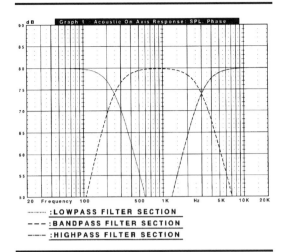

.... :LOWPASS FILTER SECTION
– – – :BANDPASS FILTER SECTION
–·–· :HIGHPASS FILTER SECTION

FIGURE 7.3

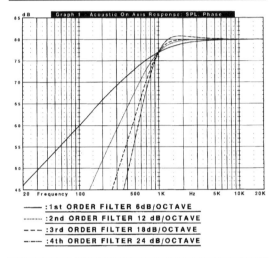

——— :1st ORDER FILTER 6dB/OCTAVE
.... :2nd ORDER FILTER 12 dB/OCTAVE
– – – :3rd ORDER FILTER 18dB/OCTAVE
–·–· :4th ORDER FILTER 24 dB/OCTAVE

FIGURE 7.4

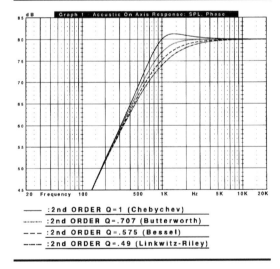

——— :2nd ORDER Q=1 (Chebychev)
.... :2nd ORDER Q=.707 (Butterworth)
– – – :2nd ORDER Q=.575 (Bessel)
–·–· :2nd ORDER Q=.49 (Linkwitz-Riley)

puter crossover optimization software. These programs open new ways to experiment and rational network design that are either impossible or terribly time-consuming and tedious to do empirically. Programs such as CALSOD, Filter Designer, and LEAP 4.0 are making network design more nearly a science and less "black magic."

7.20 CROSSOVER BASICS.

Loudspeaker crossovers can be configured as series or parallel (*Fig. 7.1*). Of the two, the parallel

filter is the overwhelming choice of the loudspeaker industry, and has the advantage of allowing each driver in a multi-way system to be treated independently. Using a series network, the component variations can affect both the high- and low-pass drivers.[9] Because the parallel filter tends to be more flexible and better-suited for the applications described in this chapter, they will be the only crossover configuration discussed. More information on series networks can be found in R. Small's article "Constant Voltage Crossover Network Design," *JAES*, January 1971.

Loudspeaker crossover networks are made up of L/C (inductance and capacitance) filter sections. Three basic filter formats are used in parallel configuration crossover design. The responses of these filter sections are illustrated in *Fig. 7.2* and are:

1. Low-pass filters—roll off the upper frequency response, generally used with woofers.

2. High-pass filters—roll off the lower frequency response of a speaker, and are mostly used with tweeters.

3. Bandpass filters—roll off both the lower and upper frequency ranges, and are typically used with midrange drivers.

L/C filters combined as crossover networks are basically frequency-dependent attenuation circuits which rely on the reactive properties of inductors and capacitors. The frequency dependent property of these components is described by the formulas for reactance (which is AC resistance):

$$X_C = \frac{1}{2\pi f C}$$

$$X_L = 2\pi f L$$

The relationships described by the formulas are clear. Capacitive reactance is inversely proportional to frequency, and capacitors provide increasing AC resistance (become more reactive) as frequency decreases. Inductive reactance is directly proportional to frequency, and inductors provide increasing AC resistance (become more reactive) as frequency increases.

Filters are generally described by three basic properties, the slope of the rolloff, the filter resonance, and the Q. Slope is commonly related in terms of the amount of attenuation per octave, or dB/octave. Depending on the circuit topology, and the way in which the L and C components are grouped, the slope of a filter can be 6, 12, 18, or 24dB of attenuation per octave of frequency change as depicted in *Fig. 7.3*. Although uncommon, the slope rate can be even greater than 24dB/octave. These attenuation rates are also referred to by the "order" of the slope: first-order for 6dB/octave, second-order for 12dB/octave, third-order for 18dB/octave, and fourth-order corresponding to 24dB/octave slope rates.

A filter circuit's resonance, which applies to filter orders greater than one, is the frequency at which the component reactances are equal, and designates the crossover frequency. The resonance of a simple second-order filter is given by:

$$f = \frac{1}{2\pi(LC)^{1/2}}$$

The product of L and C (L × C), sometimes referred to as the L/C ratio, is important, since the same resonance frequency will result if the values of L and C are varied, but the L × C product remains constant.

"Q" of a filter network relates the same relationships that described "Q" for drivers and driver/box combinations. Q, which is also called the "figure of merit," is a calculated quantity used to describe resonance and is equivalent to:

$$Q = 2 \times \frac{(energy\ stored\ at\ resonance)}{(energy\ dissipated\ at\ resonance)}$$

For a second-order L/C circuit, the formula for Q is:

$$Q = [(R^2C)]L]^{1/2}$$

Different filter Qs describe the shape of the "knee" of the rolloff response, as shown in *Fig. 7.4*. The "Q" shapes are identical to those described for closed-box woofer enclosures in *Chapter 1*. In the case of a second-order filter, the Q, and hence the shape of the response, is controlled by the L/C ratio of the components. The different filter Qs, illustrated in *Fig. 7.4*, have different properties and have usually been named for the engineer who first mathematically described the response characteristics of that shape, such as Butterworth (Q = 0.707), Bessel (Q = 0.58), or Linkwitz-Riley (Q = 0.49). Please note that this discussion of filter shapes refers to the electrical transfer function of these filters, and not necessarily to the acoustic transfer function of the loudspeaker.

7.21 COMBINED RESPONSE OF TWO-WAY CROSSOVERS— HIGH-/LOW-PASS SUMMATIONS.

The goal of any crossover network design effort is to provide a flat response transition between two different frequency range drivers being combined to produce a full-range loudspeaker system. The designer's task is to combine two independent sound sources with overlapping frequency ranges so the combination produces no new peaks or dips in the transition from one driver to the other. The way in which two independent signal sources combine depends on the phase relationship between the sources. Signals combine differently when they are in phase than when they are not. *Figure 7.5* illustrates the acoustical addition of two correlated (in-phase) and two uncorrelated (not in-phase) sound sources. When two independent sound sources are correlated, the two signals combine as simple scalar quantities. In *Fig. 7.5B*, the two drivers fed by the same signal sources are phase correlated, and sum to a level 6dB greater than the output of a single driver. When two independent sound sources are

uncorrelated, the signals combine as an RMS quantity. Looking at *Fig. 7.5C*, the two drivers fed by two different and uncorrelated signal sources sum to a level that is 3dB greater than the output of a single driver.

As in the above example, the electric or voltage

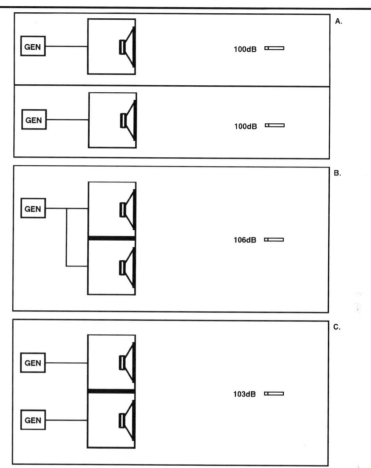

FIGURE 7.5: Combining acoustic waveforms.

summation of crossover filters works in the same manner. Filters used for loudspeaker crossover networks can be divided into two groups based on the phase relationship (correlated or uncorrelated) between the high-pass and low-pass filter sections. Phase, which is discussed in more detail in *Chapter 8*, is a function of the network slope. Different filter Qs and slopes have different phase characteristics. Odd-order Butterworth filters exhibit a high-pass and low-pass phase relationship which is constantly 90° out of phase, a phenomenon known as phase quadrature. Being 90° out of phase at all frequencies is the same as the phase being uncorrelated, and the two filter sections will sum together flat when the level of both filters is down 3dB at the crossover frequency.

All even-order networks, Butterworth, Bessel, Linkwitz-Riley, and Chebyshev, both second- and fourth-order, exhibit a high-pass and low-pass phase relationship which is in-phase. Specifically, second-order filters are 180° out of phase, but when the polarity is reversed, the phase is coincident. Fourth-order filters are actually 360° out of phase, which is effectively the same as being in-phase. When even-order filters are combined, the phase is correlated, and the two filter sections sum together flat when the level of both filters is down 6dB at the crossover frequency. *Figure 7.6* compares the summation of a third-order Butterworth crossover with a fourth-order

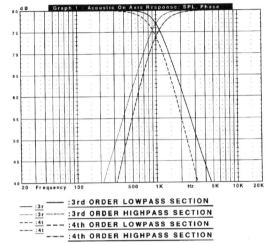

FIGURE 7.6

— — : 3r ———— : 3rd ORDER LOWPASS SECTION
· · · · · · : 3r · · · · · · : 3rd ORDER HIGHPASS SECTION
— — — : 4t — — — : 4th ORDER LOWPASS SECTION
— · — · : 4t — · — · : 4th ORDER HIGHPASS SECTION

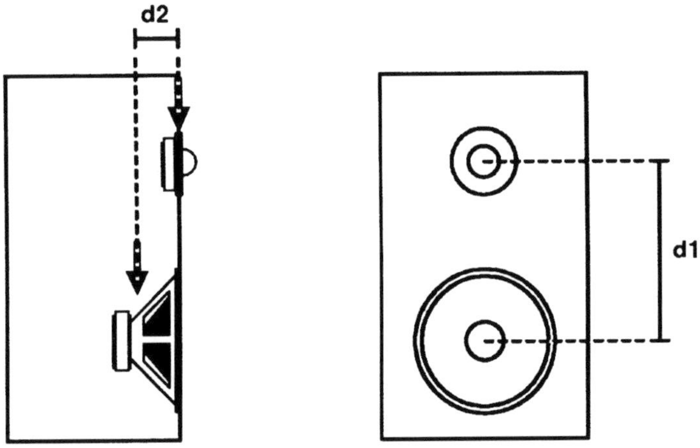

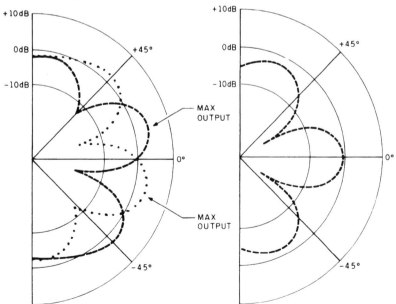

FIGURE 7.7 (above): Noncoincident driver separation.
FIGURE 7.8 (below): Vertical polar response radiation tilt.

FIRST ORDER REVERSE POLARITY NETWORK — — —   L-R FOURTH ORDER NETWORK
FIRST ORDER NORMAL POLARITY NETWORK · · ·

Linkwitz-Riley type.

This is contrary to conventional wisdom concerning the nature of even-order Butterworth filters, but is nevertheless correct. When even-order Butterworth high-pass and low-pass filters are calculated, and their phase is coincident, both HP and LP sections are 3dB down at the crossover frequency (due to the shape of the filter Q), and the summation is +3dB, which is the 6dB gain of two correlated sources (+3dB minus -3dB is 6dB). If the exact in-phase constraint is relaxed and the filter frequencies spread by a given factor (about 1.3 for second-order Butterworth filters), the filter will sum to nearly flat when both HP and LP sections are 6dB down at the crossover frequency.

7.22 ALL-PASS AND LINEAR PHASE CROSSOVERS.

We have a choice of three crossover network types: linear phase (also known as minimum phase), all-pass, and non-all-pass. Only one crossover type fits the minimum phase criterion, the first-order Butterworth. The first-order crossover produces a summed zero phase response which is flat at all frequencies and has a flat magnitude response.[10] Although the first-order crossover is phase coherent and theoretically produces no phase distortion, it generally provides insufficient attenuation for most high-pass drivers (midranges and tweeters).

All-pass crossovers have all-pass phase characteristics, which means phase-shifting, and have a flat magnitude response. Four all-pass networks are typically used as loudspeaker crossovers.[4] These are the first- and third-order Butterworth types, and the second- and fourth-order Linkwitz-Riley configurations. Although the first-order filter is minimum phase, it is included in this category. These filters are most frequently used for loudspeakers because they provide a flat magnitude response and, except for the first-order type, provide sufficient attenuation to prevent midrange and tweeter distortion.

Non-all-pass crossovers include all the remaining filters used for loudspeaker design. These also have all-pass phase characteristics, but don't sum to a flat magnitude response. This includes the second- and fourth-order Butterworth, the second- and fourth-order Bessel, and the group of asymmetrical (non-circular poles) fourth-order crossovers: the Legendre, Gaussian, and fourth-order Linear Phase filters. However, as I will show later, if we relax the in-phase criterion, these filters can be made to sum nearly flat as well.

7.23 THE ACOUSTIC SUMMATION: DRIVER ACOUSTIC CENTERS AND ZDP.

Our discussion of classic filter shapes which we use for crossover networks is about the electrical response of the filters (their transfer function) and the summation is an electrical, or voltage, summation. In order for the same phase and magnitude consequences which we've described for the networks to occur acoustically, which is our goal, we must make certain assumptions. Our most important assumption is that the radiation by the two drivers must be coincident. This means they must radiate from the exact same point in space and time. This doesn't, however, describe the way the majority of loudspeakers work.

The only loudspeaker types which have high-pass and low-pass drivers which radiate from the same point in space are coaxial types. These usually have a dome tweeter mounted on top of the woofer pole piece so that the radiating positions are nearly identical (unlike horn-loaded drivers mounted coaxially with woofers, where the horn tweeter's voice coil may actually be positioned some distance behind the woofer voice coil). Although we're seeing a recent resurgence of these designs in high-end loudspeaker manufacturing (including several from KEF and Tannoy), the vast majority of loudspeakers have noncoincident drivers separated both vertically and horizontally, like the one depicted in *Fig. 7.7*.

When noncoincident drivers are used for different frequency ranges, the radiating origins are separated in both the horizontal and vertical planes. Vertical separation has a number of consequences which have been well documented.[3,7] Lobing is the most important problem in radiation patterns (refer to *Chapter 6, Fig. 6.4*). The wider the separation the worse the lobing. The only solution is to minimize driver separation and make certain no low-pass/high-pass driver combination is separated by a distance greater than one wavelength at the crossover frequency.

Vertical separation of drivers can also cause radiation pattern tilt depending upon the type of crossover network being employed. The first- and third-order Butterworth all-pass networks have a frequency dependent tilt in the vertical polar response, as illustrated in *Fig. 7.8*. For a first-order filter, if no horizontal driver separation exists, and the vertical separation is no greater than one wavelength at the crossover frequency, the tilt is about 15° downward for a normal polarity connection, and 15° upward when the relative polarity between filter sections is reversed (the directions of the tilt are reversed for the third-order filter). The lobe axis shifting occurs because of the 90° phase difference in the high- and low-pass sections. All in-phase networks, which includes all even-order networks, do not have the same frequency dependent tilt and exhibit a symmetrical vertical polar response (assuming no horizontal offset).

The radiation pattern will also be inclined because of the horizontal displacement of two noncoincident drivers. The amount of tilt, β, depends on the ratio of vertical and horizontal displacement and can be calculated by:

$$\beta = Tan^{-1}\frac{d_1}{d_2} \ in \ degrees$$

This tilt in the on-axis radiation will be in addition to any phase-related frequency dependent lobing caused by the use of odd-order networks, and will also modify the symmetrical radiation

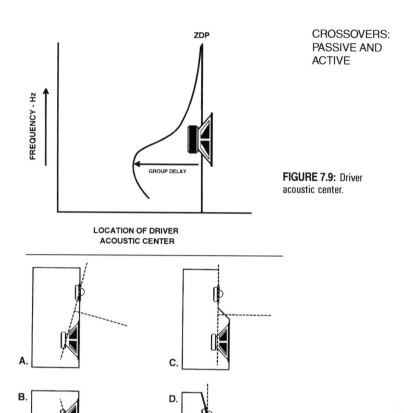

FIGURE 7.9: Driver acoustic center.

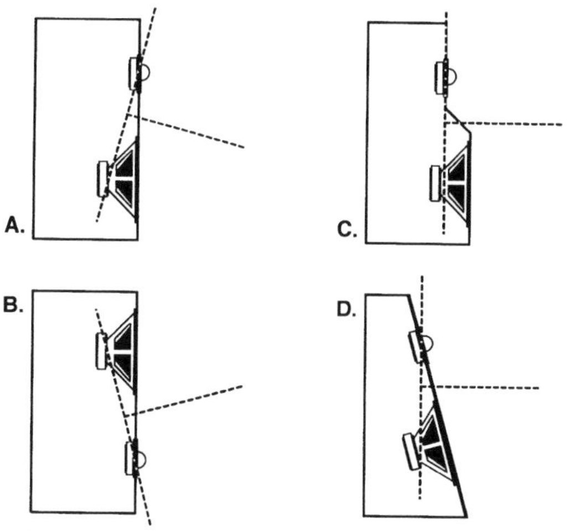

FIGURE 7.10: ZDP reference axis.

pattern of in-phase networks.

The exact location of the driver radiating center, or zero delay plane (ZDP), can be somewhat of a mystery. When calculating the horizontal offset distance (or the amount of time delay caused by that distance) for crossover design purposes, the only important factor is the relative amount of offset and not the driver acoustic centers. A loudspeaker's acoustic center varies with frequency and by definition is a function of the natural phase response of the driver.[9,11] The diagram in *Fig. 7.9* illustrates the relationship between the acoustic center of a speaker and the time delay function. At lower frequencies where the group delay (derived from the slope of the driver phase response) is greatest, the acoustic center of the speaker can be a substantial distance behind the driver ZDP. In terms of crossover design, when driver responses are shaped to match the network filter transfer function, the phase response of the driver has been accounted for, and the frequency dependent changes in the acoustic center are not relevant.

For most purposes, such as calculating the radiation tilt caused by horizontal driver offset, the radiating center may be assumed to be the center of the voice coil for either cone or dome type drivers. The voice coil center position for any driver will always be the midpoint of the driver frontplate. Using the voice coil centers will do a reasonably good job of establishing the relative difference in the ZDPs between two differ-

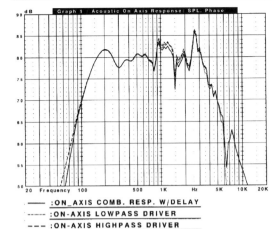

FIGURE 7.11

___ :ON_AXIS COMB. RESP. W/DELAY
.... :ON-AXIS LOWPASS DRIVER
___ _ :ON-AXIS HIGHPASS DRIVER

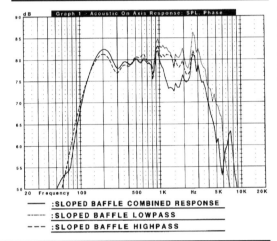

FIGURE 7.12

___ :SLOPED BAFFLE COMBINED RESPONSE
.... :SLOPED BAFFLE LOWPASS
___ _ :SLOPED BAFFLE HIGHPASS

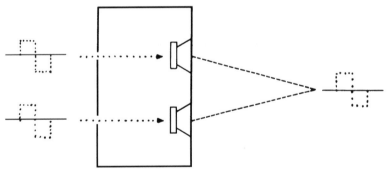

FIGURE 7.13: First-order network minimum phase response.

ent drivers. The exact acoustic position of the driver ZDP can be somewhat different than expected depending on the construction of the speaker. Determining this, however, cannot be done without the aid of sophisticated test gear. For example, cone drivers tend to radiate from a point just in front of the apex of the cone. Measuring this exactly can be done by impulse analysis using ETCs (Energy-Time-Curve) for the two drivers, and subtracting their time difference to get their relative separation distance, or by determining the time delay from the microphone to the beginning of the impulse. Another technique is to use a shaped tone burst generator[12] driven at the crossover frequency and using an oscilloscope to align the drivers. The offset is then physically measured. Often the relative distances you calculate with more sophisticated

techniques will be nearly the same as the distance you measure from the center of the gap.

You can see the effects of driver offset on the reference axis of the speaker in *Fig. 7.10*. The diagram shows the different ways that the axis of zero inter-unit time delay will be affected by the driver mounting configuration.[13] The A configuration is typical of most two-way designs with the tweeter mounted above the woofer. The ZDP axis for this configuration is off-axis toward the floor.

When measured on the 0° listening axis, the delay caused by misalignment of driver ZDPs will induce response variations in the crossover region.[14] If the distance is sufficient at the wavelength of the crossover frequency, a complete null will occur at the crossover frequency. For example, a 3kHz second-order Butterworth network will show a null for the reverse polarity connection if the drivers are offset by 2.25″. The distance is a time delay of 166μs, which is exactly one-half wavelength at 3kHz. If the drivers are coincident, the reverse polarity condition will yield a flat response. Different filter types and slopes have different sensitivities to misalignment. This will be detailed for each filter type in *Section 7.30*.

The ZDP axis in A could be returned to the 0° listening axis if the tweeter were delayed by the appropriate amount. If the horizontal separation distance were 3″, then the time delay required to bring these two drivers back into the same radiating plane would be:

$$t_g = \frac{d_2}{C}$$

where C is equal to the velocity of sound in inches. In this case (3/13560) = 221μs.

Experts often suggest the inverted driver position in B when you use odd-order-networks (normal polarity first-order, and reversed polarity third-order) in order to compensate for the downward polar tilt caused by the phase difference between the high- and low-pass sections. The radiation tilt from the driver geometry causes a +15° tilt, which, when added to the –15° tilt caused by the odd-order-filter, yields a 0° ZDP reference axis (provided the d_2 and d_1 distances have been appropriately accounted for).

Figure 7.10D illustrates another method some use for making ZDP axis the same as the 0° listening axis. For this case the ZDP has been tilted upward by sloping the front baffle, in order to be coincident with the listening axis. However, although the ZDP axis location has changed, using a sloped baffle is not equivalent to having the two driver radiating planes aligned as in C, or using electronic time delay correction to align the high-pass driver as in A or B.

To demonstrate this, I measured two identical drivers (using MLSSA and the ACO microphone) first, using time delay offset to correct for the physical offset delay; and second, measured at an off-axis position equivalent to using a sloped baffle. The horizontal separation was 4″ and the vertical separation was 10.5″. Using the

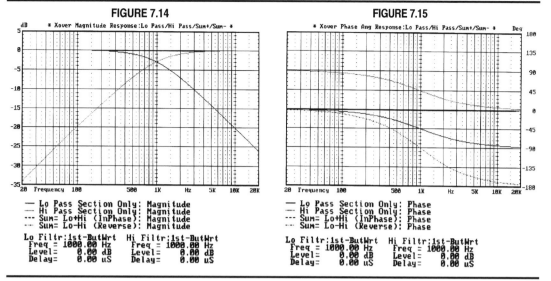

FIGURE 7.14

FIGURE 7.15

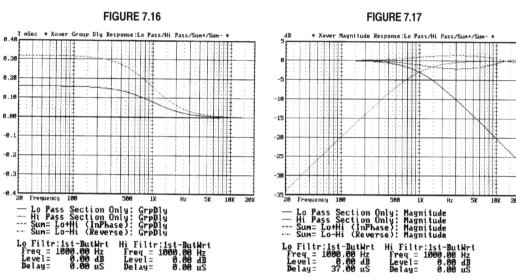

FIGURE 7.16

FIGURE 7.17

above formula, the ZDP would be about 20° from the 0° axis (Tan^{-1} 4/10.5) for the two drivers mounted on the same baffle. I simulated the crossover in the LEAP program as a fourth-order Linkwitz-Riley active network at 1.2kHz. In order to make the results easier to interpret, I switched around the same driver and used it for both high-pass and low-pass measurements, so the combined response, if correct, should be identical to the single driver response. *Figure 7.11* shows the results of the two non-coincident drivers measured on the 0° listening axis, the low-pass with a 4″ offset, and the high-pass driver corrected with 295μs of delay. The combined response is compared on the same graph with the two individual high-pass/low-pass driver responses (without the network). The summed response with the delay correction is nearly identical to the two single driver responses.

Figure 7.12 shows the same two drivers measured 20° off-axis without the corrective time delay, but using the same fourth-order active crossover filters. The combined off-axis measurement is compared with the individual driver off-axis measurements. Obviously, the phase relationships are not exactly the same off-axis as they are on-axis and

delayed, and thus, the two methods of correcting driver offset do not produce equivalent results.

The radiation tilt and phase shift consequences described for noncoincident drivers are usually only a concern at crossover frequencies above 700Hz. At higher frequencies the typical physical offsets (in relation to typical driver dimensions) will cause a phase change which is a sufficiently large percentage of the relevant wavelength at the crossover frequency to be significant. For example, a 2″ offset at 500Hz will only produce a 7% change in phase, which is relatively trivial. The same 2″ offset at 3kHz will produce a 44% change in phase, which is substantial.

7.24 CROSSOVER NETWORK POWER RESPONSE.

Power response for a loudspeaker is more or less the same thing as its off-axis response. If hypothetically you disregard directivity for the purpose of this example, the power response could be derived by integrating multiple off-axis free-field measurements. If you combined all the possible off-axis measurement points, and found their average, that would be the power response.

Conventional wisdom says the power response

151

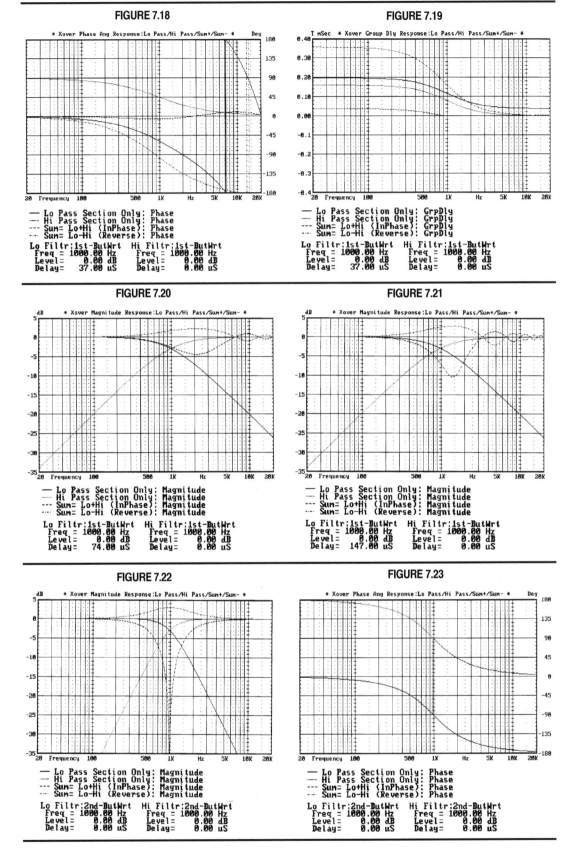

FIGURE 7.18

FIGURE 7.19

FIGURE 7.20

FIGURE 7.21

FIGURE 7.22

FIGURE 7.23

of a multiway loudspeaker with a crossover network is derived differently than the on-axis response.[4,7,13,15–17] This means all filters sum on-axis as scalar quantities with correlated phase, while the off-axis power response is calculated as an RMS quantity as if phase were uncorrelated. Although there is some disagreement as to what

importance, if any, should be given to the power response difference between even- and odd-order filters, the methodology is specific.

The main consequence of this crossover power response idea has been the recognition of the filter class known as CPCs, or Constant Power Crossovers. CPCs are supposedly the even-order

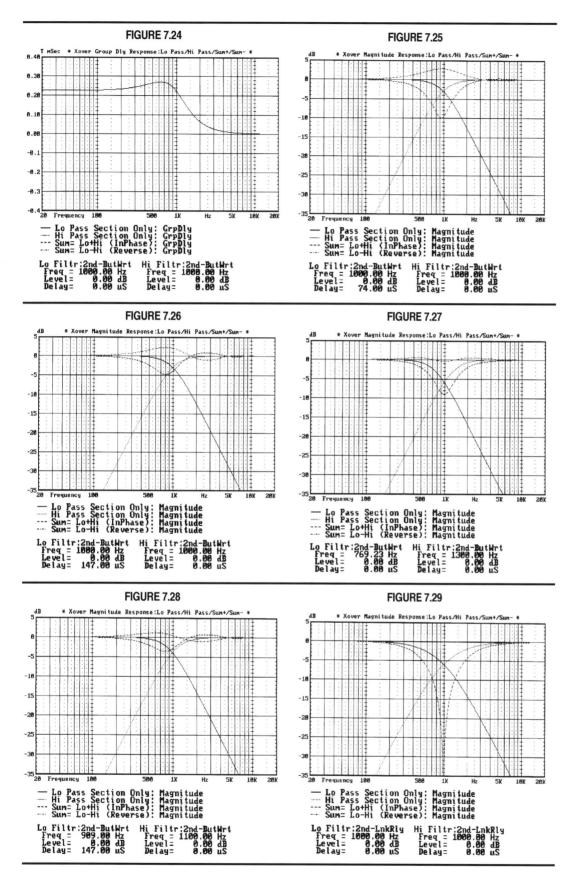

FIGURE 7.24

FIGURE 7.25

FIGURE 7.26

FIGURE 7.27

FIGURE 7.28

FIGURE 7.29

Butterworth networks, which, as I mentioned above, sum to +3dB on-axis when their high-pass and low-pass sections are phase coincident. If the power response sums as uncorrelated signals, which are 3dB less than correlated signals, it implies that the power response of an even-order Butterworth network is then flat, since its two sections are 3dB down at the crossover frequency (−3dB + 3dB = 0). This also implies that even-ordered Linkwitz-Riley networks, which sum flat when their high-pass and low-pass sections are −6dB at the crossover frequency, do not have a flat power response but have a dip of −3dB. Odd-order Butterworth filters, which already have

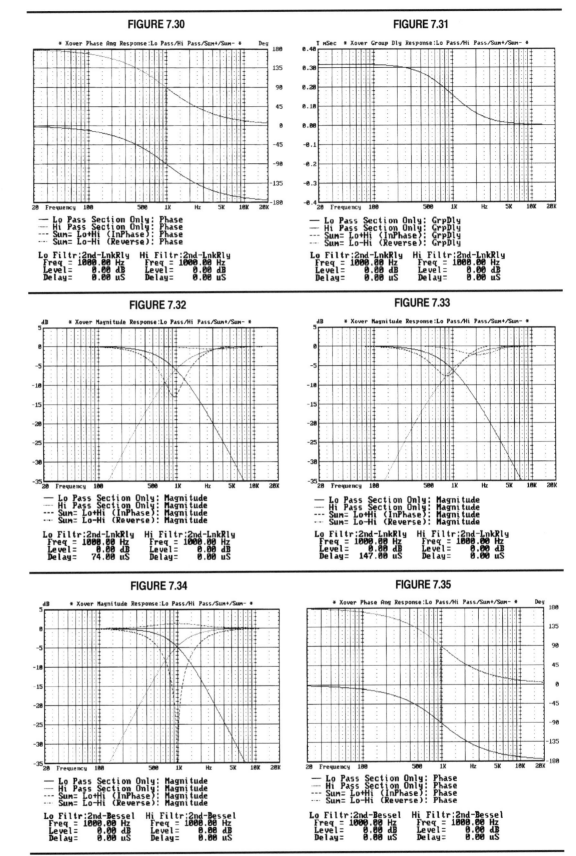

FIGURE 7.30

FIGURE 7.31

FIGURE 7.32

FIGURE 7.33

FIGURE 7.34

FIGURE 7.35

uncorrelated phase on-axis, fit neatly into this picture, and are also considered CPC and to have a flat power response.

Although it is not popular to go against the mainstream of thought in any field, this view of power response is simply not correct. The reality is much simpler. In-phase high- and low-pass filters sum as scalar quantities both on- and off-axis. Phase quadrature high-pass and low-pass filters sum as an RMS quantity both on- and off-axis. Calculating a crossover's power response as though the signals are uncorrelated is unjustified, in my view. If, for example, you are summing two speakers located on the opposite sides

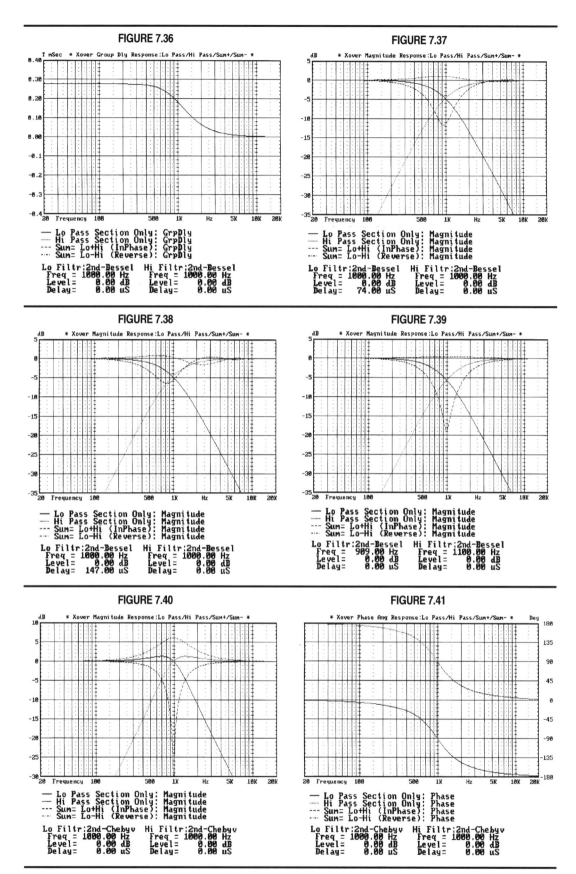

FIGURE 7.36

FIGURE 7.37

FIGURE 7.38

FIGURE 7.39

FIGURE 7.40

FIGURE 7.41

of an auditorium stage, taking their place as uncorrelated in the power domain would certainly be justified and correct. But when two drivers are mounted on the same baffle, radiating from very nearly the same plane and being fed by the same program material, there is no other way to process the two signals as being anything but phase correlated. They are phase correlated on-axis and they are phase correlated in the power domain. No difference exists between the power response of any crossover network and the on-axis response. There are no separate categories of filters, such as CPCs, which are more useful for situations dominated

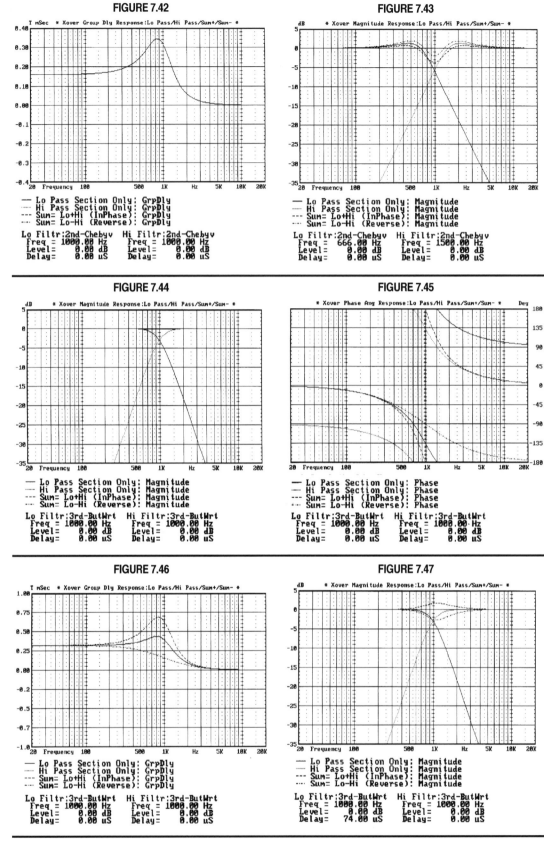

FIGURE 7.42

FIGURE 7.43

FIGURE 7.44

FIGURE 7.45

FIGURE 7.46

FIGURE 7.47

by a reverberant field than those where direct sound dominates.

7.30 TWO-WAY CROSSOVER CHARACTERISTICS.

Twelve different two-way crossover types have been applied to loudspeaker design. They are: first-order Butterworth, second-order Butterworth, second-order Linkwitz-Riley, second-order Bessel, second-order Chebyshev, third-order Butterworth, fourth-order Butterworth, fourth-order Linkwitz-Riley, fourth-order Bessel, fourth-order Legendre, fourth-order Gaussian, and fourth-order Linear Phase. Each has its own set

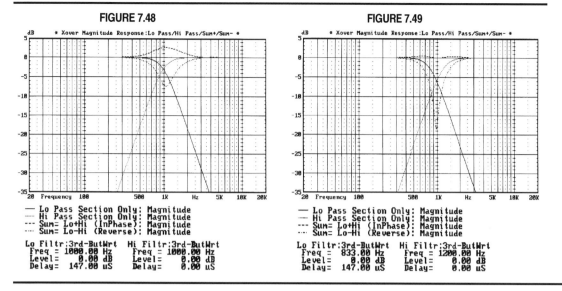

FIGURE 7.48

FIGURE 7.49

— Lo Pass Section Only: Magnitude
⋯ Hi Pass Section Only: Magnitude
--- Sum= Lo+Hi (InPhase): Magnitude
-·- Sum= Lo-Hi (Reverse): Magnitude

Lo Filtr:3rd-ButWrt		Hi Filtr:3rd-ButWrt	
Freq =	1000.00 Hz	Freq =	1000.00 Hz
Level=	0.00 dB	Level=	0.00 dB
Delay=	147.00 uS	Delay=	0.00 uS

— Lo Pass Section Only: Magnitude
⋯ Hi Pass Section Only: Magnitude
--- Sum= Lo+Hi (InPhase): Magnitude
-·- Sum= Lo-Hi (Reverse): Magnitude

Lo Filtr:3rd-ButWrt		Hi Filtr:3rd-ButWrt	
Freq =	833.00 Hz	Freq =	1200.00 Hz
Level=	0.00 dB	Level=	0.00 dB
Delay=	147.00 uS	Delay=	0.00 uS

of response, phase, and group delay characteristics as well as differing levels of sensitivity to driver horizontal offsets. The following individual descriptions of these crossover functions illustrate the operation of the networks at 1kHz. The effects of horizontal driver offsets are shown for 1″ and 2″ which amount to 74μs and 147μs of delay, respectively. The effects shown will be exactly the same for different frequencies, but with proportionately different offset delay distances. The table below gives the relationship for distance and different offsets:

Frequency	Offset Distances		
0.5kHz	1″	2″	4″
1kHz	0.5″	1″	2″
3kHz	0.17″	0.33″	0.67″
5kHz	0.1″	0.2″	0.4″

In other words, what holds true in the description of the operational character of the network at 1kHz with 1″ of horizontal driver separation will be exactly the same at 3kHz with 0.33″ of separation, and 0.2″ of separation at 5kHz, and so on.

1. First-order Butterworth

The normal polarity first-order Butterworth crossover is an all-pass type and the only network listed which has a minimum phase response. This means its output has the same frequency magnitude and same phase as the input signal, and satisfies all the criteria for a constant-voltage network.[18] At points equidistant from both drivers, the sum of the driver outputs, using first-order filters, produces zero phase distortion (*Fig. 7.13*). This describes the operational characteristic usually referred to as phase coherent or linear phase. This crossover sums flat with both high-pass and low-pass sections at -3dB at the crossover frequency, as shown in *Fig. 7.14*. The summation is flat for either normal or reversed polarity. The phase, shown in *Fig. 7.15*, is 90° apart at all frequencies for each section, and has a flat zero magnitude for the summation. Because of the 90° phase dif-

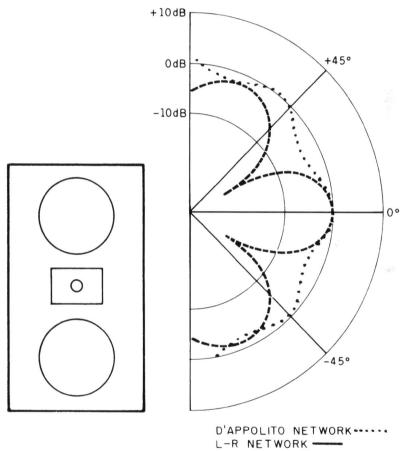

D'APPOLITO NETWORK ⋯⋯
L−R NETWORK ——

FIGURE 7.50: Vertical response comparison.

ference, a -15° tilt will occur in the vertical polar response with high-pass and low-pass driver separated by a distance of one wavelength at the crossover frequency (+15° for the reversed polarity connection). This reversed polarity connection shows all-pass phase for its summation and is not minimum phase as the normal polarity. The summed group delay curve in *Fig. 7.16* shows the normal polarity crossover group delay at zero, with both high-pass and low-pass sections at 0.16μs. The group delay for the reversed polarity network, unlike the normal polarity connection,

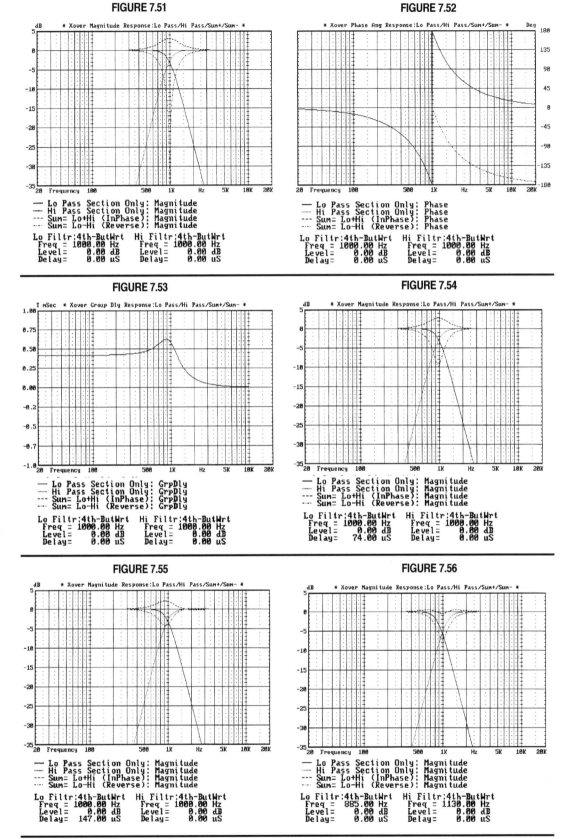

FIGURE 7.51

FIGURE 7.52

FIGURE 7.53

FIGURE 7.54

FIGURE 7.55

FIGURE 7.56

is not zero, and is about 0.32µs.

The first-order Butterworth's minimum phase characteristic exists only in a narrow window, which requires nearly exact driver alignment. The response, phase, and group delay curves shown in *Figs. 7.17–7.19* are with a 0.5″ offset. The phase and group delay are no longer mini-

mum phase and the frequency response exhibits a broad dip of nearly 2.5dB. The response picture with 1″ and 2″ offsets, illustrated in *Figs. 7.20–7.21*, show response variations up to 10dB, spread over more than six octaves. The first-order Butterworth is obviously very sensitive to driver alignment.

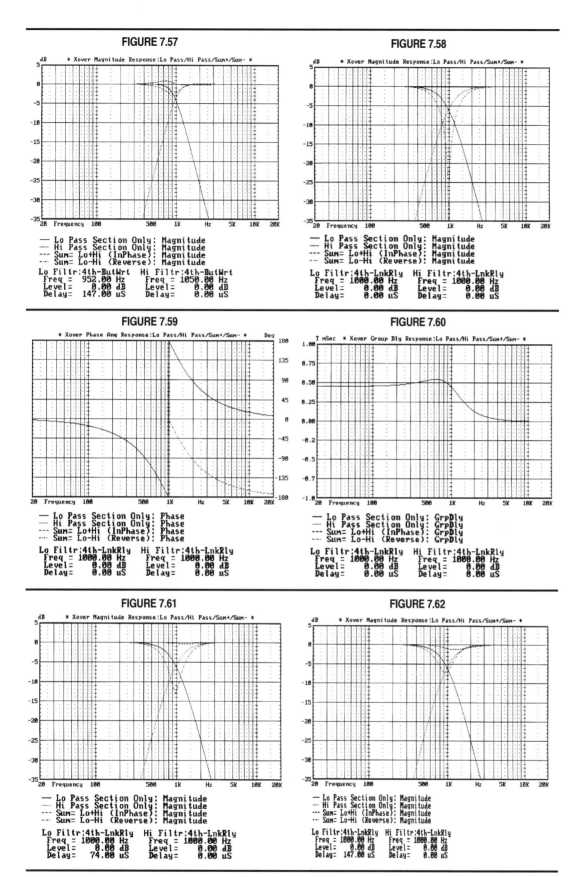

FIGURE 7.57

FIGURE 7.58

FIGURE 7.59

FIGURE 7.60

FIGURE 7.61

FIGURE 7.62

The first-order Butterworth is also very sensitive to driver resonance within the crossover stopband (attenuation range).[6] This, plus the usually insufficient attenuation rate to prevent driver distortion, and the frequency dependent polar tilt, seemingly make this simple network a poor choice for loudspeaker applications. However, this filter has enjoyed great popularity and has something near cult following in certain "audiophile" circles. Since the preference is by no means universally embraced, and the minimum phase characteristic has yet to be established as audibly superior, the choice remains a subjective one. It is

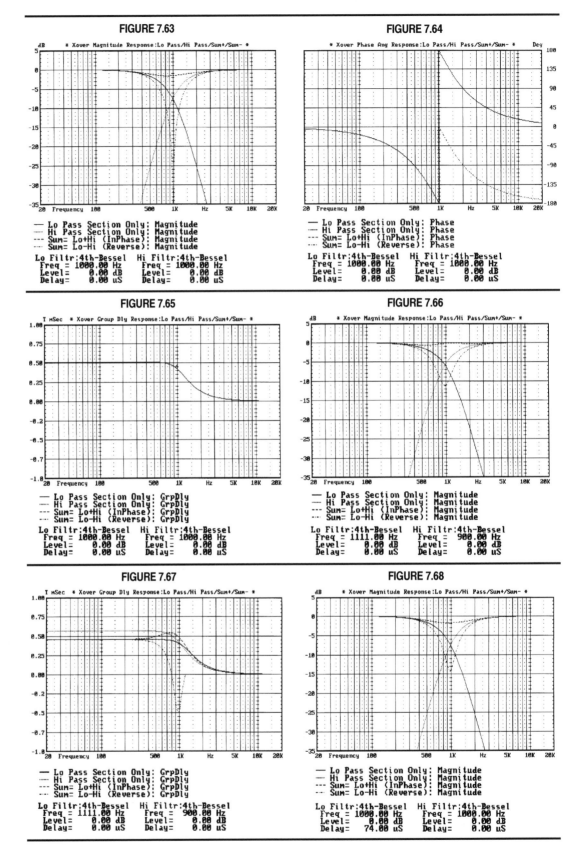

FIGURE 7.63

FIGURE 7.64

FIGURE 7.65

FIGURE 7.66

FIGURE 7.67

FIGURE 7.68

part of what some consider the "art" of loudspeaker design.

2. Second-order Butterworth

This configuration was, at one time, the network of choice for manufacturing, although it has been mostly discarded for the all-pass class of crossovers. The network frequency response,

phase, and group delay are shown in *Figs. 7.22–7.24* (no offset). The phase for the normal polarity version is 180° out and causes a virtual null in the response, while the reverse polarity connection is in-phase and the network sums to +3dB. Most sources recommend the reverse polarity connection; however, its success will depend on driver alignment. The filter Q is 0.707 and the

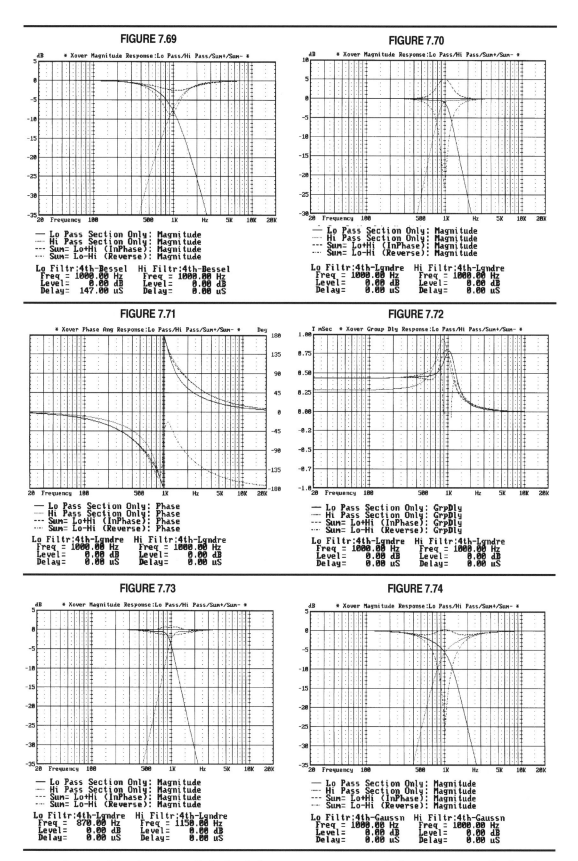

FIGURE 7.69

* Xover Magnitude Response:Lo Pass/Hi Pass/Sum+/Sum- *

— Lo Pass Section Only: Magnitude
···· Hi Pass Section Only: Magnitude
--- Sum= Lo+Hi (InPhase): Magnitude
-·- Sum= Lo-Hi (Reverse): Magnitude
Lo Filtr:4th-Bessel   Hi Filtr:4th-Bessel
 Freq = 1000.00 Hz   Freq = 1000.00 Hz
 Level=   0.00 dB   Level=   0.00 dB
 Delay= 147.00 uS   Delay=   0.00 uS

FIGURE 7.70

* Xover Magnitude Response:Lo Pass/Hi Pass/Sum+/Sum- *

— Lo Pass Section Only: Magnitude
···· Hi Pass Section Only: Magnitude
--- Sum= Lo+Hi (InPhase): Magnitude
-·- Sum= Lo-Hi (Reverse): Magnitude
Lo Filtr:4th-Lgndre   Hi Filtr:4th-Lgndre
 Freq = 1000.00 Hz   Freq = 1000.00 Hz
 Level=   0.00 dB   Level=   0.00 dB
 Delay=   0.00 uS   Delay=   0.00 uS

FIGURE 7.71

* Xover Phase Ang Response:Lo Pass/Hi Pass/Sum+/Sum- *

— Lo Pass Section Only: Phase
···· Hi Pass Section Only: Phase
--- Sum= Lo+Hi (InPhase): Phase
-·- Sum= Lo-Hi (Reverse): Phase
Lo Filtr:4th-Lgndre   Hi Filtr:4th-Lgndre
 Freq = 1000.00 Hz   Freq = 1000.00 Hz
 Level=   0.00 dB   Level=   0.00 dB
 Delay=   0.00 uS   Delay=   0.00 uS

FIGURE 7.72

T mSec * Xover Group Dly Response:Lo Pass/Hi Pass/Sum+/Sum- *

— Lo Pass Section Only: GrpDly
···· Hi Pass Section Only: GrpDly
--- Sum= Lo+Hi (InPhase): GrpDly
-·- Sum= Lo-Hi (Reverse): GrpDly
Lo Filtr:4th-Lgndre   Hi Filtr:4th-Lgndre
 Freq = 1000.00 Hz   Freq = 1000.00 Hz
 Level=   0.00 dB   Level=   0.00 dB
 Delay=   0.00 uS   Delay=   0.00 uS

FIGURE 7.73

* Xover Magnitude Response:Lo Pass/Hi Pass/Sum+/Sum- *

— Lo Pass Section Only: Magnitude
···· Hi Pass Section Only: Magnitude
--- Sum= Lo+Hi (InPhase): Magnitude
-·- Sum= Lo-Hi (Reverse): Magnitude
Lo Filtr:4th-Lgndre   Hi Filtr:4th-Lgndre
 Freq = 870.00 Hz   Freq = 1150.00 Hz
 Level=   0.00 dB   Level=   0.00 dB
 Delay=   0.00 uS   Delay=   0.00 uS

FIGURE 7.74

* Xover Magnitude Response:Lo Pass/Hi Pass/Sum+/Sum- *

— Lo Pass Section Only: Magnitude
···· Hi Pass Section Only: Magnitude
--- Sum= Lo+Hi (InPhase): Magnitude
-·- Sum= Lo-Hi (Reverse): Magnitude
Lo Filtr:4th-Gaussn   Hi Filtr:4th-Gaussn
 Freq = 1000.00 Hz   Freq = 1000.00 Hz
 Level=   0.00 dB   Level=   0.00 dB
 Delay=   0.00 uS   Delay=   0.00 uS

summed group delay shows a small rise just below the crossover frequency (group delay is the same for the high-pass, low-pass, and summed normal and reversed polarity).

All second-order filters are less sensitive to driver horizontal alignment than the first-order filter. *Figures 7.25* and *7.26* show the response changes for 1″ and 2″ offsets. At 1″ there is almost

no response variation, and no really drastic changes at 2″. If the offset were one half a wavelength at the crossover frequency, which is 6.78″ at 1kHz, the phase is reversed and the normal polarity connection becomes in-phase instead of producing a null (the reversed polarity would now produce a null).

As I observed earlier, the even-order

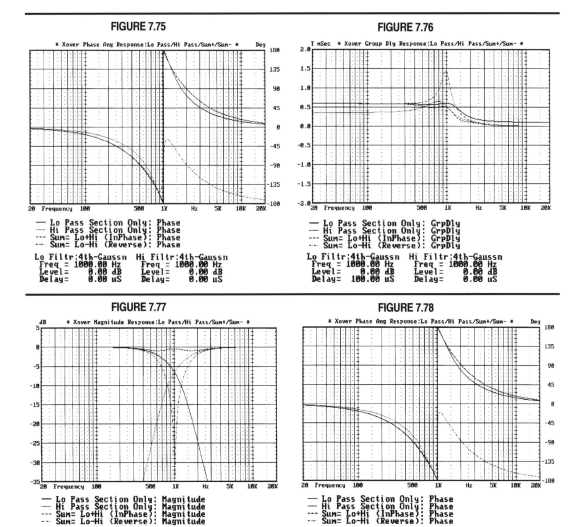

FIGURE 7.75

Xover Phase Ang Response:Lo Pass/Hi Pass/Sum+/Sum- *

— Lo Pass Section Only: Phase
···· Hi Pass Section Only: Phase
--- Sum= Lo+Hi (InPhase): Phase
-·- Sum= Lo-Hi (Reverse): Phase

Lo Filtr:4th-Gaussn   Hi Filtr:4th-Gaussn
Freq = 1000.00 Hz   Freq = 1000.00 Hz
Level=   0.00 dB   Level=   0.00 dB
Delay=   0.00 uS   Delay=   0.00 uS

FIGURE 7.76

mSec * Xover Group Dly Response:Lo Pass/Hi Pass/Sum+/Sum- *

— Lo Pass Section Only: GrpDly
···· Hi Pass Section Only: GrpDly
--- Sum= Lo+Hi (InPhase): GrpDly
-·- Sum= Lo-Hi (Reverse): GrpDly

Lo Filtr:4th-Gaussn   Hi Filtr:4th-Gaussn
Freq = 1000.00 Hz   Freq = 1000.00 Hz
Level=   0.00 dB   Level=   0.00 dB
Delay= 100.00 uS   Delay=   0.00 uS

FIGURE 7.77

dB   * Xover Magnitude Response:Lo Pass/Hi Pass/Sum+/Sum- *

— Lo Pass Section Only: Magnitude
···· Hi Pass Section Only: Magnitude
--- Sum= Lo+Hi (InPhase): Magnitude
-·- Sum= Lo-Hi (Reverse): Magnitude

Lo Filtr:4th-Linear   Hi Filtr:4th-Linear
Freq = 1000.00 Hz   Freq = 1000.00 Hz
Level=   0.00 dB   Level=   0.00 dB
Delay=   0.00 uS   Delay=   0.00 uS

FIGURE 7.78

* Xover Phase Ang Response:Lo Pass/Hi Pass/Sum+/Sum- *

— Lo Pass Section Only: Phase
···· Hi Pass Section Only: Phase
--- Sum= Lo+Hi (InPhase): Phase
-·- Sum= Lo-Hi (Reverse): Phase

Lo Filtr:4th-Linear   Hi Filtr:4th-Linear
Freq = 1000.00 Hz   Freq = 1000.00 Hz
Level=   0.00 dB   Level=   0.00 dB
Delay=   0.00 uS   Delay=   0.00 uS

FIGURE 7.79

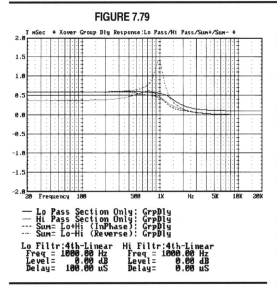

mSec * Xover Group Dly Response:Lo Pass/Hi Pass/Sum+/Sum- *

— Lo Pass Section Only: GrpDly
···· Hi Pass Section Only: GrpDly
--- Sum= Lo+Hi (InPhase): GrpDly
-·- Sum= Lo-Hi (Reverse): GrpDly

Lo Filtr:4th-Linear   Hi Filtr:4th-Linear
Freq = 1000.00 Hz   Freq = 1000.00 Hz
Level=   0.00 dB   Level=   0.00 dB
Delay= 100.00 uS   Delay=   0.00 uS

Butterworth filters will sum flat, or nearly so, when combined at –6dB for high-pass and the low-pass. *Figure 7.27* shows the effect of multiplying the high-pass crossover frequency by 1.3 and the low-pass frequency by 0.7692 (1.3^{-1}) to achieve a nearly flat response (for zero horizontal offset).

Another important observation is that offset response variations can be corrected somewhat by altering the high-pass and low-pass frequencies, in the same manner as above, for causing the second-order Butterworth to sum flat. *Figure 7.28* shows the response variations for a 2″ offset given in *Fig. 7.26*, minimized by shifting the filter section frequencies by a factor of 1.1. While using this technique is unlikely to produce an absolutely flat response, it's one way of dealing with driver offset magnitude changes without resorting to physical or electrical delay.

3. Second-order Linkwitz-Riley

The second-order L-R crossover is an all-pass configuration which sums to a flat magnitude with the high-pass and low-pass filters at –6dB at the crossover frequency. The magnitude response, phase, and group delay graphs are shown in *Figs. 7.29–7.31* (no offset). The phase and polarity relationships are the same for all second-order filter types, so what was true for the second-order Butterworth also obtains for the L-R filters. The Q of the L-R filter is 0.49 (the square of the Butterworth Q), and the summed group delay curve has a magnitude, but is flat (the group delay is the same for the high-pass, low-pass, and summed responses with both polarities).

Sensitivity to driver offset for all the second-order filters is identical. *Figures 7.32* and *7.33* show the response changes caused by 1″ and 2″ offsets. The flat magnitude response, low sensitivity to offset, and in-band driver resonances have made the L-R a popular choice among manufacturers.

4. Second-order Bessel

The second-order Bessel filter is similar to the Linkwitz-Riley filter, only it has a somewhat higher Q, 0.58, and does not sum flat, so is not an all-pass crossover. The response, phase, and group delay curves are shown in *Figs. 7.34–7.36* (no offset). The magnitude response is about +1dB for the reversed polarity connection. Phase is the same as other second-order crossovers. The summed group delay is flat like the L-R crossovers with slightly less magnitude.

Sensitivity to driver offset is the same as the other second-order filters as shown in *Figs. 7.37* and *7.38* for 1″ and 2″ offsets. This crossover will sum flat if the high-pass and low-pass frequencies are altered by a factor of 1.1, as illustrated in *Fig. 7.39*. When forced to sum as a flat magnitude, the summed group delay for this filter is no better than the L-R filter.

5. Second-order Chebyshev

The second-order Chebyshev crossover is not often used, unless the Q, which is 1 in this example, is used to combine with a low driver response Q to achieve some particular target response. The frequency response, phase, and group delay are given in *Figs. 7.40–7.42* (no offset). The response sums at +6dB and the summed group delay has a substantial "knee," which is indicative of low damping. Sensitivity to driver offset is typical of second-order filters and is not shown. The filter can be made to sum within ±2dB of flat by shifting the filter section values by a factor of 1.5, as shown in *Fig. 7.43* (with no offset).

6. Third-order Butterworth

Like the first-order Butterworths, the third-order types sum flat with each section at -3dB at the crossover frequency, and are all-pass. The response, phase, and group delay are shown in *Figs. 7.44–7.46* (no offset). The response is flat for both connection polarities; however, the summed group delay is flat with a relatively low magnitude for the reverse polarity connection, and has a sharply higher magnitude "knee" in the group delay for the normal polarity connection. The reverse polarity is normally considered preferable because of the improved group delay. Like the first-order filter, the third-order Butterworth exhibits a 15° tilt in the vertical polar response due to the 90° phase difference in the high-pass and low-pass sections. The tilt is +15° for normal polarity and -15° for reverse polarity.

Third-order Butterworth crossovers also exhibit low sensitivity to driver offset as illustrated in *Figs. 7.47* and *7.48* for 1″ and 2″ offsets. *Figure*

7.49 shows the effect of shifting the high-pass and low-pass filter section frequencies by a factor of 1.2 with a 2″ offset. The response is corrected to a nearly flat magnitude.

The third-order Butterworth has gained popularity by its use in the M-T-M (mid-tweeter-mid) driver configuration described by Joe D'Appolito. The combination of third-order filter and the driver geometry yields a fairly smooth vertical polar response, as shown in *Fig. 7.50*. The idea originally was to eliminate the lobing error normally found in even-order filters when drivers are non-

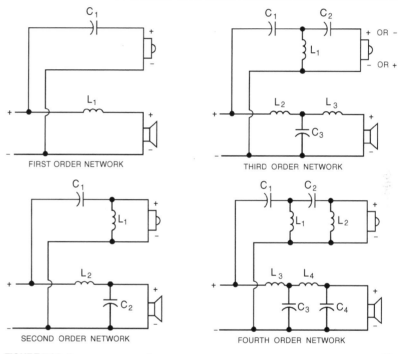

FIGURE 7.80: Two-way crossover diagrams.

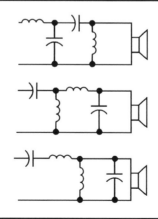

FIGURE 7.81: Possible bandpass circuit topography.

coincident. However, the lobing is relatively inoffensive for listening, and later versions of the D'Appolito designs use the same driver configuration, but with fourth-order L-R acoustic slopes.[19] The higher slope rate has its advantages, and the M-T-M configuration keeps the axis from being tilted because of the horizontal driver offsets. In this case, the drivers need not have zero offset for the polar pattern to stay on the 0° axis, and a flat response can be achieved by optimizing the crossover frequency. Aligning drivers is less important for high-quality sound than flat response and good polar behavior. At most, aligning drivers by time will force predictable polar behavior and make achieving a flat response somewhat easier. The main advantage of the M-T-M configuration is the control it gives the designer over the vertical polar response behavior.

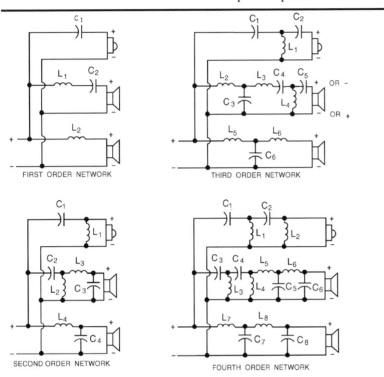

FIGURE 7.82: Three-way crossover diagrams.

7. Fourth-order Butterworth
We build fourth-order filters by cascading two second-order types. Since the Qs of the second-order sections of the fourth-order Butterworth are 1.307 and 0.541, the total Q is 0.707. The response, phase, and group delay of the fourth-order Butterworth is depicted in *Figs. 7.51–7.53* (no offset). Like the second-order Butterworth, when the phase is coincident, the filter sums to a -3dB for both high- and low-pass filters, giving a +3dB peak at the crossover frequency, but with normal polarity connection instead of reverse polarity. The out-of-phase null with fourth-order filters comes with the reverse polarity connection. The summed group delay of the filter shows a knee or peak just below the crossover frequency. Fourth-order filters have most of the attributes of second-order types, except the increased attenuation yields lower driver distortion. Also, minimum overlap means that any negative mutual radiation effects will occur only over a relatively small range. The only drawback is the possible insertion loss caused by the DCR of the two inductors in a filter section.

Because of the decreased amount of overlap, fourth-order filters are less sensitive to driver offset than second-order filters. *Figures 7.54* and *7.55* show the response variations for 1″ and 2″ offsets, which in both cases is minimal. The

fourth-order Butterworth will sum nearly flat if the high-pass and low-pass sections combine at -6dB. This can be accomplished by altering the individual filter section frequencies by a factor of 1.13, as shown in *Fig. 7.56*. If driver offset is taken into consideration, the ratio of frequency change can be adjusted to cause a flat response to occur. The graph in *Fig. 7.57* shows a 2″ offset corrected with the filter frequencies changed by a factor of 1.05.

8. Fourth-order Linkwitz-Riley
The fourth-order L-R crossovers sum to a flat magnitude and belong to the all-pass category of crossovers. Both second-order sections have a Q of 0.707, for a total Q of 0.49, which is why this filter is sometimes referred to as the squared Butterworth filter. *Figures 7.58–7.60* show the frequency response, phase, and group delay for this crossover design. The summed group delay magnitude shows a slight peak just below the crossover frequency.

Sensitivity to driver offset is low, as with other fourth-order filters. *Figures 7.61–7.62* show the response variations for 1″ and 2″ offsets, which are both minimal. The flat magnitude response, high attenuation rate, and low sensitivity to offset error makes this one of the best tweeter filters.

9. Fourth-order Bessel
These do not sum to flat magnitude and are not all-pass crossovers. *Figures 7.63–7.65* (no offset) show the response, phase, and group delay for this crossover design. The response yields a 1.5dB dip at the crossover frequency, while the summed group delay is flat.

The fourth-order Bessel can be made to sum to a nearly flat magnitude by altering the high-pass and low-pass filter frequencies by a factor of 0.9 (overlapping the filters), as shown in *Fig. 7.66*. The resulting summed group delay, given in *Fig. 7.67*, is about the same as the fourth-order L-R filter.

Sensitivity to driver horizontal offset is low, as shown in *Figs. 7.68* and *7.69*, for 1″ and 2″ offsets.

The last three fourth-order crossovers are a special class of asymmetrical networks. They are asymmetrical because of the way they are constructed. They don't have "circular poles" as do the previous filters (all except the Chebyshev, which has poles located on an elliptical plane), but rather a mutated plane which facilitates their derivation. The fourth-order Butterworth, Bessel, and Linkwitz-Riley are all formed by two cascaded sections with the same Q and corner frequency. These three asymmetrical fourth-order filters have second-order sections with dissimilar Qs and skewed corner frequencies. Thus, no second-order implementations of these filters are possible.

While some have expressed interest in using these filters in loudspeakers, I believe they have no advantage over the fourth-order L-R crossovers, and have the disadvantage of being highly sensitive to parameter variation. This makes them neither useful for manufacturing nor attractive for amateurs. I include them because of the interest some have expressed and to clear up any misconceptions about their viability.

10. Fourth-order Legendre

This type has a response similar to a Chebyshev filter. Its frequency response, phase, and summed group delay are shown in *Figs. 7.70–7.72.* The asymmetrical nature of the filter is evident from the phase plot, which shows the low-pass skewed from the high-pass. The summed group delay is also similar to a Chebyshev and shows substantial peaking. The filter sums to +5dB for the in-phase condition. The Legendre can be forced to sum to nearly flat by altering the high-pass and low-pass frequencies by a factor of 1.15, as shown in *Fig. 7.73.* Sensitivity to driver offset is low, as with any fourth-order filter.

11. Fourth-order Gaussian

The frequency response, phase, and summed group delay for the fourth-order Gaussian crossover is illustrated in *Figs. 7.74–7.76* (no offset). The crossover sums to nearly flat for the normal polarity connection, and has a summed group delay similar to a Bessel filter.

12. Fourth-order Linear-Phase

The name sounds attractive, but again the asymmetrical derivation makes this filter's suitability questionable. The frequency response, phase, and summed group delay are given in *Figs. 7.77–7.79* (no offset). The filter sums to nearly flat for the normal polarity and has a summed group delay which is fairly flat and similar to the Bessel fourth-order crossover.

7.31 TWO-WAY NETWORK DESIGN FORMULAS.

The following design formulas can be used for deriving symmetrical high-pass/low-pass crossover networks. The schematic diagrams in *Fig. 7.80* illustrate the required circuit topography. Values are in henries (L), farads (C), ohms (R), and hertz (f).

7.40 THREE-WAY CROSSOVER NETWORKS.

Two-way networks describe a nearly ideal situation. Adding an additional driver and a second crossover frequency, however, considerably complicates the situation. Because of the compromises and tradeoffs about to be described, it is a good time to point out that if you are considering a three-way speaker, you should seriously consider using an active network for the second frequency division point. Although requiring an additional amplifier as well as the electronic network, the advantages are quite substantial (*Section 7.100*).

It has been established that three-way networks cannot be satisfactorily derived by combining two two-way networks.[8] The situation is further complicated by a choice of several circuit topologies. Unfortunately, each of the three-way bandpass filter topographies illustrated in *Fig. 7.81* will pro-

FIGURE 7.83

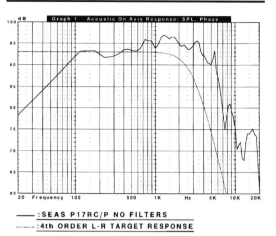

——— :SEAS P17RC/P NO FILTERS
·········· :4th ORDER L-R TARGET RESPONSE

FIGURE 7.84

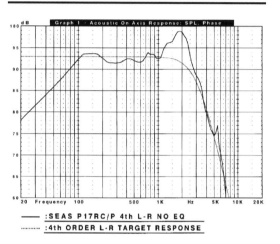

——— :SEAS P17RC/P 4th L-R NO EQ
·········· :4th ORDER L-R TARGET RESPONSE

FIGURE 7.85

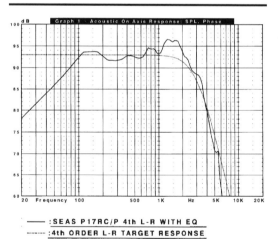

——— :SEAS P17RC/P 4th L-R WITH EQ
·········· :4th ORDER L-R TARGET RESPONSE

First-Order Networks
Butterworth

$$C_1 = \frac{.159}{R_H f} \qquad L_1 = \frac{R_L}{6.28f}$$

Second-Order Networks
Linkwitz-Riley

$$C_1 = \frac{.0796}{R_H f} \qquad L_1 = \frac{.3183R_H}{f}$$

$$C_2 = \frac{.0796}{R_H f} \qquad L_2 = \frac{.3183R_L}{f}$$

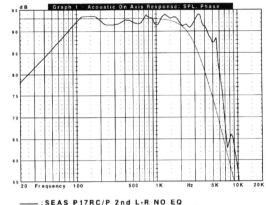

FIGURE 7.86

— :SEAS P17RC/P 2nd L-R NO EQ
······ :4th ORDER L-R TARGET RESPONSE

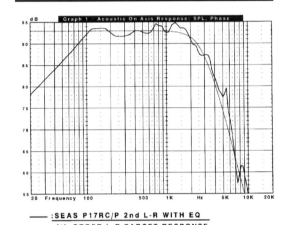

FIGURE 7.87

— :SEAS P17RC/P 2nd L-R WITH EQ
········ :4th ORDER L-R TARGET RESPONSE

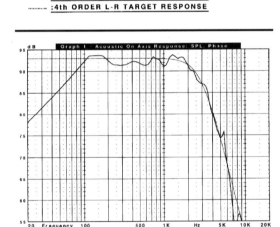

FIGURE 7.88

— :SEAS P17RC/P 2nd L-R WITH EQ
········ :4th ORDER L-R TARGET RESPONSE

Chebychev (Q = 1)

$$C_1 = \frac{.1592}{R_H f} \qquad L_1 = \frac{.1592 R_H}{f}$$

$$C_2 = \frac{.1592}{R_L f} \qquad L_2 = \frac{.1592 R_L}{f}$$

Third-Order Networks
Butterworth

$$C_1 = \frac{.1061}{R_H f} \qquad L_1 = \frac{.1194 R_H}{f}$$

$$C_2 = \frac{.3183}{R_H f} \qquad L_2 = \frac{.2387 R_L}{f}$$

$$C_3 = \frac{.2122}{R_L f} \qquad L_3 = \frac{.0796 R_L}{f}$$

Fourth-Order Networks
Linkwitz-Riley

$$C_1 = \frac{.0844}{R_H f} \qquad L_1 = \frac{.1000 R_H}{f}$$

$$C_2 = \frac{.1688}{R_H f} \qquad L_2 = \frac{.4501 R_H}{f}$$

$$C_3 = \frac{.2533}{R_L f} \qquad L_3 = \frac{.3000 R_L}{f}$$

$$C_4 = \frac{.0563}{R_L f} \qquad L_4 = \frac{.1500 R_L}{f}$$

Bessel

$$C_1 = \frac{.0702}{R_H f} \qquad L_1 = \frac{.0862 R_H}{f}$$

$$C_2 = \frac{.1719}{R_H f} \qquad L_2 = \frac{.4983 R_H}{f}$$

$$C_3 = \frac{.2336}{R_L f} \qquad L_3 = \frac{.3583 R_L}{f}$$

$$C_4 = \frac{.0504}{R_L f} \qquad L_4 = \frac{.1463 R_L}{f}$$

Bessel

$$C_1 = \frac{.0912}{R_H f} \qquad L_1 = \frac{.2756 R_H}{f}$$

$$C_2 = \frac{.0912}{R_L f} \qquad L_2 = \frac{.2756 R_L}{f}$$

Butterworth

$$C_1 = \frac{.1125}{R_H f} \qquad L_1 = \frac{.2251 R_H}{f}$$

$$C_2 = \frac{.1125}{R_L f} \qquad L_2 = \frac{.2251 R_L}{f}$$

Butterworth

$$C_1 = \frac{.1040}{R_H f} \qquad L_1 = \frac{.1009 R_H}{f}$$

$$C_2 = \frac{.1470}{R_H f} \qquad L_2 = \frac{.4159 R_H}{f}$$

$$C_3 = \frac{.2509}{R_L f} \qquad L_3 = \frac{.2437 R_L}{f}$$

$$C_4 = \frac{.0609}{R_L f} \qquad L_4 = \frac{.1723 R_L}{f}$$

Legendre

$$C_1 = \frac{.1104}{R_H\, f} \qquad L_1 = \frac{.1073R_H}{f}$$

$$C_2 = \frac{.1246}{R_H\, f} \qquad L_2 = \frac{.2783R_H}{f}$$

$$C_3 = \frac{.2365}{R_L\, f} \qquad L_3 = \frac{.2294R_L}{f}$$

$$C_4 = \frac{.091}{R_L\, f} \qquad L_4 = \frac{.2034R_L}{f}$$

Gaussian

$$C_1 = \frac{.0767}{R_H\, f} \qquad L_1 = \frac{.1116R_H}{f}$$

$$C_2 = \frac{.1491}{R_H\, f} \qquad L_2 = \frac{.3251R_H}{f}$$

$$C_3 = \frac{.2235}{R_L\, f} \qquad L_3 = \frac{.3253R_L}{f}$$

$$C_4 = \frac{.0768}{R_L\, f} \qquad L_4 = \frac{.1674R_L}{f}$$

Linear-Phase

$$C_1 = \frac{.0741}{R_H\, f} \qquad L_1 = \frac{.1079R_H}{f}$$

$$C_2 = \frac{.1524}{R_H\, f} \qquad L_2 = \frac{.3853R_H}{f}$$

$$C_3 = \frac{.2255}{R_L\, f} \qquad L_3 = \frac{.3285R_L}{f}$$

$$C_4 = \frac{.0632}{R_L\, f} \qquad L_4 = \frac{.1578R_L}{f}$$

duce a different response shape. Clarifying all of this, Robert Bullock offers a complex derivation which considers mutual loading problems caused by the cascaded filters in the bandpass section. This information is provided in his *JAES* presentation "Passive Three-Way All-Pass Crossover Networks,"[8] along with some practical realizations given in his "Passive Crossover Networks" article in *Speaker Builder* 2/85. A "T" bandpass topology is also an alternative for three-way networks presented in *Speaker Builder* 2/87. This more complex circuit design has less input impedance sensitivity than the standard type, but for the circuits described below, this should be of no consequence.

7.41 CHARACTERISTICS OF THREE-WAY NETWORKS.

Unlike two-way networks, no three-way networks are both APC and CPC. Most of what was said about polar response and power response, however, is more or less the same. The exception is the odd-order APC, which has a response dip of 1–2dB.

Probably the most important characteristic to consider is the effects of the spread between the crossover frequencies. Generally speaking, the further apart the two crosspoints are, the better the combined response of the drivers will be (three octaves is a good starting point).[19] Crosspoints closer together than the three-octave ideal will suffer from complicated, undesirable interference patterns.

The design formulas in *Section 7.42* (see *Fig. 7.82*) represent what I consider to be the best network type and crossover frequencies for minimum interdriver interference. They are also practical for speaker formats most likely to be followed by speaker builders and limited-production manufacturers.

7.42 THREE-WAY APC NETWORK DESIGN FORMULAS.

The design formulas (*Fig. 7.82*) are configured for two basic pairs of crossover frequencies most often used in three-way loudspeaker systems. Each pair represents a different frequency spread between the midrange-to-tweeter crosspoint (f_H) and the woofer-to-midrange crosspoint (f_L). The two frequency "spreads" chosen represent (A) 3.4 octaves ($f_H/f_L=10$), and (B) 3 octaves ($f_H/f_L=8$). The (A) crossover formulas can be applied to 3kHz/300Hz, useful with woofer/mid-woofer/dome-tweeter formats, and 5kHz/500Hz, useful with woofer/small canister or dome midrange/tweeter formats. The (B) crossover formulas can be applied to 3kHz/375Hz, useful with woofer/mid-woofer/tweeter formats, as well as 5kHz/625Hz and 6kHz/750Hz, both useful with woofer/small canister or dome midrange/tweeter formats. For other crossover frequency spreads or network types, refer to Bullock[19] and be prepared to do a lot of calculating.

When you use crossover frequencies in the 300Hz or lower range, inductor values can be large and have substantial insertion losses. You can compare inductor losses by using the following:
where

$$L_L = 20\,log_{10}\left(\frac{R_m}{R_s + R_m}\right)$$

L_L = inductor losses in dB
R_m = driver impedance
R_s = total measured inductor resistance (DCR)

Using the three-way networks will result in a bandpass filter stage gain increase. You can use this figure, provided for each network example, when you compare driver efficiencies. The midrange driver in a three-way system would have a total efficiency in dB equal to:

Total Driver Gain = (driver ref. eff.) + (bandpass gain) - (inductor loss).

For all the design equations presented,

$$fm = (f_H \times f_L)^{1/2}.$$

First-Order APC
(A)

$$C_1 = \frac{.1590}{R_H \, f_H} \qquad L_1 = \frac{.0458R_M}{f_M}$$

$$C_2 = \frac{.5540}{R_M \, f_M} \qquad L_2 = \frac{.1592R_L}{f_L}$$

(B)

$$C_1 = \frac{.1590}{R_H \, f_H} \qquad L_1 = \frac{.0500R_M}{f_M}$$

$$C_2 = \frac{.5070}{R_M \, f_M} \qquad L_2 = \frac{.1592R_L}{f_L}$$

Second-Order APC
(reverse polarity bandpass)
(A)

$$C_1 = \frac{.0791}{R_H \, f_H} \qquad L_1 = \frac{.3202R_H}{f_H}$$

$$C_2 = \frac{.3236}{R_M \, f_M} \qquad L_2 = \frac{1.0291R_M}{f_M}$$

$$C_3 = \frac{.0227}{R_m \, f_M} \qquad L_3 = \frac{.0837R_M}{f_M}$$

$$C_4 = \frac{.0791}{R_L \, f_L} \qquad L_4 = \frac{.3202R_L}{f_L}$$

Bandpass Gain = 2.08dB
(B)

$$C_1 = \frac{.0788}{R_H \, f_H} \qquad L_1 = \frac{.3217R_H}{f_H}$$

$$C_2 = \frac{.3046}{R_M \, f_M} \qquad L_2 = \frac{.9320R_M}{f_M}$$

$$C_3 = \frac{.0248}{R_M \, f_M} \qquad L_3 = \frac{.0913R_M}{f_M}$$

$$C_4 = \frac{.0788}{R_L \, f_L} \qquad L_4 = \frac{.3217R_L}{f_L}$$

Bandpass Gain = 2.45dB

Third-Order APC
(reverse polarity bandpass)
(A)

$$C_1 = \frac{.0995}{R_H \, f_H} \qquad L_1 = \frac{.1191R_H}{f_H}$$

$$C_2 = \frac{.3402}{R_H \, f_H} \qquad L_2 = \frac{.0665R_M}{f_M}$$

$$C_3 = - \frac{.0683}{R_H \, f_H} \qquad L_3 = \frac{.0233R_M}{f_M}$$

$$C_4 = \frac{.3128}{R_M \, f_M} \qquad L_4 = \frac{.4285R_M}{f_M}$$

$$C_5 = \frac{1.148}{R_M \, f_M} \qquad L_5 = \frac{.2546R_L}{f_L}$$

$$C_6 = \frac{.2126}{R_L \, f_L} \qquad L_6 = \frac{.0745R_L}{f_L}$$

Bandpass Gain = 1.6dB
(B)

$$C_1 = \frac{.0980}{R_H f_H} \qquad L_1 = \frac{.1190R_H}{f_H}$$

$$C_2 = \frac{.3459}{R_H f_H} \qquad L_2 = \frac{.0711R_M}{f_M}$$

$$C_3 = \frac{.0768}{R_M f_M} \qquad L_3 = \frac{.0254R_M}{f_M}$$

$$C_4 = \frac{.2793}{R_M f_M} \qquad L_4 = \frac{.3951R_M}{f_M}$$

$$C_5 = \frac{1.061}{R_M f_M} \qquad L_5 = \frac{.2586R_L}{f_L}$$

$$C_6 = \frac{.2129}{R_L f_L} \qquad L_6 = \frac{.0732R_L}{f_L}$$

Bandpass Gain = 2.1dB
(normal polarity bandpass)
(A)

$$C_1 = \frac{.1138}{R_H f_H} \qquad L_1 = \frac{.1191R_H}{f_H}$$

$$C_2 = \frac{.2976}{R_H f_H} \qquad L_2 = \frac{.0598R_M}{f_M}$$

$$C_3 = \frac{.0765}{R_M f_M} \qquad L_3 = \frac{.0253R_M}{f_M}$$

$$C_4 = \frac{.3475}{R_M f_M} \qquad L_4 = \frac{.3789R_M}{f_M}$$

$$C_5 = \frac{1.068}{R_M f_M} \qquad L_5 = \frac{.2227R_L}{f_L}$$

$$C_6 = \frac{.2127}{R_L f_L} \qquad L_6 = \frac{.0852R_L}{f_L}$$

Bandpass Gain = .85dB
(B)

$$C_1 = \frac{.1158}{R_H f_H} \qquad L_1 = \frac{.1189R_H}{f_H}$$

$$C_2 = \frac{.2927}{R_H f_H} \qquad L_2 = \frac{.0634R_M}{f_M}$$

$$C_3 = \frac{.0884}{R_M f_M} \qquad L_3 = \frac{.0284R_M}{f_M}$$

$$C_4 = \frac{.3112}{R_M f_M} \qquad L_4 = \frac{.3395R_M}{f_M}$$

$$C_5 = \frac{.9667}{R_M f_M} \qquad L_5 = \frac{.2187R_L}{f_L}$$

$$C_6 = \frac{.2130}{R_L f_L} \qquad L_6 = \frac{.0866 R1}{f_L}$$

Bandpass Gain = .99dB
Fourth-Order APC
(A)

$$C_1 = \frac{.0848}{R_H f_H} \qquad L_1 = \frac{.1004 R_H}{f_H}$$

$$C_2 = \frac{.1686}{R_H f_H} \qquad L_2 = \frac{.4469 R_H}{f_H}$$

$$C_3 = \frac{.3843}{R_M f_M} \qquad L_3 = \frac{.2617 R_M}{f_M}$$

$$C_4 = \frac{.5834}{R_M f_M} \qquad L_4 = \frac{1.423 R_M}{f_M}$$

$$C_5 = \frac{.0728}{R_M f_M} \qquad L_5 = \frac{.0939 R_M}{f_M}$$

$$C_6 = \frac{.0162}{R_M f_M} \qquad L_6 = \frac{.04 f_L}{f_M}$$

$$C_7 = \frac{.2523}{R_L f_L} \qquad L_7 = \frac{.2987 R_L}{f_L}$$

$$C_8 = \frac{.0567}{R_L f_L} \qquad L_8 = \frac{.1502 R_L}{f_L}$$

Bandpass Gain = 2.28dB
(B)

$$C_1 = \frac{.0849}{R_H f_H} \qquad L_1 = \frac{.1007 R_H}{f_H}$$

$$C_2 = \frac{.1685}{R_H f_H} \qquad L_2 = \frac{.4450 R_H}{f_H}$$

$$C_3 = \frac{.3774}{R_M f_M} \qquad L_3 = \frac{.2224 R_M}{f_M}$$

$$C_4 = \frac{.5332}{R_M f_M} \qquad L_4 = \frac{1.273 R_M}{f_M}$$

$$C_5 = \frac{.0799}{R_M f_M} \qquad L_5 = \frac{.1040 R_M}{f_M}$$

$$C_6 = \frac{.0178}{R_M f_M} \qquad L_6 = \frac{.0490 R_M}{f_M}$$

$$C_7 = \frac{.2515}{R_L f_L} \qquad L_7 = \frac{.2983 R_L}{f_L}$$

$$C_8 = \frac{.0569}{R_L f_L} \qquad L_8 = \frac{.1503 R_L}{f_L}$$

Bandpass Gain = 2.84dB

7.43 APPLICATION OF CROSSOVER NETWORK FORMULAS.

The formulas presented here for crossover design will yield the acoustic response indicated by the characteristics of the filter type only if these conditions are met:

1. The filter is terminated by a flat magnitude, zero phase impedance.

2. The response of the driver extends 1.5–2 octaves from the crossover frequency through the network stopband (up in frequency for low-pass filters, and down in frequency for high-pass filters) with a reasonably flat response.

3. Both high-pass and low-pass drivers are radiating from the same plane. If you fall short on any of these conditions in doing your crossover design, your results won't match the predictions. Only two methods will be useful in this case. The first is simple trial and error. With this method the formulas can only give you a ballpark range of what you need. You measure the response, adjust a network value, re-measure, adjust again, and so on until you get the response you wish. This is, in reality, the most common design technique manufacturers use because the variables you must deal with leave you little choice. The key is to devise a rapid or real-time measurement system which gives you nearly immediate feedback, such as a high-speed FFT analyzer, successive sine-wave sweeps, or a high-quality real-time analyzer (RTA).

The other method is to use a computer-based circuit optimizer program. These are becoming increasingly popular and effective. Computer-aided engineering (CAE) programs for loudspeakers, including network optimization software, are discussed in detail in *Chapter 9*.

Your design problems can be minimized if you do the following:

A. Regarding the first rule, you needn't have a flat impedance to achieve the response you desire from a speaker, but it's easier if you do. Other benefits accrue in the quality of the amplifier/speaker interface, as well as overall damping of the driver, which occurs if the load impedance is flat. Changing the impedance of a driver from the typical load into a reasonably flat magnitude requires the use of conjugate filters tuned to the speaker's particular parameters. This is discussed in *Sections 5.50–5.52*. Unless you are willing to completely rely on trial and error procedures, start your network design project by correcting the impedance of the drivers.

B. About rule 2, having the appropriate response extension is not always possible, but is mostly dependent on driver directivity, low frequency extension, the selected crossover frequency, and choice of crossover slope. The first consideration is that the extension amount can be less critical depending on the slope of the crossover's target response. Fourth-order filters require less extension to achieve a flat on- and off-axis response than do first-order filters.

The directivity problem may be solved by combining drivers having directivity and low frequency extensions which complement each other. A 12″ woofer is a poor choice for a two-way system if your dome tweeter can't be crossed below 2kHz. The 12″ driver's directivity will cause a huge hole in the off-axis response. Using the same 12″ woofer with a 6″ mid-bass, 3″ mid-

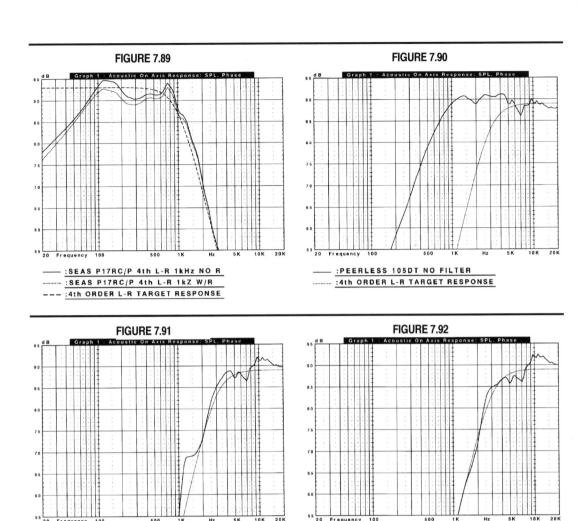

FIGURE 7.89

— :SEAS P17RC/P 4th L-R 1kHz NO R
······ :SEAS P17RC/P 4th L-R 1kZ W/R
— — :4th ORDER L-R TARGET RESPONSE

FIGURE 7.90

— :PEERLESS 105DT NO FILTER
······ :4th ORDER L-R TARGET RESPONSE

FIGURE 7.91

— :PEERLESS 105DT 4th L-R NO EQ
······ :4th ORDER L-R TARGET RESPONSE

FIGURE 7.92

— :PEERLESS 105DT 4th L-R WITH EQ
······ :4th ORDER L-R TARGET RESPONSE

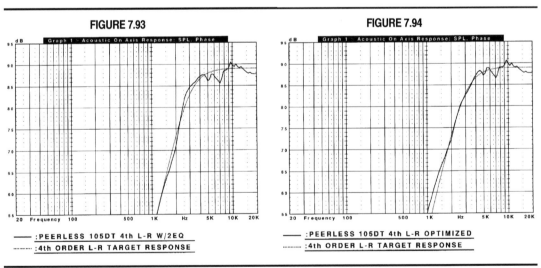

FIGURE 7.93

— :PEERLESS 105DT 4th L-R W/2EQ
······ :4th ORDER L-R TARGET RESPONSE

FIGURE 7.94

— :PEERLESS 105DT 4th L-R OPTIMIZED
······ :4th ORDER L-R TARGET RESPONSE

dome, and a 0.75″ tweeter would work. Ten-inch woofer two-ways are usually marginal, even though they have been quite successful commercially. Ten-inch woofers combined with 4″ midranges and 1″ tweeters do work well, crossed at 750Hz–1kHz and 4–5kHz. Having 1.5–2 octaves of extension can be achieved with three- or four-way designs if components and crossover frequencies are appropriate.

With two-way designs, however, avoiding crossing drivers in a changing frequency range is gen-erally impossible, either due to directivity or to low-pass rolloff. Given the high frequency rolloff and directivity of the drivers typically used in two-way speakers (4″ to 8″ woofers, and 0.5″–1.5″ tweeters), it is unrealistic to expect driver respons-es which will extend 1.5–2 octaves above or below a crossover frequency of 2 or 3kHz. If the crossover operates in a range where the driver response is changing significantly both on- and off-axis, you will have no alternative but to resort to either trial and error, or use crossover comput-

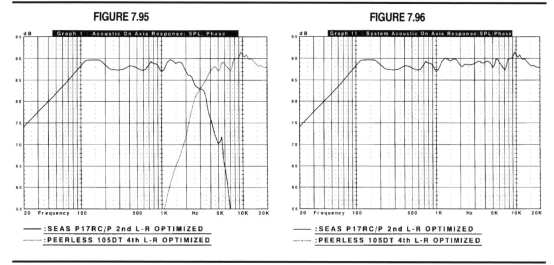

FIGURE 7.95

FIGURE 7.96

— :SEAS P17RC/P 2nd L-R OPTIMIZED
······ :PEERLESS 105DT 4th L-R OPTIMIZED

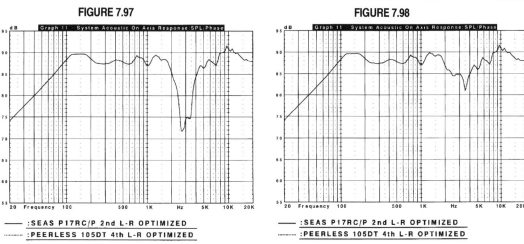

FIGURE 7.97

FIGURE 7.98

— :SEAS P17RC/P 2nd L-R OPTIMIZED
······ :PEERLESS 105DT 4th L-R OPTIMIZED

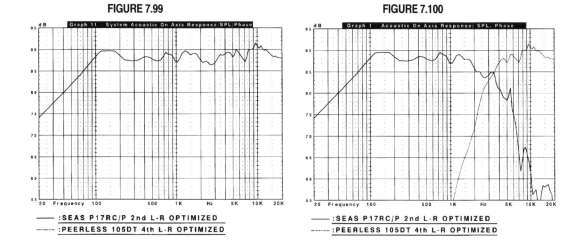

FIGURE 7.99

FIGURE 7.100

— :SEAS P17RC/P 2nd L-R OPTIMIZED
······ :PEERLESS 105DT 4th L-R OPTIMIZED

er optimization tools to solve the design problem.

For example, if a high-pass filter is being made for a tweeter at 2kHz, where the response is starting to roll off around 1.2kHz, the transfer function of the filter network must be designed to combine with the transfer function of the driver to achieve the target response. The trick is that a flat response can always be accomplished in a variety of ways. Every network problem has multiple solutions, some of which will be acceptable and some not. Using computer optimizer soft-

ware, you'll find it's possible to program a second-order filter topology using a simple C/R (capacitance/ resistance) conjugate filter, and optimize the values for a second-, third-, or fourth-order response, all with the same topology. Obviously, you need not use the same order topology as the target slope of the response. In fact, this is almost never the case when dealing with two-way loudspeakers. Optimizers also open the door to experimenting with non-standard topologies, such as using first-order low-pass fil-

FIGURE 7.101

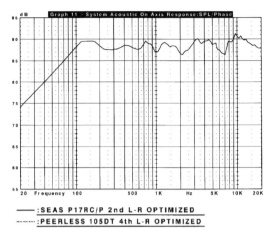

————— :SEAS P17RC/P 2nd L-R OPTIMIZED
·········· :PEERLESS 105DT 4th L-R OPTIMIZED

FIGURE 7.102

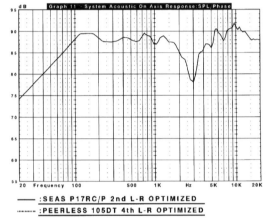

————— :SEAS P17RC/P 2nd L-R OPTIMIZED
·········· :PEERLESS 105DT 4th L-R OPTIMIZED

ters in conjunction with parallel-connected L/C/R trap filters placed an octave above the crossover frequency.

When the driver's transfer function is part of the equation, you can develop crossovers which appear to be asymmetrical. Two-way systems are often designed with a first-order low-pass, and a third-order high-pass. In this case the first-order low-pass combines with the second-order woofer response to create a third-order acoustic response, while the tweeter may be crossed over high enough to require a third-order topology to create a third-order response.

C. The response problems caused by driver horizontal offset (item 3) are usually something less than ideal that you either specifically correct or minimize by using higher order filters. One useful trick which will help you discover whether your final network design has been adjusted appropriately, and the high- and low-pass sections are properly in phase: merely reverse the high-pass section's polarity and measure the response again. If the network is correctly designed, that is, if the response is reasonably flat, and the driver phase and roll-off magnitudes appropriately adjusted, you will see a symmetrical null at the crossover frequency. Sometimes even though you have adjusted filter values to achieve an apparently flat response, the network may still be improperly designed. If so you may hear evidence of

erratic polar behavior and a poor image.

In addition to the criteria of load, directivity, and driver alignment, other considerations are also determinative in deciding which crossover frequencies will give you optimum performance. Keeping networks out of certain frequency ranges seems to contribute to successful designs. While there are exceptions to every rule, three-way crossovers seem to work best with the low- to mid-crosspoint at 200–350Hz, and the mid- to high-crosspoint at 2–2.5kHz. In other words, avoid networks operating in the frequency range between 350Hz–1.5kHz. Low-frequency drivers (15″–10″) tend to cause the male voice to sound too full if allowed to play above the 200–350Hz range (dependent on slope, of course), and it is difficult to find cone or dome type tweeters which will operate with low enough distortion below 2kHz.

7.44 CROSSOVER APPLICATION EXAMPLES.

You may better understand the "rules" discussed in *Section 7.43* through graphic examples. What follows are computer-simulated examples of each type of design dilemma encountered in a two-way loudspeaker project employing a fourth-order Linkwitz-Riley crossover:

The low-pass filter

The woofer is a SEAS P17RC/P 6″ whose impedance I measured using the Audio Precision's System 1 and the voltage divider method. Response measurements, for simplicity made at 1M on-axis, were done with the DRA MLSSA, ACO Pacific's 7012 mike, and the collected data exported into LEAP 4.0. My target response is a fourth-order topology with the standard L-R calculated values. I did so to show the effects of using a classic filter shape without an impedance-correcting conjugate network. The result is depicted in *Figure 7.84*. While the overall slope from 3–6kHz doesn't look bad, some serious interaction is obviously causing the 4dB peak at the rolloff point.

The next logical step is to add a conjugate filter, a simple series C/R (capacitance/resistance) circuit in parallel with the driver. The effect is shown in *Fig. 7.85*. While this is certainly an improvement, the rolloff rate is about 27dB/octave between 3kHz and 6kHz. Since our target is supposed to be 24dB/octave, evidently the combination of the driver's natural rolloff added to the network's produces something higher than the fourth-order response we need.

Since the fourth-order topology seems to attenuate too much, we will try a lower order filter. *Figure 7.86* shows the response of a calculated second-order L-R filter with the driver, but without impedance compensation. This does not come as close as the original fourth-order response, but if we apply impedance EQ, as shown in *Fig. 7.87*, the response becomes the best to date.

So far, all we needed was a little intuition, and a minimum amount of adjusting, to get fairly close to the target response. *Figure 7.88* shows the same network after being optimized by the computer, which is an even closer fit.

Our exercise ignores the effects of "parasitic" resistance in our series inductors. If we wish to explore the consequences of inductor DC resistance (DCR), another example will help us see the effects more easily. *Figure 7.89* shows the same driver with a fourth-order L-R network and impedance EQ at 1kHz. The two inductors measure 2.4mH and 1.2mH, respectively. If these inductors were air-core types, wound with 18-gauge wire, their DCRs would be 1Ω and 0.7Ω, for a total series DCR of 1.7Ω. The top curve shows the response without the effects of inductor parasitic resistance, while the lower one includes it. The series inductors' attenuation is greater at lower frequencies, and diminishes as we approach the crossover frequency, where its effect is swamped by filter action. The other consideration, of course, is the effect of series resistance on driver Q.

The high-pass filter

Our driver is a Peerless 105DT fabric dome tweeter. I chose it because it has a normal resonance and does not employ Ferrofluid (a special viscous fluid which transfers heat from the voice coil to the magnet). I did the measurements using the same equipment and methods detailed above. My target response is a fourth-order L-R high-pass crossover at 3kHz. *Figure 7.90* shows the unfiltered driver response overlaid by the target.

The first step is to try a fourth-order topology with standard L-R calculated values. Again this shows us how the classic filter shape looks without any resonance trap conjugate network. The result is depicted in *Fig. 7.91*. The driver's resonance is causing the response glitch at about 1kHz. Note also that a peak has resulted in the 8–15kHz region caused by the circuit's interaction with the tweeter voice coil.

If we add a series LCR resonance trap tuned for 1kHz resonance to the same filter network, *Fig. 7.92* makes it evident the 1–2kHz anomaly has disappeared, but the 8–15kHz peak is still with us.

The cure is adding a simple CR impedance-correcting network whose effect is clear in *Fig. 7.93*. In this case, leveling out the reactive rise of the tweeter was enough to get rid of the 8–15kHz peak problem. Now the response really doesn't look that bad. *Figure 7.94*, however, shows how helpful computer optimization can be on the same circuit, bringing it a bit closer to the target.

High-pass/Low-pass Summation

Figure 7.95 shows the optimized responses of the two drivers. This illustrates a typical L-R crossover setup, with the response of both drivers 6dB down at the crossover frequency of 3kHz. The summation of these two driver responses is given in *Fig. 7.96*. As expected, the summation achieves a fairly flat response. If we

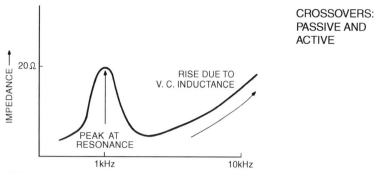

FIGURE 7.103: Dome tweeter impedance curve.

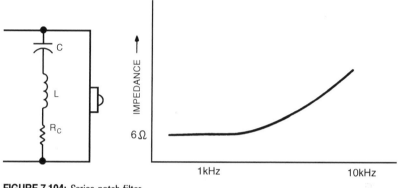

FIGURE 7.104: Series notch filter.

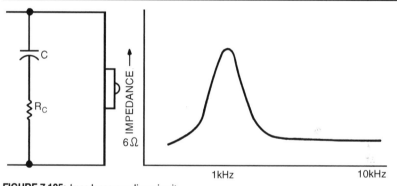

FIGURE 7.105: Impedance equalizer circuit.

evaluate this network by reversing the phase of one of the drivers, results are evident in *Fig. 7.97*. The out-of-phase null is closely centered on the 3kHz crossover frequency and reasonably symmetrical.

Unfortunately, our example until now has assumed the two drivers were mounted on the same radiating plane, and were physically staggered for alignment by time. If we mount the drivers one above the other on a single flat baffle as are 99% of all drivers, we will get a delay of about 115µs, corresponding to 1.56″ distance between the woofer and the tweeter radiating planes. If we take this into account in the summation, the effect is shown in *Fig. 7.98*. We now have a substantial dip in the response at the crossover frequency.

Since the dip is fairly deep and not far from the 3kHz crosspoint, reversing the relative phase of the drivers might possibly return the response to a reasonable shape. *Figure 7.99* shows this would indeed be the effect, and the response is again acceptably flat.

Another way to realign the phase is to alter

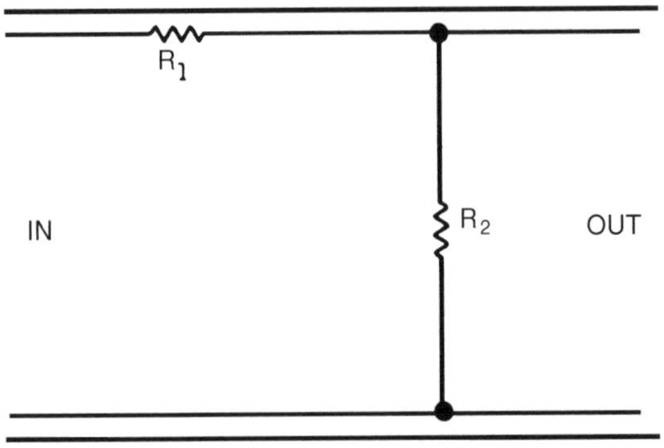

FIGURE 7.106:
Attenuation circuit.

either the low-pass or the high-pass filter shape to restore the response to a satisfactory level. The LEAP optimizer was used to accomplish this. In this instance, I left the high-pass filter values as they were, and allowed the low-pass values to change. The altered low-pass response is shown along with the previous high-pass response in *Fig. 7.100*. The summation of the new optimized response is given in *Fig. 7.101*. Reversing the phase, to check the network alignment, is shown in *Fig. 7.102*. Although the shape is not as flat as the original with the drivers time adjusted, it does show a fairly symmetrical null at the 3kHz crossover frequency.

Which one sounds best? The answer is entirely up to the listener. It is always possible that one combination may sound better with certain program material with a spectral bias based on recording techniques and microphone choice. But, this variation is the essence of the art form.

7.45 MAKING RESPONSE MEASUREMENTS FOR CROSSOVER DESIGN.

The techniques for making frequency response measurements are discussed in detail in *Chapter 8*. However, whatever the measurement domain, other criteria will also be helpful.

Traditionally, design work data is taken at the standard distance of one meter, on-axis, to the listening position. The height can be considered to be about 38″ above the floor, and the design axis as the midpoint between the high-pass and low-pass drivers (between the mid and tweeter in a three-way design). However, others[13] have suggested taking measurements between 2–3 meters from the enclosure, which is nearer the average listener-to-speaker setup in most listening rooms. This alternative requires a more anechoic-like environment if the experimenter is to get reflection-free signals down to a usable frequency. However, such a location is reasonably useful.

Good power response is also important. Flat on-axis response and flat power response are the common characteristics in speakers consistently judged to be high quality in surveys conducted by Floyd Toole at Canada's National Research Council.[20] If a designer chooses a crossover point in a range where the driver's frequency response is changing rapidly off-axis, the off-axis response will have large response anomalies. Large variations in the off-axis response degrade the power response the listener perceives. Reflected and reverberant responses will be significantly differ-

ent from the on-axis response, and generally devalue the overall subjective quality.

Poor off-axis response may be avoided in a crossover stopband by using one of two techniques. First, avoid locating a driver's crossover in a region of changing off-axis response. The designer can't always avoid this, especially in two-way designs. The other technique is to design the network off-axis, and accept whatever consequences develop on-axis as a result. If most listeners will be hearing the speaker off-axis, which is frequently the case, say, with Nearfield or Studio Monitors, designing at 15–30° from the horizontal 0° axis can result in a better overall design.

7.50 DRIVER LOAD COMPENSATING CIRCUITRY.

Although you may use trial and error or computer optimization to design a crossover which compensates for the typical loads presented by a driver, a load-leveling conjugate filter is also a practical solution. Even if you are using computer optimization, conjugate load filters are still added in many commercial designs. The two troublesome impedance anomalies are the driver resonance peak and the reactive rise due to voice coil inductance[21,22] (*Fig. 7.103*). Resonance peaks can be corrected using series LCR filters, described in *Section 7.51*. These can be designed for any driver: woofer, midrange, or tweeter. But if you are trapping low-frequency resonances you will need some large capacitors and inductors. We often see such LCR resonance filters in the 80–100Hz region of the crossover for the satellites of three-piece systems.

If you need compensating circuitry for a voice coil's inductive nature, use a simple CR filter, as described in *Section 7.52*. These are normally applied to low-pass woofer and midrange filters, but are also used with tweeters to provide flat amplifier loads out to 100kHz, and to alter the high-frequency response of the tweeter a bit.

7.51 SERIES NOTCH FILTERS.

The primary function of the circuit illustrated in *Fig. 7.104* is to damp and eliminate the effects of driver resonance on crossover networks. If the driver has been treated with Ferrofluid, however, the resonance has already been mechanically damped and will probably not benefit from the application of this circuit. Assuming, however, the driver has an undamped resonance peak, and the peak is located less than two octaves from a high-pass crosspoint, this circuit will greatly improve driver performance. It is particularly useful on tweeter domes, midrange domes, and cone type midrange drivers whose enclosure resonance is above 200Hz. It is possible to use the device on resonances in the lower octaves, but it usually calls for some extremely large value inductors.

Design Formulas

$$C = \frac{.1592}{R_E Q_{es} f_s}$$

$$L = \frac{.1592\left(Q_{es}R_E\right)}{f_s}$$

$$R_c = R_E + \frac{Q_{es}R_E}{Q_{ms}}$$

Impedance magnitude:

$$Z = R_c{}^2 + \left(6.2832\,fL - \frac{.1592}{fC}\right)^2$$

Phase Angle:

$$\theta\circ = Tan^{-1}\left(\frac{6.2832L - \dfrac{1}{6.2832C}}{R_C}\right)$$

Note: R_c should be calculated to include the measured resistances of L, or alternatively, you can wind L from small gauge wire so that the DCR equals R_c.

A shortcut method, which does not require driver Q parameters, is given by:

$$C = \frac{.03003}{f}$$

$$L = \frac{.02252}{f^2 C}$$

R_c = approx. rated impedance of driver.

Check the circuits described above by running an impedance curve of the driver and circuit combination. If the flattened impedance is not quite level, try increasing R_c in increments of 0.5Ω until you achieve a satisfactory impedance.

7.52 IMPEDANCE EQUALIZATION.

All voice coil type drivers exhibit a rising impedance caused by the voice coil inductive reactance. In order for midrange and woofer low-pass crossover circuits to operate properly, you can equalize this rising impedance by using the CR circuit shown in *Fig. 7.105*. You can also use this circuit for tweeter domes, not to facilitate network operation, but to help eliminate harshness and to assure the accurate application of L-type shelving networks.

Design Formulas

$$C = \frac{L_e}{R_c{}^2}$$

L_e = driver voice coil inductance in henries.

$R_c = 1.25 \times R_E$

The values of Rc and C are only approximate. You should adjust them experimentally for a flat measured impedance curve.

7.60 DRIVER ATTENUATION CIRCUITS.

Midrange and tweeter drivers are typically more efficient and play louder in their bandwidth than most woofers. In order to achieve a uniform overall frequency response from a given loudspeaker, you must adjust the different mid and high-frequency driver output levels to match the nominal woofer SPL. There are several ways the resistance can be inserted into a filter circuit to accomplish this task, each one having a different effect upon the overall frequency response, the impedance of the circuit/driver combination, and the transfer function of the circuit. There are four basic methodologies to consider: a series resistance placed on the "amplifier side" of the circuit (near the circuit input prior to any positive to ground elements); a series resistance placed on the "driver side" of the circuit (between the driver and the positive to ground circuit elements); a combination of equal amp side and driver side resistance (as well as unequal resistances); and last, a balanced impedance L-pad type attenuation circuit (*Fig. 7.106*). *Figures 7.107–7.114* illustrate these four methods of driver attenuation for typical tweeter high-pass filters (a third-order topography) and midrange bandpass circuits.

How these different approaches at level attenuation affect the overall functioning of a driver actually varies from driver to driver, as each transducer has a more or less individual reactive nature. However, this brief tutorial will give you an idea of

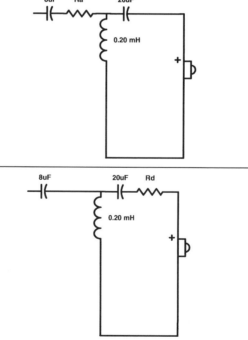
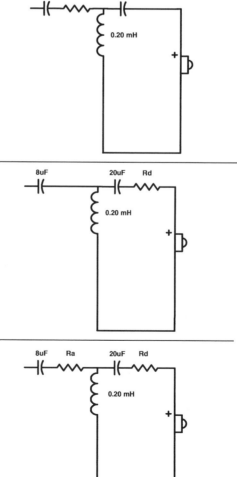
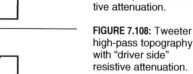
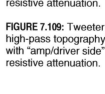

FIGURE 7.107: Tweeter high-pass topography with "amp side" resistive attenuation.

FIGURE 7.108: Tweeter high-pass topography with "driver side" resistive attenuation.

FIGURE 7.109: Tweeter high-pass topography with "amp/driver side" resistive attenuation.

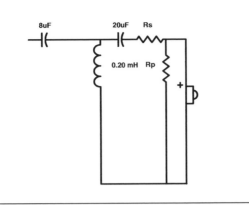

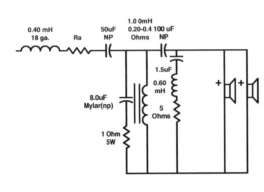

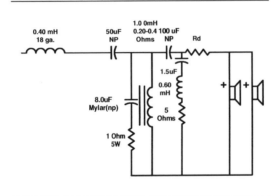

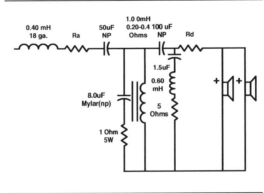

FIGURE 7.110: Tweeter high-pass topography with L-pad resistive attenuation.

FIGURE 7.111: Midrange bandpass topography with "amp side" resistive attenuation.

FIGURE 7.112: Midrange bandpass topography with "driver side" resistive attenuation.

FIGURE 7.113: Midrange bandpass topography with "amp/driver side" resistive attenuation.

FIGURE 7.114: Midrange bandpass topography with L-pad resistive attenuation.

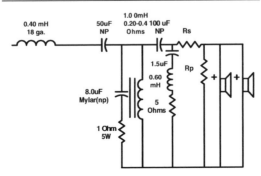

how to apply these different methods and how each type tends to interact with the transducer.

Amplifier side attenuation:

1) Tweeter high-pass circuits—*Figure 7.115* shows the SPL changes for an amp side resistor from 1–6Ω. *Figures 7.116* and *7.117* give the accompanying changes in impedance and transfer function, respectively. This method results in a fairly shallow and damped transfer function, the damping increasing logically as the resistor value increases. Attenuation is fairly even across the bandwidth, but somewhat frequency dependent as is seen in *Fig. 7.118* where the attenuated response (with 6Ω in series) was offset to the same level as the unattenuated response. Overall, this method is the one I employ almost universally in designs I do for manufacturers in my consulting business. It provides a well-damped filter function and has only moderate frequency dependent changes in the response. Using a filter optimizer, such as the one in the LinearX Leap software, it is a simple matter of fixing the value of the attenuation resistor and letting the optimizer vary the other circuit elements to achieve a target response.

2) Midrange bandpass circuits—Bandpass circuits are much more difficult to deal with because of the proximity of two independent filters to each other, the problem worsening the closer the high-pass and low-pass frequencies are. *Figure 7.119* gives the SPL changes for an amp side resistor from 1–6Ω, with the impedance shown in *Fig. 7.120* and the accompanying transfer functions in *Fig. 7.121*. Like the tweeter circuit with this type of attenuation, the filter function is well damped, but also has even more frequency dependent SPL changes. *Figure 7.122* depicts the unattenuated response with the 6Ω attenuated response offset for comparison. The result is a sort of shelving of the high-pass response with an increase in the low-pass roll-off frequency. This is more severe than with a simple tweeter high-pass filter, and it means that the entire bandpass has to be re-optimized to achieve a flat target response. This is a very time consuming and tedious procedure for cut-and-try designing, but relatively quick and painless using a circuit optimizer.

Driver side attenuation:

1) Tweeter high-pass circuits—With the resistor on the other side of the ground leg inductor, the behavior of the circuit is entirely different. *Figure 7.123* shows the result of this type of resistor placement from 1–6Ω, with the impedance and transfer function consequences illustrated in *Figs. 7.124* and *7.125*. It's obvious from *Fig. 7.123* that the attenuation is very frequency dependent with resistances over 2Ω. In a two-woofer two-way speaker where the tweeter requires only a dB or so to match the woofer levels, this could work out, but it is also apparent that the circuit impedance declining at the filter resonance and in the case of the 6Ω example, results in an impedance minimum of 2.5Ω, which is really too low. As the impedance drops at this frequency, the transfer

function becomes progressively less damped, the opposite of amplifier side attenuation. *Figure 7.126* depicts the unattenuated response with the level of the 6Ω example offset to the same level, making the nature of the frequency dependent changes obvious. You really need to watch both the impedance and transfer function closely when using this method.

2) Midrange bandpass circuits—Response changes for driver side resistance applied to band-pass filters, shown in *Fig. 7.127*, in a manner similar to the tweeter high-pass with shelving of the response above the high-pass corner frequency, but with the additional low-pass filter the result is a "sag" in the midband that gets worse as the resistance increases. The impedance and transfer function changes are also similar (*Figs. 7.128* and *7.129*) with the impedance getting very low (at least for this driver which happens to be a set of Bravox 3.5" woofers) at the high-pass crossover frequency and the filter knee getting progressively sharper. *Figure 7.130* compares the unattenuated driver response with the 6Ω example offset to the same SPL, again showing the large sag in the response that would require at least an attempt to

re-optimize both high- and low-pass filters.

Combination equal amp side and driver side attenuation:

1) Tweeter high-pass circuits—Placing equal amounts of resistance is just to give you an idea about this approach. In practice, this technique can be applied when low-*impedance* problems occur in a circuit to try to bring up too low minimum impedance. For this particular example, it's not a problem, but it does illustrate the method. *Figure 7.131* illustrates the SPL changes for equal resistances on both sides of the circuit from 1–6Ω. It's fairly similar to the amp side SPL changes shown in *Fig. 7.115*, with similarities also in the impedance and transfer function changes (*Figs. 7.132* and *7.133*) without the negative effects of driver side attenuation. *Figure 7.134* compares the unattenuated response with the offset equal 6Ω amp/driver side resistance, again showing a frequency dependent nature that can probably be optimized for a more flat response as long as it doesn't raise the Q of the circuit too much (tests have shown that ringing associated with high-Q circuits is somewhat audible[23]).

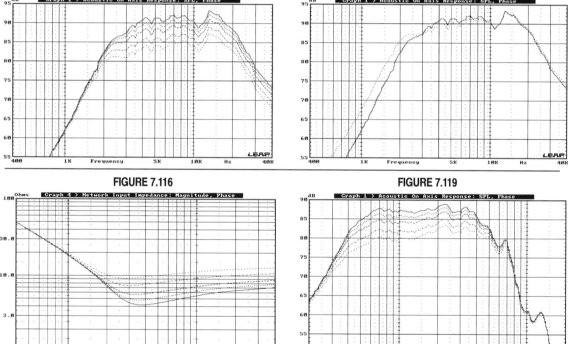

FIGURE 7.115 FIGURE 7.118
FIGURE 7.116 FIGURE 7.119
FIGURE 7.117 FIGURE 7.120

FIGURE 7.115: SPL for high-pass with "amp side" attenuation (solid = no attenuation; dot = 1Ω; dash = 2Ω; dash/dot = 4Ω; short dash = 6Ω).

FIGURE 7.116: Impedance for high-pass with "amp side" attenuation (solid = no attenuation; dot = 1Ω; dash = 2Ω; dash/dot = 4Ω; short dash = 6Ω).

FIGURE 7.117: Transfer function for high-pass with "amp side" attenuation (solid = no attenuation; dot = 1Ω; dash = 2Ω; dash/dot = 4Ω; short dash = 6Ω).

FIGURE 7.118: Comparison of no attenuation and 6Ω attenuation (offset) in *Fig. 7.115* (solid = no attenuation; short dash = 6Ω).

FIGURE 7.119: SPL for bandpass with "amp side" attenuation (solid = no attenuation; dot = 1Ω; dash = 2Ω; dash/dot = 4Ω; short dash = 6Ω).

FIGURE 7.120: Impedance for band-pass with "amp side" attenuation (solid = no attenuation; dot = 1Ω; dash = 2Ω; dash/dot = 4Ω; short dash = 6Ω).

2) Midrange bandpass circuits—The SPL changes caused by the use of amp/driver side resistance show similarities to both amp and driver side single resistance SPL (*Fig. 7.135*). The midband sag is less severe, and the shelving on the high-pass side of the circuit is more severe. The impedance and transfer function changes shown in *Figs. 7.136* and *7.137* are also similar to both types, with maybe more in common with the amp side resistance method. *Figure 7.138* compares the unattenuated response with the dual 6Ω response being offset. Again, this technique can be used to balance minimum impedance and adjust transfer function damping, as well as SPL manipulation, however, all of these methods generally require overall circuit re-optimization to compensate for the frequency dependent effects of the attenuation circuits.

Combination unequal amp/driver side attenuation:

1) Tweeter high-pass circuits—This comparison is different in that it affects the effects of a total 6Ω on the circuit, but placed in different places. The comparison is of the unattenuated high-pass, 6Ω placed on the amp side of the circuit, unequal resistance of 2Ω on the amp side and 4Ω on the driver side, the reverse of this, 4Ω on the amp side and 2Ω on the driver side, and for comparative purposes, equal 3Ω on both amp/driver sides of the circuit. *Figure 7.139* gives the comparison of all four circuits, with the impedance and transfer functions shown in *Figs. 7.140* and *7.141*. It's obvious that these techniques could be used as a tool in contouring a response shape, along with adjusting filter values. This in combination with re-optimizing the tweeter circuit gives you some idea of the tools available for response shaping and at the same time manipulating both the impedance and transfer functions. All three are important in network design. *Figure 7.142* compares all four attenuation examples offset to the unattenuated response. Again, the amp side alone technique yields the shallowest damped transfer function and is almost always my choice in design unless I need to compensate for an

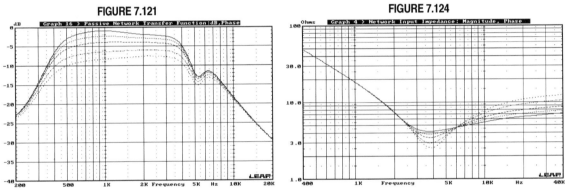

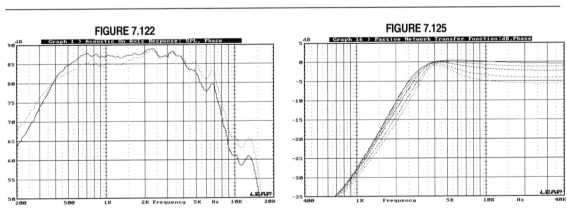

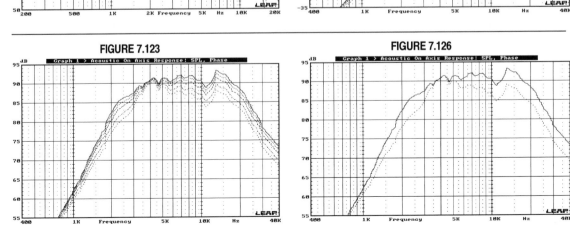

impedance *anomaly* of some sort.

2) Midrange bandpass circuits—The effect on this type of circuit is perhaps similar to the tweeter high-pass. However, with a bandpass filter with unequal both side attenuation (*Fig. 7.143*), it is possible to manipulate the knee at both high-pass and low-pass sections, with the consequential impedance and transfer function changes depicted in *Figs. 7.144* and *7.145*. However, all the attenuation methods result in some sort of shelving or sag in the midband (*Fig. 7.146*) accompanied by an increase in the low-pass output and would require re-optimization no matter which method was employed.

L-Pad attenuation:

1) Tweeter high-pass circuits—L-pads are inserted as the last element in the circuit prior to the driver. The concept is to provide a constant load for the network while at the same time attenuating the response of the device, which means that an L-pad maintains a constant impedance for the filter circuit, unlike series type resistance methods.

Figure 7.147 illustrates different levels of attenuation using an L resistance circuit. The response shows none of the frequency dependent behavior associated with the series resistance circuits, which is the primary benefit for this type of circuit. The impedance is maintained at a constant value, as shown in *Fig. 7.148*, for different levels of attenuation. Likewise, the shape of the transfer function (*Fig. 7.149*) stays constant, although not as well damped as the amp side series resistor method. *Figure. 7.150* compares the unattenuated response to the –12dB curve offset to the same level. As can be seen, no frequency dependent shift occurs with an L-pad.

2) Midrange bandpass circuits—L-pads aren't as benign on bandpass filters and do exhibit some frequency dependent anomalies. *Figure 7.151* shows the changes in response with changing levels of attenuation, with an obvious decrease in midband output. Impedance and transfer functions (*Figs. 7.152* and *7.153*) also show changes as attenuation increases. *Figure 7.154* compares the –12dB (5/2Ω) example offset to the unattenuated

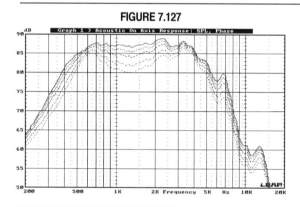

FIGURE 7.127

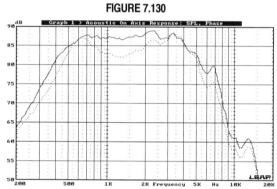

FIGURE 7.130

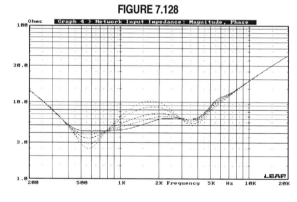

FIGURE 7.128

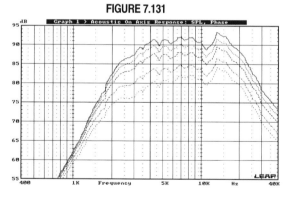

FIGURE 7.131

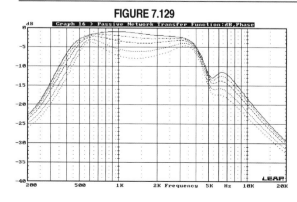

FIGURE 7.129

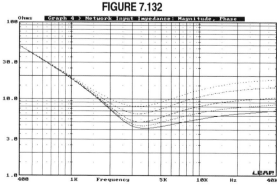

FIGURE 7.132

FIGURE 7.127: SPL for bandpass with "driver side" attenuation (solid = no attenuation; dot = 1Ω; dash = 2Ω; dash/dot = 4Ω; short dash = 6Ω).

FIGURE 7.128: Impedance for bandpass with "driver side" attenuation (solid = no attenuation; dot = 1Ω; dash = 2Ω; dash/dot = 4Ω; short dash = 6Ω).

FIGURE 7.129: Transfer function for bandpass with "driver side" attenuation (solid = no attenuation; dot = 1Ω; dash = 2Ω; dash/dot = 4Ω; short dash = 6Ω).

FIGURE 7.130: Comparison of no attenuation and 6Ω attenuation (offset) in *Fig. 7.127* (solid = no attenuation; short dash = 6Ω).

FIGURE 7.131: SPL for high-pass with equal "amp/driver side" attenuation (solid = no attenuation; dot = 1/1Ω; dash = 2/2Ω; dash/dot = 4/4Ω; short dash = 6/6Ω).

FIGURE 7.132: Impedance for high-pass with equal "amp/ driver side" attenuation (solid = no attenuation; dot = 1/1Ω; dash = 2/2Ω; dash/dot = 4/4Ω; short dash = 6/6Ω).

response, showing the response variations. As with all the attenuation methods applied to bandpass circuits, the high-pass/bandpass filters need to be re-optimized.

7.65 DESIGNING L-PADS.

You must obtain or calculate the following data:

1. the rated sensitivity in dB of each driver (D_S)*
2. total insertion losses in dB, equal to (amplifier source resistance) + (total inductor resistances), where:

$$R_t = 20\log_{10}\frac{R_m + (R_G + R_L)}{R_m}$$

R_t = total insertion loss in dB
R_m = effective driver impedance inclusive of all load compensating circuitry
R_G = amp source resistance (refer to *Chapter 8*)
R_L = measured inductor series resistance

3. bandpass gain, if applicable, in dB (B_G).

Then total driver sensitivity (D_{ts}) equals:

$$D_{ts} = D_s + B_G - R_t$$

With this information, the sensitivity differences between the woofer and the midrange and tweeter drivers are readily apparent.

Now that you know the amount of attenuation you need, you can use two types of attenuation circuits: either a simple series resistor, or an L attenuation network like the one illustrated in *Fig. 7.106.* A series resistance is adequate so long as you refigure the crossover components to account for the increase in total driver impedance. This, however, increases inductor sizes and creates additional insertion loss. An L-type attenuation circuit will provide attenuation while maintaining minimum driver impedance, as long as you use proper load circuits to adjust the driver impedance to a constant level.[24]

Design Formulas

$$R_2 = \frac{10^{(A/20)} \times Z}{1 - 10^{(A/20)}}$$

FIGURE 7.133: Transfer function for high-pass with equal "amp/driver side" attenuation (solid = no attenuation; dot = 1/1Ω; dash = 2/2Ω; dash/dot = 4/4Ω; short dash = 6/6Ω).

FIGURE 7.134: Comparison of no attenuation and dual 6Ω attenuation (offset) in *Fig. 7.131* (solid = no attenuation; short dash = 6/6Ω).

FIGURE 7.135: SPL for bandpass with equal "amp/driver side" attenuation (solid = no attenuation; dot = 1/1Ω; dash = 2/2Ω; dash/dot = 4/4Ω; short dash = 6/6Ω).

FIGURE 7.136: Impedance for bandpass with equal "amp/driver side" attenuation (solid = no attenuation; dot = 1/1Ω; dash = 2/2Ω; dash/dot = 4/4Ω; short dash = 6/6Ω).

FIGURE 7.137: Transfer function for bandpass with equal "amp/driver side" attenuation (solid = no attenuation; dot = 1/1Ω; dash = 2/2Ω; dash/dot = 4/4Ω; short dash= 6/6Ω).

FIGURE 7.138: Comparison of no attenuation and dual 6Ω attenuation (offset) in *Fig. 7.135* (solid = no attenuation; short dash = 6/6Ω).

FIGURE 7.133

FIGURE 7.136

FIGURE 7.134

FIGURE 7.137

FIGURE 7.135

FIGURE 7.138

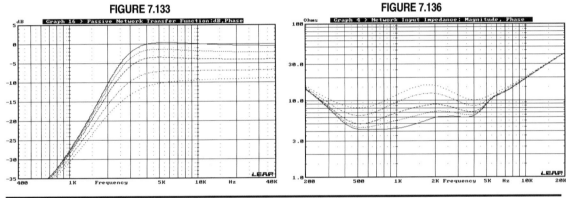

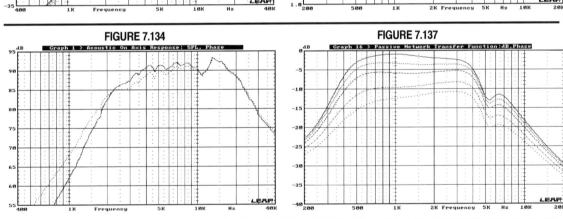

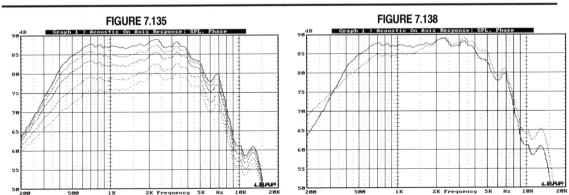

$$R_1 = Z - \left(\frac{1}{R_2} + \frac{1}{Z} \right)^{-1}$$

Z = total driver impedance.
A = amount of required attenuation in -dB (negative decibels, e.g., -3dB).

*Note: manufacturers' driver sensitivity ratings are 1W/1m broadband measurements, and because of response anomalies (such as the rising response near rolloff typical of most two-layer voice coil woofers), can give a distorted picture of sensitivity. A more accurate and relevant method is to ascertain sensitivity from the appropriate portion of the driver bandwidth using a 1W/1m frequency response curve.

7.70 RESPONSE SHAPING CIRCUITRY.
Two circuits are useful for modifying driver response in addition to standard networks and load-leveling filters. Both circuits, the contour network and the parallel trap filter, generally require trial and error for proper execution, but the following information will help you find usable starting values.

(1). Contour Networks. *Figure 7.155* illustrates the two simple RC and RL circuits you can use to modify rising frequency response tendencies in situations where a) the response is rising with increasing frequency, and b) the response is rising with decreasing frequency (*Fig. 7.156*).

Design Sequence

(1). Find the component value at the frequency where the attenuation should be minimum (the point at which the rise in response begins). The reactance of the component (L or C) should be 1Ω at this point, where:
A) For L

$$L = \frac{.15916}{f}$$

L = inductance in henries
f = frequency of minimum

FIGURE 7.139

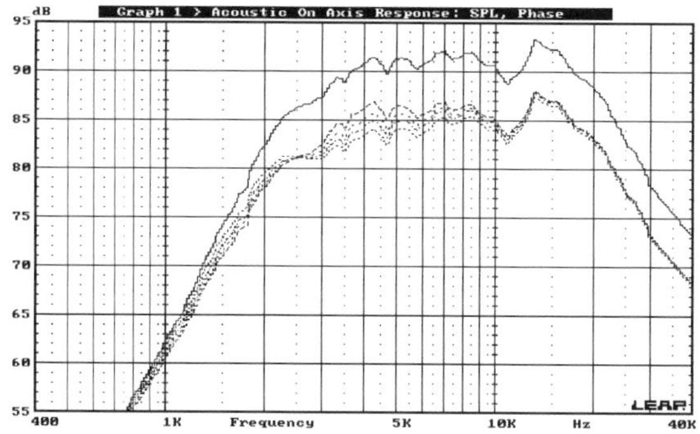
Graph 1 > Acoustic On Axis Response: SPL, Phase
dB
Frequency
Hz
LEAP

FIGURE 7.142

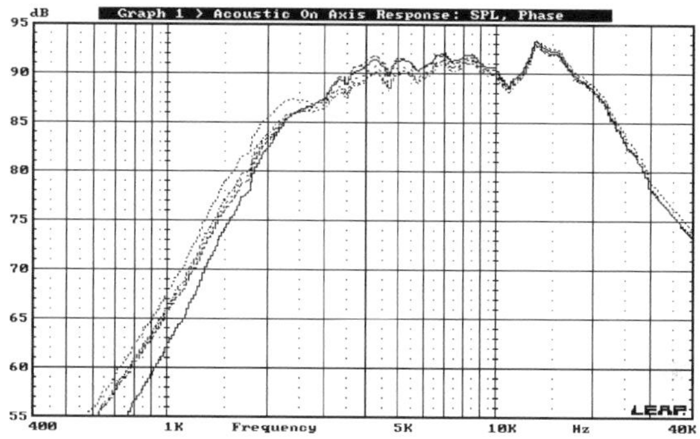
Graph 1 > Acoustic On Axis Response: SPL, Phase
dB
Frequency
Hz
LEAP

FIGURE 7.140

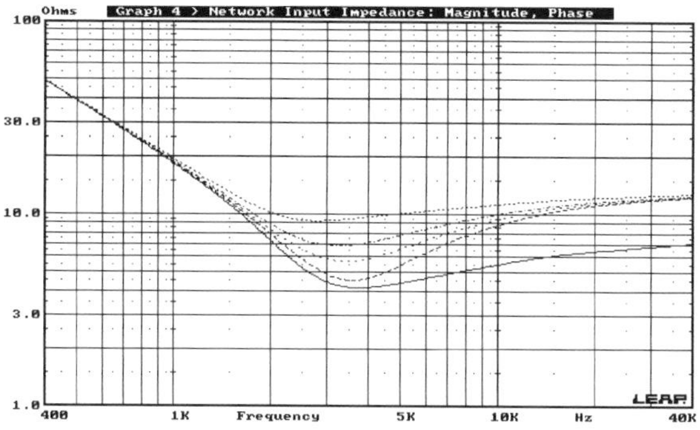
Graph 4 > Network Input Impedance: Magnitude, Phase
Ohms
Frequency
Hz
LEAP

FIGURE 7.143

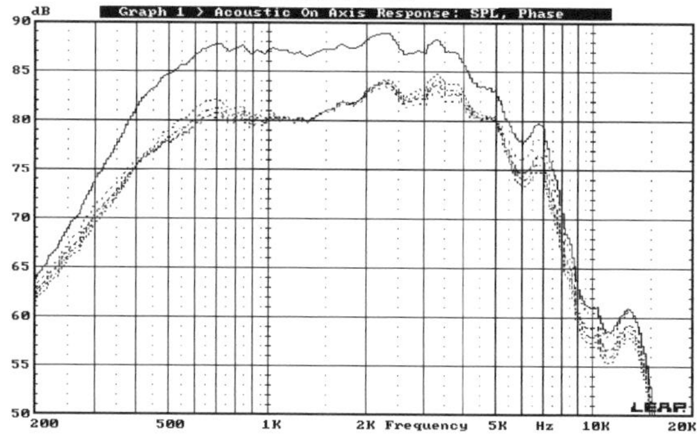
Graph 1 > Acoustic On Axis Response: SPL, Phase
dB
Frequency
Hz
LEAP

FIGURE 7.141

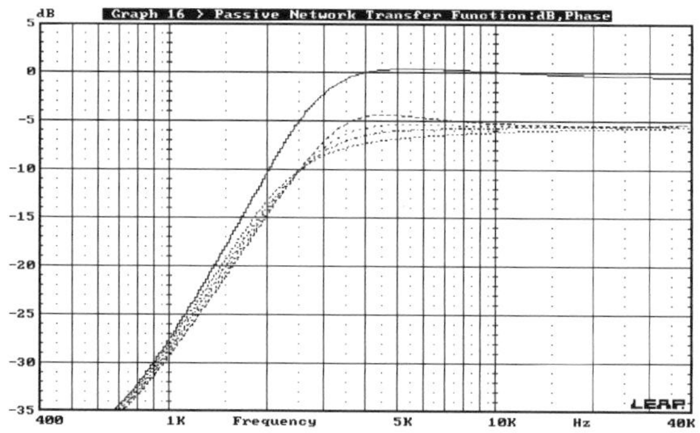
Graph 16 > Passive Network Transfer Function:dB,Phase
dB
Frequency
Hz
LEAP

FIGURE 7.144

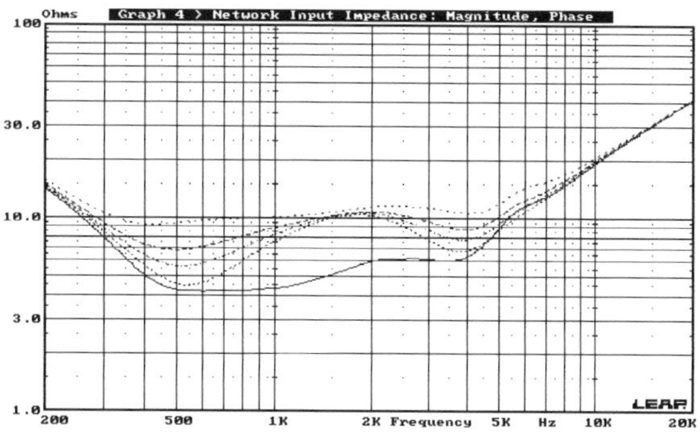
Graph 4 > Network Input Impedance: Magnitude, Phase
Ohms
Frequency
Hz
LEAP

FIGURE 7.139: SPL for high-pass with unequal "amp/driver side" attenuation (solid = no attenuation; dot = 2/4Ω; dash = 4/2Ω; dash/dot = 3/3Ω; short dash = 6Ω).

FIGURE 7.140: Impedance for high-pass with unequal "amp/driver side" attenuation (solid = no attenuation; dot = 2/4Ω; dash = 4/2Ω; dash/dot = 3/3Ω; short dash = 6Ω).

FIGURE 7.141: Transfer function for high-pass with unequal "amp/driver side" attenuation (solid = no attenuation; dot = 2/4Ω; dash = 4/2Ω; dash/dot = 3/3Ω; short dash = 6Ω).

FIGURE 7.142: Comparison of no attenuation and all four unequal attenuation examples (offset) in *Fig. 7.140* (solid = no attenuation; dot = 2/4Ω; dash = 4/2Ω; dash/dot = 3/3Ω; short dash = 6Ω).

FIGURE 7.143: SPL for bandpass with unequal "amp/driver side" attenuation (solid = no attenuation; dot = 2/4Ω; dash = 4/2Ω; dash/dot = 3/3Ω; short dash = 6Ω).

FIGURE 7.144: Impedance for bandpass with unequal "amp/driver side" attenuation (solid = no attenuation; dot = 2/4Ω; dash = 4/2Ω; dash/dot = 3/3Ω; short dash = 6Ω).

reactance

$$C = \frac{.15916}{f}$$

(B) For C

$C =$ capacitance in farads

$f =$ frequency of minimum reactance.

(2). R is selected so the combined impedance of the total circuit equals the amount of maximum-needed attenuation (or average if the rise is not well-defined), where:

$$Z = \frac{RX}{\left(R^2 + X^2\right)^{1/2}}$$

$Z =$ total circuit impedance

$X =$ component reactance at frequency of maximum attenuation (see *Section 7.20* for reactance formulas).

(3). The amount of attenuation in decibels is then:

$$A_t = 20 \log_{10} \frac{R_d + Z}{R_d}$$

$A_t =$ attenuation in dB

$R_d =$ total driver impedance, including load circuits.

An example is using circuit A for a rising response problem, beginning about 250Hz and rising to 10dB at 5kHz.

$$L = \frac{.15916}{250}$$

$$= .6\text{mH}$$

0.63mH has a reactance of 20Ω at 5kHz. *Table 7.1* gives values of attenuation for different R values.

It is difficult to establish hard and fast rules for these types of filters, so again, trial and error play an important part. It is important to remember to include the contour network with the driver and load compensating circuitry when you measure the impedance for calculating the crossover network.

(2). Parallel Notch Circuits. You can remove broad peaks by using the circuit in *Fig. 7.157*. The frequency response plot shows a typical situation

FIGURE 7.145: Transfer function for bandpass with unequal "amp/driver side" attenuation (solid = no attenuation; dot = 2/4Ω; dash = 4/2Ω; dash/dot = 3/3Ω; short dash = 6Ω).

FIGURE 7.146: Comparison of no attenuation and dual 6Ω attenuation (offset) in *Fig. 7.143* (solid = no attenuation; dot = 2/4Ω; dash = 4/2Ω; dash/dot = 3/3Ω; short dash = 6Ω).

FIGURE 7.147: SPL for high-pass with "L-pad" attenuation (solid = no attenuation; dot = 1S/100P ohms; dash = 2S/20P ohms; dash/dot = 4S/4P ohms; short dash = 5S/2P ohms).

FIGURE 7.148: Impedance for high-pass with "L-pad" attenuation (solid = no attenuation; dot = 1S/100P ohms; dash = 2S/20P ohms; dash/dot = 4S/4P ohms; short dash = 5S/2P ohms).

FIGURE 7.149: Transfer function for high-pass with "L-pad" attenuation (solid = no attenuation; dot = 1S/100P ohms; dash = 2S/20P ohms; dash/dot = 4S/4P ohms; short dash = 5S/2P ohms).

FIGURE 7.150: Comparison of no attenuation and −12dB L-pad attenuation (offset) in *Fig. 7.147* (solid = no attenuation; short dash = 5S/2P ohms).

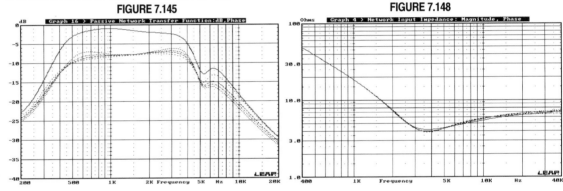

FIGURE 7.145

FIGURE 7.148

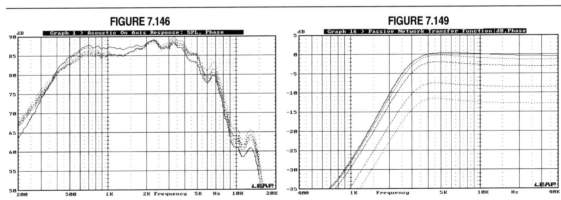

FIGURE 7.146

FIGURE 7.149

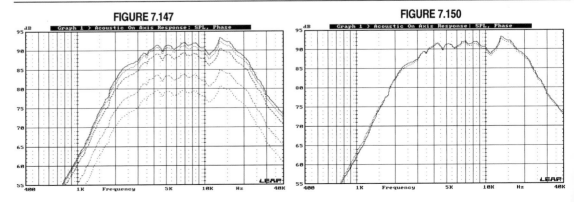

FIGURE 7.147

FIGURE 7.150

for this type of filter:

Design Sequence

Find f, the midpoint frequency of the peak, and its magnitude in dB. Also locate the -3dB frequencies f_1 and f_2, then:

$$C = \frac{.03003}{f} \text{ in farads}$$

$$L = \frac{.02252}{f^2C} \text{ in henries}$$

$$R = \frac{1}{6.2832CB}$$

$$B = \text{-3dB bandwidth } (f_1 - f_2)$$

The impedance circuit is given by:

$$Z = \frac{1}{\left[\frac{1}{R^2} + \left(\frac{1}{6.2832fC} - 6.2832fC\right)^2\right]^{1/2}}$$

Total attenuation in dB is given by:

$$A_t = 20\log_{10}\frac{R_d + Z}{R_d}$$

The phase angle in degrees is given by:

$$\theta\circ = Tan^{-1}R\left(\frac{1}{6.2832fL} - 6.2832fC\right)$$

Unfortunately, the parallel notch filter is not as easy to generate from design formulas as the series notch filter. The DC resistance in the inductor portion of the circuit at once turns the parallel filter into a series-parallel filter. Also, the different dissipation factors of various types of capacitors will also affect final performance (Mylar and polypropylene capacitors will yield different results than electrolytic capacitors). Since this is unavoidable, the best way to proceed is to start with the calculated values, making certain to minimize the inductor DCR, and then selectively increase the value of R until the desired effect is achieved (increasing the value of R will increase the depth or Q of the notch). The L/C ratio of the above formulas creates a fairly narrow filter shape (hi-Q) which should work for most "peak" situations. If a wider filter shape is desired to accommodate a peak which spans in excess of two octaves, use smaller values of C and proportionately larger values of L. As long as the product of $L \times C$ is the same number as that derived from the original formula, the circuit resonance will remain the same. Conversely, if a more narrow shape is required, increase the value of C and decrease L. David Weems' article "Notch Filters," *SB* 2/86, does an excellent job of graphically depicting this situation. The charts and tables should, however, only be taken as a relative indicator, since the original inductor DCR and capacitor dissipation factors are not stated.

7.80 CROSSOVER NETWORK INDUCTORS.

Two types of inductors are commonly used in loudspeaker crossovers: air-core and metal-core. Metal-core inductors, using transformer laminate or the various types of ferrite, are normally employed only where large values of inductance are required with a low series resistance not obtainable with air-core type inductors. Because of

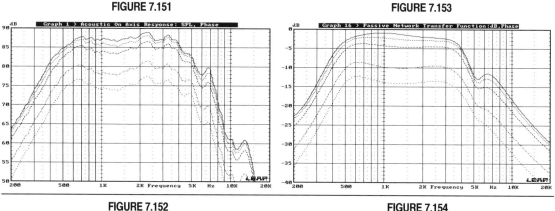

FIGURE 7.151

FIGURE 7.153

FIGURE 7.152

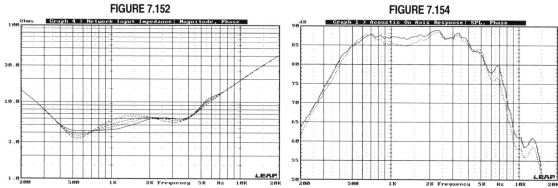

FIGURE 7.154

FIGURE 7.151: SPL for bandpass with "L-pad" attenuation (solid = no attenuation; dot = 1S/100P ohms; dash = 2S/20P ohms; dash/dot= 4S/4P ohms; short dash = 5S/2P ohms).

FIGURE 7.152: Impedance for bandpass with "L-pad" attenuation (solid = no attenuation; dot = 1S/100P ohms; dash = 2S/20P ohms; dash/dot = 4S/4P ohms; short dash = 5S/2P ohms).

FIGURE 7.153: Transfer function for bandpass with "L-pad" attenuation (solid = no attenuation; dot = 1S/100P ohms; dash = 2S/20P ohms; dash/dot = 4S/4P ohms; short dash = 5S/2P ohms).

FIGURE 7.154: Comparison of no attenuation and −12dB L-pad attenuation (offset) in *Fig. 7.143* (solid = no attenuation; short dash = 5S/2P ohms).

the propensity of metal-cored inductors to saturate and cause distortion at higher operating levels, they represent a compromise. Bi-amping with subwoofer crossovers becomes an attractive alternative when faced with the occasional requirement of large inductance values. As will be shown below, however, you may use large low resistance air-cored inductors if a large physical size can be tolerated.

If proper tools are available, such as a coil winding machine (preferably motorized) and an impedance bridge, air-core inductors are more or less practical for speaker builders. Without these tools, you are probably better off purchasing prewound and measured inductors. Having manufactured thousands of inductors, the thought of handwinding is discouraging to me. But if you have the patience, it is certainly possible without a winding machine. The inductors, however, do need to be measured (as all crossover components should be, before assembly).

The design methodology described by Thiele[25,26] establishes the physical size of the inductor in relationship to the reactive time constant at the crossover frequency, rather than only upon the inductance alone. The result is an inductor often larger than what you normally see in production loudspeakers, but an inductor that has its DCR optimized to cause as small a change as possible in driver Q. A good rule of "thumb" is to keep DCR and 1/10 of the voice coil DCR (i.e., no more than 0.7Ω inductor DCR for a voice coil DCR of 7Ω)[27].

For the calculations below, the DCR of the inductors is usually arbitrarily set to 1/20 of the value of the driver load impedance (including impedance compensating circuitry) at the crossover frequency. Thiele reasoned this is the allowable amount which will not adversely affect the driver Q. For an 8Ω driver this would be about 0.4Ω, for a 4Ω driver this would be about 2Ω, and so on.

Two inductor shapes are useful, depending on the circumstances. Type A is normally good for any application except small value higher frequency crossover inductors, where an extremely small core is sometimes called for. Type B is proportioned to make usable core sizes for small value inductors. *Figure 7.158* illustrates the basic configuration used to derive both types. For the following:

R	=	the desired DCR of the inductor
H	=	core height in inches
d	=	wire diameter
L	=	inductance in microhenries (6.5mH = 6500µH)
N	=	number of turns required
r	=	core radius in inches

$$(A)$$
$$r = H$$

$$H = \left(\frac{L}{5590R}\right)^{1/2}$$

$$N = 3.94\left(\frac{L}{H}\right)^{1/2}$$

$$d = \frac{.841H}{\sqrt{N}}$$

$$(B)$$
$$r = 2H$$

$$H = \left(\frac{L}{6170R}\right)^{1/2}$$

$$N = 2.61\left(\frac{L}{H}\right)^{1/2}$$

$$d = \frac{.738H}{\sqrt{N}}$$

Use *Table 7.2* to convert wire diameters into wire gauges. Choose the wire gauge closest to the calculated wire diameter. Incidentally, if you have never purchased copper wire for the purpose of coil winding, you can buy bulk quantities from supply houses which sell electric motor winding parts.

When you round the value of r to find a practical core size (r changed to 0.5″ for a calculated r = 0.46″), the value of the number of required turns will change. If you wish to round to a relatively close next size, use the value of N as a target figure. Wind the inductor, measure its value, and adjust by adding or subtracting turns as needed.

In cases where large gauge wire is called for, such as 12 or 10 gauge, you can wind the inductor from multiple smaller gauges of wire. *Table*

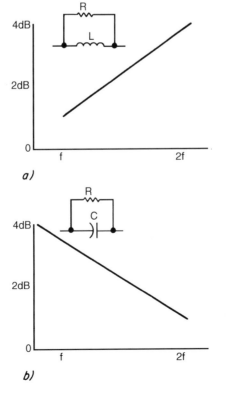

FIGURE 7.155: R_L (a) and R_C (b) contour networks.

7.3 gives approximate equivalent wire gauges for multiple windings of smaller gauge wire.

As an example, a 4mH inductor with a 0.4Ω DCR would have the following parameters:

H = 1.33″
r = 1.33″
N = 216 turns
d = 0.07663 or 12 ga., or 4 strands 18 ga.

Inductors should be mechanically bound with nylon wire ties and stabilized by dipping them in wire varnish (such as Glyptol).

In making physical placement decisions about inductors on a crossover circuit board, place them as far apart as possible and mount them at 90° angles to each other (see *Fig. 7.159*). If you maintain at least 3″ of separation and right-angle mounting, you will avoid magnetic coupling between different sections of the crossover.[28]

Since really large high or even medium Q air core inductors with acceptably low DCRs are not always possible due to limitations of physical size, cored inductors become more desirable and really the only practical solution. While air core inductors do not saturate and have very low distortion, virtually all cored type inductors do saturate and do produce higher levels of distortion and are to some degree nonlinear with respect to voltage input[27,29,30,31]. However, there is considerable variation among different types of cores. Of the commonly available core types the following is a relative ranking of their linearity, distortion, and Q:

MPP toroids
Iron bobbin (many variables in this group—some are quite good, others not)
Laminate core
Ferrite core (slugs and bobbins)

Of this group MPP toroids are the most linear with voltage changes, also the most costly and usually only available custom made. Iron bobbins are next in terms of linearity and distortion, but there are a number of different compounds used for core materials with substantially varied results. High Q laminate core types are not the best, but overall these types work well and are widely used in manufactured loudspeakers. Typical formats used by manufacturers utilize lower Q coil formats that have lengths that are several times greater than the winding height in order to make a coil that lays low on a crossover circuit board. These inductors can be greatly improved using optimal Q coil windings (equal depth and winding height) and laminate bars that extend beyond the winding at least the height of the core. I built this type of laminate bar inductor into a SRA (Speaker Research Associates) speaker I exhibited at CES in 1978. For this speaker, a three-way using a 10″ woofer with a 100Hz passive crossover that used 16-gauge wire for the coil and transformer "Is" for the core. This crossover sounded extremely good, and compared well with the same system bi-

TABLE 7.1
CONTOUR EXAMPLE

RΩ	A_t
25	10.6dB
20	10.0dB
15	9.0dB
10	7.5dB
5	4.8dB

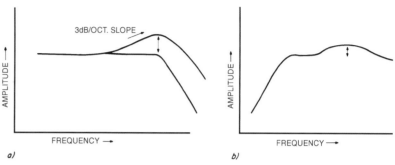

FIGURE 7.156: Frequency response (a) rising with increasing frequency and (b) rising with decreasing frequency.

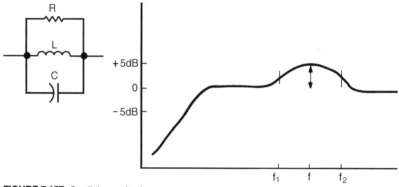

FIGURE 7.157: Parallel trap circuit.

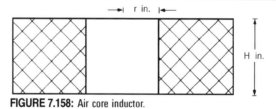

FIGURE 7.158: Air core inductor.

amped. I have also observed the use of these format laminate core inductors (max Q coil winding) in manufacturer products, such as the PSB Mini Stratus I reviewed in *SB* 3/93. The last group, ferrite bobbins and slugs, generally have much higher distortion and tend to be less graceful when they saturate. While I would never suggest using them to one of my manufacturing customers in my consulting business, they were for years a very popular inductor and used in numerous British loudspeakers (the Rogers LS35A used ferrite slug inductors).

7.85 CAPACITORS IN CROSSOVER NETWORKS.

Choice of capacitor type is usually determined by consideration of loss factor and costs. This generally means using plastic film capacitors (such as Mylar and polypropylene) for values of 20μF or less in midrange and tweeter sections, and nonpolar electrolytics in most other applications. This is not a hard-and-fast rule, as certain high-quality, nonpolar capacitors are quite adequate for most applications. This does not always apply, however, since the majority of readily available nonpolar capacitors intended for loudspeaker design are only moderate-quality Asian products. One way to improve the "sound" quality of these cheap, nonpolar capacitors is to bypass them with small-value (0.1–1μF) polypropylene types. Another technique is to parallel multiple smaller values rather than use a single large value of capacitance. This way, the resistive and inductive components are also paralleled together, and the net resistive and inductive effects of such a component will be smaller than would be the case with one large capacitor.

Although nonpolar electrolytic capacitors tend to be less stable with age than solid film capacitors,[32] it is important to note that capacitors (and inductors) should be measured at the frequency they will be functioning. If an analyzer such as the LinearX LMS or an impedance bridge with selectable oscillator frequencies (as opposed to the standard fixed 1kHz type bridge) is available, crossover components should be measured at the specific high-pass or low-pass network corner frequencies (or discrete operating frequency of notch filters) and at one octave into the network stopband.

7.86 DRIVER BANDWIDTH AND CROSSOVER FREQUENCY.

Crossover frequencies are primarily determined by the usable bandwidth of the drivers you intend to use. For midrange and tweeters, the lower limit is set by the resonance. Consequently, do not cross either type of driver any lower than a minimum of one octave above resonance.[6] Two octaves, when practical, is preferable. This is to avoid phase disturbance in the stopband of the driver high-pass filter. If you are not comfortable thinking in octaves, the relationship is simple: twice the frequency is one octave higher, and half the frequency is one octave lower. Therefore, a 1kHz resonance tweeter should not be crossed any lower than 2kHz; 3–4kHz is ideal.

The upper crossover limit for midrange drivers (and for woofers) is determined by the driver horizontal polar response. As frequency increases, and the wavelength of sound becomes the same size or smaller than the diameter of the driver, the radiation pattern narrows. Two generally accepted criteria will help you determine the upper frequency limit to on-axis beaming. These criteria relate to the amount of allowable attenuation, or beaming, at the ±45° off-axis listening points. The most commonly used acceptance figure is up to -6dB attenuation at 45° off-axis. A more stringent criterion, if your driver bandwidth is sufficiently wide to allow it, is to set the upper frequency limit to -3dB of attenuation at the 45° point (*Fig. 7.160*). *Table 7.4* depicts the crossover frequency versus diameter information

TABLE 7.2
GAUGE SIZE OF COPPER WIRE

Wire Dia.	Gauge	Ω/1k′
0.10190	10	0.9989
0.09074	11	1.260
0.08081	12	1.588
0.07196	13	2.003
0.06408	14	2.525
0.05707	15	3.184
0.05082	16	4.016
0.04526	17	5.064
0.04030	18	6.385
0.03589	19	8.051
0.03196	20	10.150
0.02845	21	12.800
0.02535	22	16.140
0.02257	23	20.360
0.02100	24	25.670
0.01790	25	32.370
0.01594	26	40.810

TABLE 7.3
EQUIVALENT GAUGES FOR MULTIPLE WINDINGS

# of Wires	Gauge	Equivalent Gauge
2	18	15
3	18	13
4	18	12
2	16	13
3	16	11
4	16	10

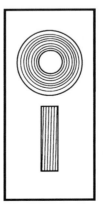

FIGURE 7.159: Proper placement of inductors.

TABLE 7.4
HORIZONTAL POLAR RESPONSE CRITERIA FOR DETERMINING THE UPPER LIMIT FOR LOW-PASS CROSSOVER FREQUENCIES.

Driver Dia.″	Frequency	
	−3dB/Hz	−6dB/Hz
15	661	1043Hz
12	912	1427Hz
10	1065	1674Hz
8	1302	2055Hz
7	1540	2421Hz
5	2051	3229Hz
4	2687	4238Hz

for low-pass filters used in conjunction with midrange and woofers.

The goal of determining the upper limit for a high-pass crossover is to produce as flat a response as possible off-axis, which in turn yields a more flat power response. While the -3/-6dB criterion is the general wisdom for making this choice of what frequency to use for a woofer or midrange low-pass filter, a criteria that I have used in my own design work and one I have frequently noted in the driver reviews that I have been doing in *Voice Coil* magazine over the years does not require looking at a polar plot. If you measure the on-axis response and the 30° off-axis response of a woofer and midrange, then a good choice for the upper limit of a low-pass crosspoint is –3dB at 30° with respect to the on-axis curve. If you look at the 0° and 30° curves in *Fig. 7.161*, -3dB at 30° off-axis with respect to the on-axis curve occurs at 3.7kHz for this 5.25″ diameter woofer. This would be considered the highest frequency you would consider for a low-pass filter for this driver, and lower is always better provided you do not induce modulation distortion (excessive excursion) in the low-pass driver you cross over to.

7.87 TWO-WAY VS. THREE-WAY LOUDSPEAKER FORMATS

For sound quality, two-way designs have no inherent superiority over three-way (or more) designs, and vice versa. History records successful examples of both formats. It is difficult, however, to resist suggesting "the simpler the design, the better." Unless you are a loudspeaker system engineering professional, fewer cross points and drivers tend to result in fewer problems and design errors.

Two-way designs are relatively simple, while three-way designs tend to separate the pros from the amateurs. The major advantage of any three-way design is greater power-handling because there are more transducers sharing the frequency spectrum. A well-executed three-way design with crossover frequencies in the range of 250–300Hz from the woofer to the midrange and 2.5–3.0kHz from the midrange to the tweeter is an outstanding format and hard to beat.

7.88 BOUNDARY RESPONSE AND CROSSOVER FREQUENCY.

If you mount the drivers close to your listening room boundaries (walls and ceiling), a substantial dip in loudspeaker power response will take place. This occurs because of the interference pattern caused by boundary reflections. Depending on the amount of distance-caused delay due to the driver location, boundary-induced interference patterns will cause peaks or dips in the loudspeaker power response. This is only a problem in the lower-frequency ranges where the distance of the driver to the wall or floor is less than 0.75l.[33] For midranges and woofers mounted 0.4–0.6 meters (16″–24″) from any room surface, the dip in frequency response will occur between 120–160Hz.[34] The magnitude

of this depression can be from 3–10dB in the overall power response, depending on your room, fixtures, exact location, and so on. In terms of loudspeaker response to a short dura-

tion pulse (i.e., 10 microseconds), floor reflections occur approximately 2mS after arrival of the initial wave front.[35]

Several solutions, or partial solutions, can resolve this problem for you. One, suggested by Colloms, is to keep midrange crossover frequencies in the 200–300Hz region. This places the dip within the midrange stopband. If you happen to be using a mid-woofer with a low resonance (around 75Hz in a small cavity), it will place the crosspoint at the desirable spot of at least two octaves above the driver resonance frequency. It is presumed you can actively equalize the dip in woofer response.

A second criterion, suggested by Allison,[2] and applied over the years by several companies (including Acoustic Research and Allison Acoustics), is to place the woofer at the intersec-

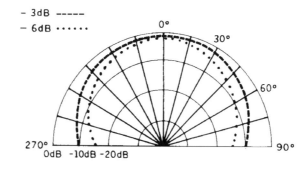

Figure 7.160: Horizontal polar plot showing bandwidth versus directivity.

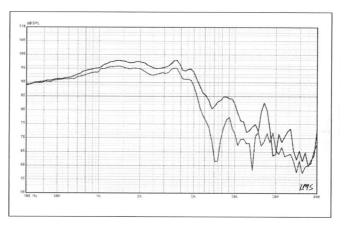

Figure 7.161: Using on- and off-axis response to determine a possible crossover frequency.

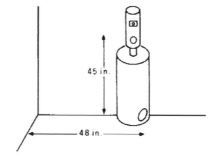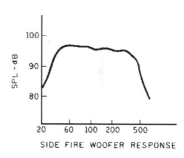

Figure 7.162: Boundary intersection woofer response.

tion of two room boundaries (the back wall and the floor). Then locate the midrange and high-frequency drivers at a distance from the boundary intersection that is 0.75 of the wavelength at the 300Hz crossover frequency, or 45″. The side "fired" woofer will exhibit a near-flat power response and will avoid the 120–160Hz response dip, while the midrange will be safely out of range of this effect (*Fig. 7.162*).

Figure 7.163: An example of active equalization.

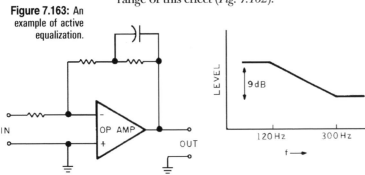

The third remedy for room boundary effect and diffraction problems is active equalization. This can take the form of adjustable frequency equalizers or simple integrated circuit filters, such as the one illustrated in *Fig. 7.163*. These represent a sort of "shotgun" approach, but they generally provide satisfactory results. Good examples of the application of this type of circuit can be found in:

1. S. Linkwitz, "A Three-Enclosure Loudspeaker System," *Speaker Builder*, 4/80.

2. J. D'Appolito, "A High-Power Satellite Speaker," *Speaker Builder*, 4/84.

7.89 ADVANCED PASSIVE NETWORK DESIGN TOPICS.

A more detailed examination of advanced passive network design techniques is beyond the scope of this book, but can be found in its companion volume, *Loudspeaker Recipes, Book I: Four Two-Way Systems*. *Loudspeaker Recipes* is a crossover design tutorial that uses four example two-way loudspeakers to explicate various aspects of network design including such concepts as compensating for interdriver time delay with network manipulation and the importance of in-phase network design. This book is available from Old Colony Sound Laboratory.

7.90 LOUDSPEAKER VOICING

"Voicing" is a term generally used in music. For piano technicians, voicing a piano means adjusting the hardness or softness of the hammers after you have tuned the piano. That hardness or softness changes the timbre of the notes and gives a different "voice" to the sound of the piano. Even when the piano tuning is perfect, you can still give it a different "sound" quality. An extreme example of piano voicing is tack piano, an old musicians' trick in which you put thumbtacks on all the hammers of an upright piano, giving the piano a tinny or chiming quali-

ty while still maintaining the same tuning.

The other musical use of the term voicing is in chord construction. Chord voicing generally refers to the harmonic order of notes in a chord. A simple type of chord voicing is a concept known as chord inversions. For example, a three-note triad C chord on a piano has the notes C, E, and G. When you play these three notes with the C note on the left, C-E-G, this is called the "root" position. By changing the order of the notes but still playing the same three notes, you get a slightly different sound, and these are called inversions. A 2nd inversion of the C chord is E-G-C and a 3rd inversion is G-C-E.

So how does the term "voicing" relate to loudspeaker design? First, because loudspeakers are supposed to just reproduce music and not add anything to the original recording, there should be no such term as voicing applied to a loudspeaker. However, while loudspeakers are indeed reproduction devices, they are sadly very imperfect ones and no matter what, *all* loudspeakers add some coloration to the original event and make things even more complicated. The room the loudspeaker is placed in also adds its own coloration to the original event!

Whether you are a loudspeaker manufacturer or a hobbyist building his own "dream" speaker design, the ultimate goal is for your speaker to sound "musical," which is another way of saying you want your speaker to sound as much as possible like the original acoustic event. This is obviously a universal goal for speaker design and "voicing" is a commonly used term among manufacturers in the loudspeaker industry used to achieve that goal. Ultimately it all comes down to some sort of final adjustment or tuning process that renders the design finished and ready for listening.

While the *Loudspeaker Design Cookbook* is a précis of the objective technology you can use to build a great-sounding loudspeaker, voicing is the act of applying a final subjective judgment that may indeed override this applied technology. Knowing how to manipulate both the objective and the subjective aspects of a design is indeed the "art" and "Zen" of loudspeaker engineering. This section is offered as a guideline for this process and will discuss some of the techniques you can use to be successful at voicing your own speakers and also offer some specific detail on what to adjust and what to leave alone. Please note that these comments primarily apply to high-order parallel crossover networks, but other than specific component part recommendations, can also be applied to other formats such as low-order networks and series networks.

Before you set up to voice your new design, the following set of assumptions is useful:

1. The SPL of the speaker should have minimal variation and be as "flat" as possible. While being "flat" does not guarantee a successful great-sounding speaker (there are countless examples of ruler flat designs that did not sound

good or sell well), it really is the best place to start. You should always strive to achieve a measured SPL that is at least ±2dB or less on-axis. In addition, the off-axis response at 30° should be as close to the same ±2dB as the on-axis response (see *Fig. 7.164* for a good example of this type of on- and off-axis SPL for a 3-way design).

2. Along with a ±2dB response, another criterion I apply is to minimize the "Q" of all the crossover filter transfer functions. You can read more about this subject in *Loudspeaker Recipes 1*, available from www.audioxpress.com.

3. Whenever possible, keep the drivers in-phase in the octave above and below the crossover frequency. With a 2-way design, this means the tweeter normally is connected in the same polarity as the woofer and when the tweeter polarity is reversed, the system summation reveals a deep 10dB or more null centered at the crossover frequency (*Fig. 7.165*). For a three-way design, the midrange and tweeter should all be connected with the same polarity as the woofer such that when the polarity of the midrange is reversed, the system summation reveals a deep 10dB or more null centered on both crossover frequencies (*Fig. 7.166*). This subject is also covered in great detail in *Loudspeaker Recipes 1*.

A. A/B Comparative Listening: References and Rooms

This is the point in the discussion where the warning label appears. Subjective adjustment of any design is difficult at best and loudspeaker systems have been probably made to sound worse with the "golden ears" approach just as often as they have been improved. While there have been many instances of manufacturers who did not rely upon objective measurement at all and been successful, there have also been just as many whose loudspeakers were not successful for the same reasons.

The two factors, besides just bad judgment and poor hearing ability, that contribute to failure with the subjective listening process have to do with microphones and rooms. If you have listened to a lot of different loudspeakers, it's obvious that there is much variability in sound quality, and the same thing is true with microphones used in recording studios. Your favorite choice of program material, especially if you habitually listen to the same recording (or a series of recordings by the same mix engineer), has been recorded with a microphone that has a particular timbre or sound quality. So the caution here is to beware of using just a single recording as your sonic yardstick.

The next issue to be aware of is the problem with rooms. While loudspeakers may all sound somewhat different, their variability is much greater when placed in different locations within the same room. A loudspeaker's sound quality changes more in different room positions and has greater sonic differences than are found from one brand of loudspeaker to another brand[36,37,38] listened to in the exact same posi-

tion! While the prospect of finding a neutral-sounding room is difficult to unlikely, you tend to just make do with what you have. Taking the room that you have available and applying all the possible acoustic treatments to remedy its mode and reflection problems is not only expensive, but requires some very specific knowledge, and if you have the time and money for it, it is certainly worthwhile.

But without going to that extreme, the best advice is to use the largest room available for the project. The smaller the room, the more difficulty there is trying to minimize its influence. If you have a choice between a larger room with higher ceilings or a much smaller room with lower ceilings, the large room is the one you want to use, all other things being equal. Once you choose a room to work with, move a speaker around the room to different locations and both listen to it and if possible use a real-time analyzer and check its response (the TrueAudio TrueRTA listed in *Chapter Nine* is an inexpensive computer software analyzer that works extremely well for this). My preference is to place the speaker out into the room aimed diagonally across the room so that the lateral reflections off the walls are at a 45° angle instead of 90°.

The majority of professionals in the loudspeak-

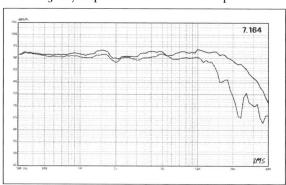

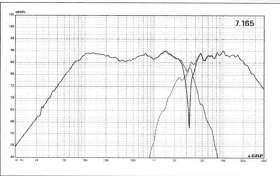

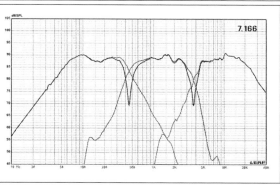

Figure 7.164: Example of acceptable on- and off-axis frequency response.

Figure 7.165: Null response with tweeter polarity reversed for a 2-way design.

Figure 7.166: Null response with midrange polarity reversed for a 3-way design.

er industry voice their products in mono and not in stereo. Trying to minimize room influence in stereo produces twice the complexity. If you are specifically working with a 2-channel system and are trying to tweak the stereo image, I still recommend doing most of the voicing in mono and then try to make small changes in stereo mode to affect the image quality. Because home theater speakers don't rely on a phantom image, it is not really as critical a consideration.

After you have the room sorted out and are satisfied you have done your best to resolve a good listening location, the next most important thing is some kind of reference. Voicing blind without using a suitable reference loudspeaker is difficult if not impossible. It is important to choose a reference speaker that you are familiar with that is close to the same category speaker you are designing. This means if you are trying to finalize a 2-way speaker with a 6.5″ poly cone woofer and a 1″ soft dome, your best reference will be a speaker that you know to sound good, or at least sounds good to you (beauty is both in the eye and the ear of the beholder!), that also has a 6.5″ woofer with a poly cone and a 1″ soft dome. This way the timbre of the two speakers will at least be similar. Trying to voice a poly cone woofer and soft dome against a magnesium cone woofer and an aluminum dome is going to be difficult as the two different loudspeakers will always have a perceptual difference that cannot be accounted for with crossover design changes.

Once you accept the need for a reference, the situation becomes much more complex because now you must switch back and forth between two sound sources located in the room. The very best you can do is to place the two speakers at the same height and as close as possible, which means literally right up against each other. Even when almost in the same acoustic space in your room, there can still be mode differentials to affect the sound quality. To check for this, A/B carefully listening to specific instruments, then swap the two speaker positions and put the left speaker on the right and the right speaker on the left. If the sound changes in this comparison, then relocate the pair until you find a place where switching the positions doesn't seem to make any difference. An alternative to this is to

build a test fixture like the one described in *Chapter 6, section 6.50* titled Subjective Evaluation of Diffraction. This rapid A/B device is inexpensive to build and very effective, but should also be combined with side-by-side comparisons.

A final, but important issue is listening level. Voicing should be accomplished at levels between 90–100dB where your hearing is more linear. This increased linearity at higher SPL can be seen in the original Bell Labs 1933 Fletcher-Munson equal loudness contour curves shown in *Fig. 7.167* (now considered more accurate and a better representation of hearing than the later Robinson-Dadson study done in 1953). Having a voicing partner who has a good set of ears and perhaps a final panel of educated listeners to preview your design is also invaluable to the voicing process.

B. Network Manipulation for Voicing
Voicing is about making changes in a crossover network that you produced using objective measurement techniques and executed by either manual iteration or computer optimization. If you have done your job well, the speaker will have appropriate SPL profile and system impedance that will not cause any amplifier issues by dropping below 3.2Ω or present too high a reactive load. Knowing what elements in the network will affect the sonic changes you desire to make and how to avoid producing problems at the same time is indeed the trick. Doing this well is a matter of experience and while not really difficult, it helps if you have done it more than once!

What follows are some guidelines for voicing 2- and 3-way speakers. While your choices of which crossover parts to change are not limited to the ones given, experimenting beyond these limits can produce problems that need to be resolved, so it is important to repeat SPL and impedance measurements following your voicing sessions.

2-Way Loudspeaker Voicing—a typical network design for a 2-way speaker is illustrated in *Fig. 7.168*. Voicing a 2-way design is mostly a tweeter issue. If you have designed a speaker that measures within a ±2dB window, with the network in phase and the Q of the network transfer functions as shallow as possible, about all that is left is adjusting the spectral balance of the speaker and dialing in the tweeter's presentation. Assuming you have chosen a crossover frequency that produces the kind of on- and off-axis response shown in *Fig. 7.164*, it's not likely that you would need to be adjusting the blend between the woofer and the tweeter.

The biggest subjective issue usually deals with tweeter level. If the tweeter level is too high, the bottom end of the speaker can sound weak and the top end too prominent. If the tweeter level is too low, high harmonics will sound recessed and the speaker can lack a "live" sound quality. Adjusting the attenuation resistor (component 1 in the schematic) can be done in 0.5–1Ω increments above and below where you set it in the initial design.

Figure 7.167:
Fletcher-Munson
Equal Loudness
Contour graph.

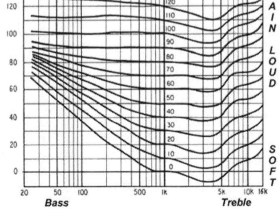

Once you have determined a new tweeter attenuation value (don't be surprised if your original resistance choice was either close or required additional resistance) and you still think the tweeter is overpowering the speaker (i.e., if the high hats and symbols image are in front of the vocalist visually), then the top end of the tweeter is probably too "hot." You may then want to experiment using a CR (capacitance/resistance) conjugate circuit shown in the diagram with the capacitor labeled as component 2. Start with low values of C = 0.5–1μF and increase it in 0.5μF increments until the tweeter sounds more "musical" in the overall presentation. Once you have "salted" the new speaker to your musical taste, go back and measure the on- and off-axis frequency response and the system impedance again to confirm what was done.

While you can adjust the other woofer and tweeter crossover components as well, start with this routine while comparing your design to a reference. If you can't resolve the sound you are looking for, then experiment with small changes in the woofer inductor L1 and tweeter capacitor C1 to see whether you can make further improvements, following up with both on- and off-axis SPL and impedance measurements to confirm the changes.

3-Way Loudspeaker Voicing—a typical network design for a 3-way speaker is given in *Fig. 7.169*. Like a 2-way design, voicing a 3-way loudspeaker is mostly a tweeter issue but is further complicated by the midrange transducer. Again, if you have designed a speaker that measures within a ±2dB or less window, with the network in phase and low-Q network transfer functions plus good off-axis performance, all that remains is adjusting the spectral balance and finalizing the tweeter presentation. The biggest subjective task is adjusting the midrange and tweeter levels so they work well together. Your goal is to obtain levels with a sonic presentation such that you can set up level switching that becomes brighter if the levels are increased, and more recessed if the levels are decreased.

Once you have determined final values for resistor components 1 and 3 (this is obviously somewhat more complicated if you choose to use L-pads), and the speaker still seems too "bright," try experiments with the same CR tweeter circuit shown for 2-way designs. As before, start with low values for component 2 with C = 0.5–1μF and increase it in 0.5μF increments until the tweeter sounds more "musical" in the overall presentation. If you can't resolve the sound you are looking for by level adjustments for the midrange and tweeter circuits or adding a tweeter conjugate circuit, then experimenting with components in the tweeter and midrange circuit is possible.

Adjusting the C1 capacitor in the tweeter circuit can increase or decrease the emphasis the tweeter is presenting, but be careful changing components in a bandpass midrange network.

This type of filter tends to have very high component sensitivity (small changes in component values can make large SPL and impedance changes). Generally, if you are using a similar circuit, you can make small changes in L1 and C3, but again, be certain to go back and make on- and off-axis SPL and impedance measurements to verify what changes were made.

A final, but important issue is listening level. Voicing should be accomplished at levels between 90-100dB where your hearing is more linear. This increased linearity at higher SPL can be seen in the original Bell Labs 1933 Fletcher-

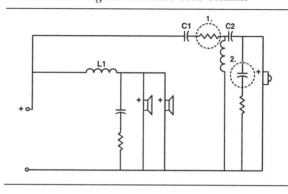

Figure 7.168: Example of typical 2-way crossover design.

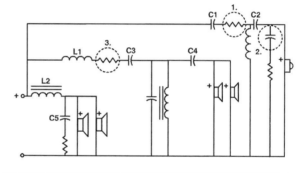

Figure 7.169: Example of typical 3-way crossover design.

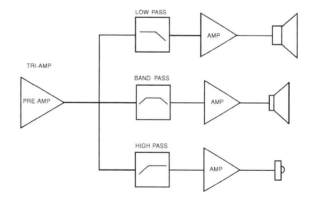

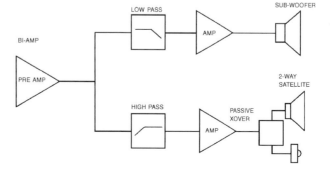

FIGURE 7.170: Low-level active networks.

Munson equal loudness contour curves shown in *Fig. 7.167* (now considered more accurate and a better representation of hearing than the later Robinson-Dadson study done in 1953). Not only is the absolute SPL important for your listening testing, but it is also extremely important that the SPL of the reference speaker and the SPL of the speaker being voiced be very closely matched. This can easily be achieved using a pink noise generator and an SPL meter. While there are a number of switching methods that can be used for this purpose, a good tool for this type of A/B listening is to use an integrated stereo amplifier that has separate channel volume controls and A/B speaker switching. Run the output of one channel of a CD player (or preamplifier output) into the single female RCA input side of a Y cord adapter and plug the two male RCA outputs into the left and right channel inputs of the amplifier. Hook the output of the left channel A speaker system to one speaker and the right channel B speaker system to the other speaker. Doing this will allow you to set the levels for each speaker independently and you can then use the A/B speaker switch literally as your A/B comparison switch. Last, it is absolutely essential to have a voicing partner who also has a good set of ears. This is the second reference in the process. Remember a successful loudspeaker is one that has agreement among a number of listeners, so as a final step in speaker development, using a panel of educated listeners to preview your design is also an outstanding tool.

7.100 ACTIVE CROSSOVER NETWORKS.

Low-level crossover networks are composed of active electronic high- and low-pass filter sections, used between the preamplifier and amplifier of a bi- or tri-amp sound system (as illustrated in *Fig. 7.170*). Bi- and tri-amplification have been used for many years in high output stage systems for live musical performance. Because an active crossover network is more complex and expensive than single amplifier/passive crossover systems, it has never been popular in commercial home stereos. Its attributes and benefits, however, make its use extremely attractive. The following are its major advantages:

(1). Lower IM distortion[18] due to amplifier operation over a more narrow bandwidth. Also, clipping caused by low frequency overload is reduced, being limited to only one driver within a multi-driver system.

(2). Increased dynamic range. One 60W and one 30W amplifier in a bi-amp setup will clip at about the same levels as one 175W amplifier with a passive crossover.[39]

(3). Improved transient performance.[40]

(4). Better amplifier/speaker coupling for woofers, and avoidance of passive crossover-induced tweeter resonances.

(5). Better crossover performance working into a constant impedance load.

(6). Better subjective sound quality than high-level networks.[41]

(7). Easier control over driver sensitivity differences.

(8). Easier manipulation of phase, time delay, resonance, and various kinds of shaping, contouring, and equalization.[3]

Two formats, as suggested by *Fig. 7.170*, are usable in a home music system. The bi-amp format usually consists of a single subwoofer enclosure and separate satellite speakers with a 100–300Hz 18–24dB/octave active crossover. The satellite speakers take the form of a small two-way speaker with a 2–3kHz passive network. (Incidentally, the single subwoofer system, in terms of image quality, is inferior to the two-channel subwoofer and was originally a costly engineering innovation.) The tri-amp format is basically just an extension of the bi-amp system with the passive network removed, replaced by an additional electronic network and amplifier. The format provides the maximum flexibility in terms of tailoring driver phase and frequency response.

Since it is beyond the scope of this book to discuss the construction details of active filter circuits, the following are offered as highly recommended resources for active filter design:

(1). S. Linkwitz, "A Three Enclosure Loudspeaker System," *Speaker Builder*, 2, 3, 4/80.

(2). R. Ballard, "An Active Crossover Filter with Phase Correctors," *Speaker Builder*, 3, 4/82.

(3). Ed Dell, "Electronic Crossovers Revisited," *Speaker Builder*, 3/82 and 3/85.

(4). R. Bullock, "Passive Crossover Networks— Active Realizations of Two-Way Designs," *Speaker Builder*, 3/85.

(5). W. G. Jung, various articles, *Audio Amateur*, 1, 2, 4/75, and 2/76.

(6). P. Hillman, "Symmetrical Speaker System with Dual Transmission Lines, Part II," *Speaker Builder*, 6/89.

(7). R. Parker, "A Tri-amplified Modular System," *Speaker Builder*, 5/90.

Most references above are supported by circuit boards and parts available from Old Colony Sound Lab, PO Box 876, Peterborough, NH 03458.

REFERENCES

1. C. P. Boegli, "Interference Effects with Crossover Networks," *Audio*, November 1956.

2. J. Ashley and L. Henne, "Operational Amplifier Implementation of Ideal Electronic Crossover Networks," *JAES*, January 1971.

3. S. Linkwitz, "Active Crossover Networks for Noncoincident Drivers," *JAES*, Jan/Feb 1976; "Passive Crossover Networks for Noncoincident Drivers," *JAES*, March 1978.

4. P. Garde, "All-Pass Crossover Systems," *JAES*, September 1980.

5. R. Gonzalez and J. D'Appolito, "Electroacoustic Model," *Speaker Builder*, 4/89.

6. W. M. Leach, "Loudspeaker Driver Phase Response: The Neglected Factor in Crossover Design," *JAES*, June 1980.

7. R. M. Bullock, III, "Loudspeaker Crossover Systems: An Optimal Crossover Choice," *JAES*, July/August 1982.

8. R. M. Bullock, III, "Passive Three-Way All-Pass Crossover Networks," *JAES*, September 1984.

9. W. F. Harms, "Series or Parallel?," *Hi-Fi News & Record Review*, December 1980.

10. J. Vanderkooy and S. Lipshitz, "Is Phase Linearization of Loudspeaker Crossover Network Possible by Time Offset and Equalization?," *JAES*, December 1984.

11. R. C. Heyser, "Loudspeaker Phase Characteristics and Time Delay Distortion: Part I," *JAES*, January 1969.

12. S. Linkwitz, "Shaped Tone-Burst Testing," *JAES*, April 1980.

13. L. Fincham, "Multiple Driver Loudspeakers," Chapter 4, *Loudspeaker and Headphone Handbook*, edited by John Borwick, published by Butterworth & Co.

14. D. Fink, "Time Offset and Crossover Design," *JAES*, September 1980.

15. R. M. Bullock, "Satisfying Loudspeaker Crossover Constraints with Conventional Networks—Old and New Designs," *JAES*, July/August 1983.

16. J. Vanderkooy and S. Lipshitz, "Power Response of Loudspeakers with Noncoincident Drivers—The Influence of Crossover Design," *JAES*, April 1986.

17. J. Backman, "Design Criteria for Smooth Energy Response," presented at the 90th AES Convention, February 1991, preprint no. 3047.

18. R. Small, "Crossover Networks and Modulation Distortion," *JAES*, January 1971.

19. J. D'Appolito, letter reply to "Going Off the Deep End," *Speaker Builder*, 3/90, p. 90.

20. F. E. Toole, "Loudspeaker Measurements and Their Relationship to Listener Preferences: Parts I and II," *JAES*, April, May 1986.

21. Swanson and Tichy, "Antiresonant Compensation of Tweeter Frequency Response Peaks," presented at the 79th AES Convention, October 1985, preprint no. 2235.

22. E. Zaustinsky, "Measuring and Equalizing Dynamic Driver Complex Impedance," presented at the 79th AES Convention, October 1985, preprint no. 2235.

23. S. Lipshitz and J. Vanderkooy, "A Family of Linear-Phase Crossover Networks of High Slope Derived by Time Delay," *JAES*, Jan/Feb 1983.

24. M. Knittel, "Microcomputer-Aided Driver Attenuation," *Speaker Builder*, 1/85.

25. A. Thiele, "Air Cored Inductors for Audio," *JAES*, June 1976.

26. A. Thiele, "Air Cored Inductors for Audio—A Postscript," *JAES*, December 1976.

27. R. Honeycutt, "Components for Passive Crossovers, Part 2," *Voice Coil*, December 1996.

28. M. Sanfilipo, "Inductor Coil Crosstalk," *Speaker Builder*, 7/94, p. 14.

29. G. R. Koonce, "A Technique to Measure Inductance," *Speaker Builder*, 6/89.

30. J. Fourdraine, "Choosing the Best Filter Coils," *Speaker Builder*, 4/96.

31. R. Russell, "Quality Issues in Iron-Core Coils," *Speaker Builder*, 6/96.

32. R. Honeycutt, "Caps for Passive Crossovers," *Speaker Builder*, 3/92, p. 34.

33. R.F. Allison, "The Influence of Room Boundaries on Loudspeaker Power Output," *JAES*, June 1974.

34. M. Colloms, *High Performance Loudspeakers*, 1978, 1985, Pentech Press.

35. Kantor and Koster, "A Psychoacoustically Optimized Loudspeaker, *JAES*, December 1986.

36. Olive, Schuck, Sally and Bonneville, "The Effects of Loudspeaker Placement on Listener Preference Ratings," *JAES*, September 1994.

37. Toole and Olive, "Hearing is Believing vs. Believing is Hearing: Blind vs. Sighted Listening Tests, and Other Interesting Things," presented at the 97th AES Convention, November 1994, preprint no. 3894.

38. Olive, Schuck, Sally and Bonneville, "The Variability of Loudspeaker Sound Quality Among Four Domestic-Sized Rooms," presented at the 99th AES Convention, October 1995, preprint no. 4092.

39. Lovda and Muchow, "Bi-Amplification—Power vs. Program Material," *Audio*, September 1975.

40. A. P. Smith, "Electronic Crossover Networks and Their Contribution to Improved Loudspeaker Transient Response," *JAES*, September 1971.

41. D. C. Read, "Using a Single Bass Speaker in a Stereo System," *Wireless World*, November 1974.

CHAPTER EIGHT

LOUDSPEAKER TESTING

8.00 INTRODUCTION.

This chapter describes a variety of test procedures which will supply you with the needed parameters to complete the calculations described in the body of the text. Most of these test procedures are relatively simple and require only the use of a good scientific calculator, a signal generator, a frequency counter (to ensure accuracy), a couple of AC voltmeters and a little patience. Given the fact that the average audio technician could easily use these directions and, for a fee, supply you with the appropriate numbers, all you really need is a scientific calculator and maybe a "C" in high school algebra.

To be honest, doing the job right can take a lot of time and patience. I spent 11 months, full-time, developing a single loudspeaker just in time for the winter Consumer Electronics Show. And I had help. So I don't want to make it seem too easy to design a really good loudspeaker. As I have commented previously, loudspeaker design is still largely an art. Engineering will guide the way, but it is no substitute for good taste, good sense, and musical sensitivity. If engineering were all it took, there would be no bad or even mediocre loudspeakers to think about.

8.10 DEFINITION OF TERMS.

BL driver motor strength given in tesla meters
f_s driver free air resonance
f_{sa} driver resonance with mass M_a attached
f_{ct} resonance of driver in box (test box)
M_{md} mass or weight of speaker cone assembly
M_{mr} radiation (air) mass load of the cone
M_{ms} total cone assembly mass including radiation mass
M_a test mass (usually a measured amount of clay)
C_{mb} compliance of box (test box)
C_{ms} driver mechanical compliance
Q_{es} driver electrical Q
Q_{ms} driver mechanical Q
Q_{ts} driver total Q
Q_{ect} Q_{es} of driver in test box
R_c substitute resistor equal to R_{evc}
R_{es} R at f_s less the R_{evc}

R_{evc} DC resistance of the voice coil
R_x substitute resistor (variable)
I_e current through R_{evc}
I_c current through R_c
V_{as} volume of air equal to the driver compliance
V_{ab} volume of unlined, air tight box (test box)
L_{evc} driver voice coil inductance

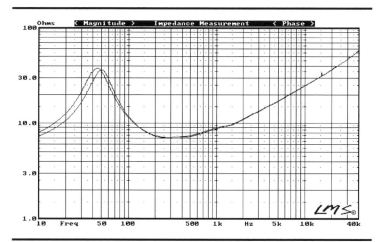

FIGURE 8.1: Woofer resonance before and after break-in.

8.20 BREAK-IN.

Prior to testing, all cone speakers should be broken in. However, the reason for doing this is not as obvious as you might imagine. While the majority of woofers will undergo a "loosening" of the suspension system after five to ten hours of play, this has very little effect on the Thiele/Small parameters used for developing box volumes. *Figure 8.1* shows a free-air impedance measurement comparison of a 6.5" Peerless woofer right out of the box to the same woofer after 12 hours of break-in using a sine wave generator (at 25Hz) and amplifier. Importing this data (along with the delta compliance curves also made before and after break-in) into the LinearX Leap Software yielded the parameter summary given in *Table 8.1*.

At first glance, it appears that there has been a substantial shift in parameters with at least an 11% decrease in the driver's resonance frequency. However, when these parameters are used to create box simulations the answer is obvious. *Figure 8.2* shows the comparison of the before and after parameter sets used to create both a sealed and a vented box computer simulation. Differences in these box simulations are shown in *Table 8.2*.

As can be seen, the changes in box performance are trivial. The reason this occurs is that

TABLE 8.1

	BEFORE BREAK-IN	AFTER BREAK-IN
F_0	49.9Hz	44.5Hz
Q_{ms}	2.11	1.97
Q_{es}	0.44	0.39
Q_{ts}	0.37	0.33
V_{as}	16.8 ltr	21.6 ltr

the F_s/Q_{ts} ratios remain constant before and after break-in. In the case of this woofer, F_s/Q_{ts} before break-in was 136.79 and after break-in

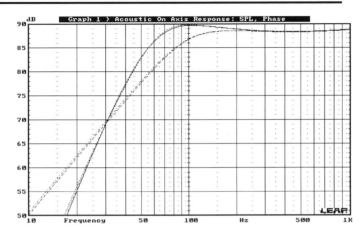

FIGURE 8.2: Computer Simulation comparison before and after break-in.

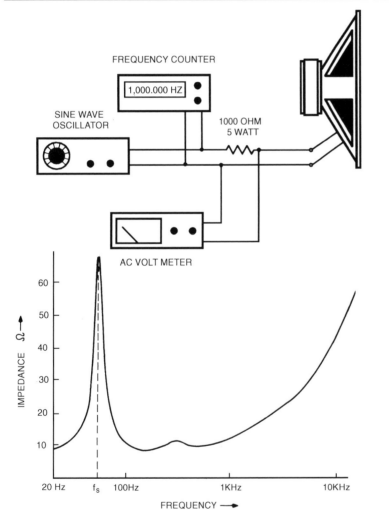

FIGURE 8.3: Impedance measurement—voltage divider method.

136.72, nearly the same. This is true for all woofers. Many times different samples of the same model will "appear" to have entirely different T/S parameter sets, when in reality they will provide identical box performance. If you suspect that two samples are very different because the

parameter set is not identical or close in the various parameter values, check the F_s/Q_{ts} ratios or perform a computer box simulation in the same box volume with the different data sets. This will immediately tell you if the woofers really are the same or if something important has changed.

So why bother to break-in drivers prior to testing? To assure that a valid test sample is being used. If a bad voice coil rub or poorly glued surround or spider are going to be a problem, banging the driver around with a reasonable amount of voltage at 25Hz for 12 hours should reveal the flaw. There is no sense in proceeding to design a project if the woofer is not a representative sample.

8.30 MEASURING RESONANCE (f_s).

The voltage divider setup in *Fig. 8.3* or *8.4* can be used to determine the driver resonance frequency f_s. This measurement should be made at the lowest nominal voltage level at which your equipment will function. Less than 1V would be preferable. As the drive level increases, especially when measuring small diameter drivers, the resonance frequency also moves upward. The measurements being made are for use with small-signal mathematical models, so the measurement voltage should be "small."

Tradition dictates that the driver be suspended, by cable or chain, in the air at least three feet from nearby objects or boundaries. However, the cone motion produces an opposite reaction such that the frame will be moving during the measurement. You'll get slightly more accurate results by clamping the driver to a test jig. The jig can be two 1×2 lumber slats placed horizontally three feet or so off the ground, with the driver placed between the two boards and clamped in place.

Make the measurement by varying the generator frequency between 10Hz and 100Hz (for woofers) until you locate the maximum voltage. Since midranges and tweeters have higher resonances, the manual "sweep" range will change appropriately. The frequency at the voltage peak is the driver free-air resonance and should be measured to the nearest tenth of a cycle (Hz).

F_{sb} is the driver resonance accounting for the loading effects of an enclosure baffle. To measure f_{sb} use the same procedure, but mount the driver on a flat baffle the same size as the enclosure baffle.

Alternatively, you may use the current source voltage setup in *Fig. 8.5*. The same procedures are followed except that the resonance frequency is determined by locating the frequency of the minimum indicated voltage.

8.31 MEASURING IMPEDANCE.

Impedance is AC resistance. The test setups shown in *Figs. 8.3* and *8.4* are examples of the voltage divider (constant current) method for measuring impedance. Using one of these setups is the easiest method for generating reasonably accurate calibrated impedance curves. The impedance curves generated with this technique, however, are not a true impedance. A major consequence of using the voltage divider type of circuit is that the higher the speaker's impedance, the less absolutely accurate the measurement will be. The difference is not great, but it is something to be aware of. *Figure 8.6* shows the comparison of a true impedance and a non-true impedance of the same driver measured with the Audio Precision System 1 using the voltage divider method. With the System 1, using the voltage divider method amounts to using the sweep generator's 600Ω source impedance for the voltage dividing resistance. The dotted impedance curve is the System 1 data imported into the LinearX LEAP software without conversion, and the solid curve represents the same data converted to true impedance (using the LEAP impedance import conversion routine). The impedance difference is about 2.7Ω at the resonance frequency, and about 1Ω at frequencies greater than 5kHz. Both the LinearX LMS and Audio Precision System 1 and 2 analyzers will measure true impedance.

The measurement using the *Fig. 8.3/8.4* setup is made by varying the oscillator frequency between 10Hz and 20kHz and manually recording the voltage value at frequency intervals. Calibrating the setup will allow you to read the voltage as an ohms equivalent. Do this by replacing the driver with a ±1% 10Ω resistor and adjusting the generator level to read 1V on the voltmeter or DVM. The voltage reading will now represent the equivalent impedance value in ohms, where 1V = 10Ω, 0.8V = 8Ω, 2V = 20Ω, and so on. Keeping the generator setting the same, replace the calibration resistor with the driver and take enough readings to produce a smooth curve. Use semi-log graph paper to record the data (semi-log graphs have a logarithmic horizontal axis, and a linear vertical axis).

The same procedure can be used to tune vented and passive-radiator enclosures to f_b. Because the flat shape of the impedance in the vicinity of the f_b frequency sometimes produces only a small voltage change and is often difficult to locate, the setup in *Fig. 8.4* can be useful. F_b is located when the scope curve becomes a flat 45° line, as shown in the diagram[1].

The current source method illustrated in *Fig. 8.5* will also yield true impedance, but has several distinct advantages over the constant voltage method. The most important advantage is that the speaker is connected directly to an amplifier (as it will be when in actual use) and not to a generator source through a 600–1000Ω resistor. The voltage divider method provides no damping for the driver other than its own mechanical system, while the current source method provides the damping factor of the amplifier to help control the woofer. Also, it is difficult to get more than a few millivolts through a 600–1000Ω resistor using most generators. This means that the measurement is pretty close to the noise floor and can easily be disturbed by ambient room noise. Using the current source method, you can not only exam-

ine impedance changes with different voltage levels to examine dynamic impedance changes, but use a minimum voltage that is above the noise environment. Applying 1V using the current source method will still yield appropriate small signal data while providing a robust enough signal to give good results. The only downside is that this type of impedance mea-

TABLE 8.2

	SEALED BOX SIMULATION	VENTED BOX SIMULATION
Before Break-in	F3 = 86.8Hz Qtc = 0.69	F3 = 61.6Hz
After Break-in	F3 = 85.5Hz Qtc = 0.68	F3 = 61.6Hz

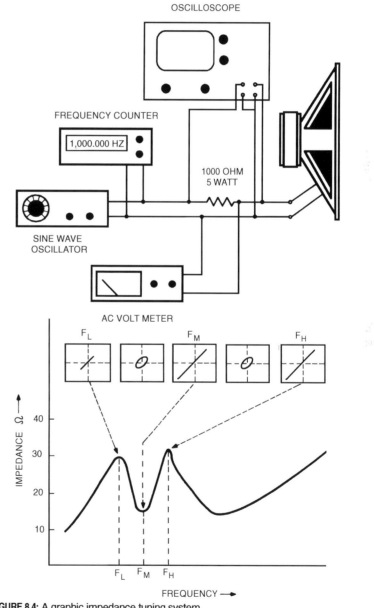

FIGURE 8.4: A graphic impedance tuning system.

surement is not as convenient as the voltage divider technique commonly used by most practitioners.

There are two ways you can perform constant current (admittance) type measurements for the purpose of creating impedance curves.

The manual method depicted in *Fig. 8.5* works well and requires a minimum of equipment to accomplish. To use this technique, use either a calibrated resistance substitution box (one that can handle sufficient current so you don't burn out the resistors) or a 10W potentiometer. Find the impedance value at any frequency by noting the voltage at that frequency with the woofer connected. Remove the woofer and substitute RX and vary its value until the same voltage is obtained. RX is then removed and measured with an ohmmeter to get the impedance value (or by reading the value off the resistance substitution box). Again, a good voltage for making impedance curves for the purpose of generating T/S parameters is 1V. If you own one of several

TABLE 8.3

Diameter	S_d (M²)	Mmr (grams)
18″	0.1300	27.0
15″	0.0890	15.3
12″	0.0530	7.0
10″	0.0330	3.5
8″	0.0220	1.9
6.5″	0.0165	1.2
6″	0.0125	0.8
5.25″	0.0089	0.5
4.5″	0.0055	0.2
3″	0.0038	0.1

computer based analyzers that have the ability to divide measurement curves such as the LinearX LMS, Audio Precision System I/II or CLIO, you can use the same method. LinearX makes a convenient test jig for this purpose called the VIBox[2]. This allows the user to make rapid voltage and current (admittance) measurements for plotting true impedance curves. The VI Box also has switching for both 0dB and –40dB attenuation for low or high voltage measurements. Using this method, two curves are produced, a voltage curve taken at the woofer terminals (no substitute resistor is used with this method) and current curves measured across the series ground leg 0.1Ω resistor. An example of these two V RMS measurements made using the LinearX LMS analyzer is shown in *Fig. 8.7*. The current or admittance curve looks like an upside down impedance curve. Using Ohms law, you then divide the voltage curve by the current curve to get the impedance curve (V/I = Z). Since both of your measurements are in volts RMS, the impedance curve is also in V RMS. The final procedure is to take this curve and convert it from volts to ohms. Using LMS, the procedure is to use the log to linear conversion routine twice (V to VdB to ohms) to produce the final result shown in *Fig. 8.8*.

8.32 MEASURING COMPLEX IMPEDANCE.

The only way to understand the reactive consequence a loudspeaker load presents to an amplifier is by measuring or calculating the complex impedance. The impedance magnitude measurement described in *Section 8.31* is a composite measurement comprised of two elements, resistance and reactance. When using the voltage divider or current source methods to measure impedance, the reactive and resistive elements are being consolidated to get a general picture of the speaker load. This is a simple measurement and is a convenient tool to reveal and avoid excessively low loads (below 3Ω). Knowing the reactive nature of this load, however, can give you a more detailed portrayal of the load's characteristics.

The composite of reactance and resistance is expressed mathematically by:

$$Z = (R^2 + X^2)^{1/2}$$

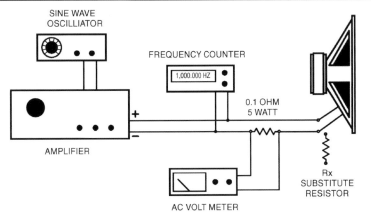

FIGURE 8.5: Impedance measurement—current source method.

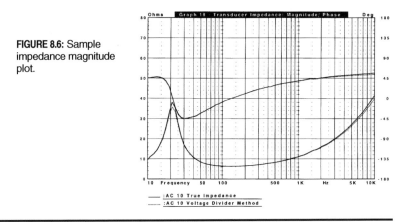

FIGURE 8.6: Sample impedance magnitude plot.

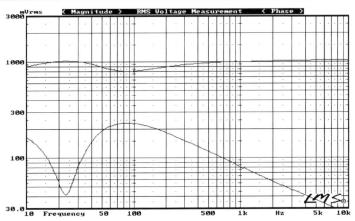

FIGURE 8.7: Voltage and current curves for impedance measurements.

where Z is the impedance magnitude (as measured in *Section 8.31*), R is the resistive component, and X is the reactive component. The resistive element in the equation can, to some extent, be observed in the impedance magnitude plot, like those shown in *Fig. 8.6*. The lowest point in the trough just following the resonance peak is almost purely resistive, and is more or less equal to the DC resistance of the driver. Also, the impedance at the resonance frequency is purely resistive (at least if the phase angle is at zero, which is generally, but not always, the case). Understanding which part of the curve is reactive is not readily apparent.

The reactance portion of the loudspeaker impedance can be either capacitive (voltage lagging current) or inductive (voltage leading current). The extent to which the load is either capacitively or inductively reactive at different frequencies determines how happy, or unhappy, an amplifier will be driving a particular loudspeaker. Determining which is which, resistive, capacitive, or inductive, is actually fairly easy. You need only the impedance magnitude, measured as shown in *Section 8.31*, plus the impedance phase. Obtaining phase requires the use of either a phase meter or a computer program which can derive phase from the impedance magnitude, such as Peter Schuck's XOPT crossover program, or LEAP 4.0. Phase is basically calculated from the slope of a magnitude curve and is explained in greater detail in *Section 8.83 Measuring Phase*.

The procedure for generating a complex impedance curve begins by finding the resistive and reactive value for each frequency point on the impedance magnitude curve. To find the resistive value at any given frequency, multiply the magnitude value (Z) by the cosine of the phase angle (θ), expressed as:

$$R = Z \times (\text{Cos } \theta)$$

To find the reactive value at any given frequency, multiply the magnitude value by the sine of the phase angle. Phase, whether measured or derived, is given in degrees positive or degrees negative. Taking the sine of a positive phase angle, a positive answer results, which signifies inductive reactance (X_L). Likewise, taking the sine of a negative phase angle, a negative answer results and this indicates capacitive reactance (X_c). Inductive or capacitive reactance is expressed as:

$$X_L = Z \times (\text{Sin } \theta)$$

$$X_c = Z \times (\text{Sin } \theta)$$

R is the real component of impedance, and X_L and X_c are the imaginary components of the impedance. The imaginary components are correctly labeled using the imaginary number j operator, and are expressed as $+j\Omega$ for inductive reactance and $-j\Omega$ for capacitive reactance.

Once the data for each frequency point has been "split" into real and imaginary components, it is graphed on a polar plot commonly referred to as a Nyquist plot. *Figure 8.9* shows a measured complex impedance plot (Nyquist) using the DRA MLSSA FFT analyzer. *Figure 8.10* gives the impedance magnitude and phase for the same measurement, which is the impedance of a vent-

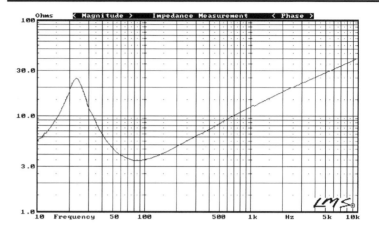

FIGURE 8.8: Impedance measurement resulting from V/I.

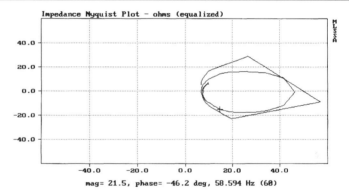

FIGURE 8.9: Impedance Nyquist—ohms (equalized).

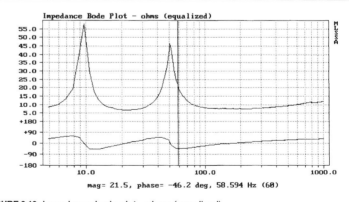

FIGURE 8.10: Impedance bode plot—ohms (equalized).

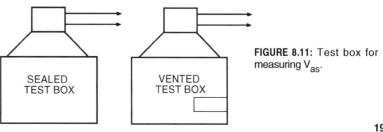

FIGURE 8.11: Test box for measuring V_{as}.

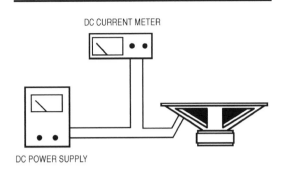

DC CURRENT METER

DC POWER SUPPLY

FIGURE 8.12: BL measurement.

ed woofer. The horizontal axis is labeled in ohms and represents the resistive/real portion of the plot. The MLSSA plot has the horizontal axis scaled in positive and negative resistive values, but plots for loudspeakers need only the right side, or positive values for R. If you are plotting complex impedance by hand, label the horizontal axis from zero to some positive value only. The vertical axis represents the imaginary values and is scaled to zero at the center of the axis and positive/inductive, $+j\Omega$, at the top half of the scale, and negative/capacitive, $-j\Omega$, at the bottom half of the scale. Both axes should use the same scale factor.

Interpreting complex impedance curves is fairly straightforward. The curves start at the left and rotate clockwise to the right. A resonance is represented by a complete circle, so in the case of the plot in *Fig. 8.9*, the two resonances of the vented enclosure have two rotations. If the enclosure is well braced, these circles will be round and symmetrical. If the enclosure is flexing to any extent, the circles will be deformed and asymmetrical (the

FIGURE 8.13: The BL meter.

MLSSA screen graphics do not maintain a correct aspect ratio so the plots will not look as described). Small loops and pigtails may show up in the mid to high frequencies of a full range speaker and indicate acoustic coupling between drivers caused by excessive overlap of driver operating ranges.

The load the speaker presents to the amplifier can be evaluated by noting the frequency of maximum capacitive reactance. Determine this by drawing a line from the origin (0,0 coordi-

nate) tangent to the capacitive side of the bass resonance circle. This frequency represents the maximum capacitive phase angle for the speaker, which in *Figure 8.9* is 46° at 58Hz. Phase angles in excess of 40° at low frequencies, and the same at frequencies about 1–2kHz, can be considered as a somewhat difficult load for an amplifier to drive. The lowest resistance point value is also important. Values below 3Ω as the curve crosses the horizontal axis at very low or very high frequencies can also present a difficult load.[3]

8.40 CALCULATING DRIVER AIR-MASS LOAD.

The air has mass and exerts a pressure on the surface of a cone which needs to be accounted for in cone assembly mass measurements. The amount of radiation air mass load depends upon the total surface area of the cone and is calculated by:

$$M_{mr} = 0.575 \times S_d^{1.5}$$

Table 8.3 gives the typical free air radiation mass load for different diameter drivers.[4]

8.41 MEASURING DRIVER MASS.

Three methods are available for ascertaining the mass, or weight, of the driver's cone assembly. The first one is obvious, just ask the manufacturer to provide the data. This is the most accurate way of obtaining the information and certainly saves time. The assembly mass, M_{md}, can then be combined with the driver radiation mass load, M_{mr}, to get the total cone mass, M_{ms}:

$$M_{ms} = M_{md} + M_{mr}$$

The other two methods, Delta Mass and Delta Compliance, must be done with test equipment.

A. Delta Mass (added Mass) Method.

Using the test setup in *Figs. 8.3, 8.4,* or *8.5*, attach a small, accurately measured amount of clay (or other material measured to 0.1 gram), M_a to the driver cone, pressing it snugly in place symmetrically around the cone and dust cap junction. With M_a in place, determine f_{sa}, the resonance of the cone plus mass, in the same manner as measuring the driver free air resonance. For accuracy, the measurement should be made with the driver clamped to a rigid, suspended surface. This measurement should be within the nearest tenth of a cycle, or 0.1Hz. The added mass (M_a) should be of sufficient weight to produce a minimum of 25–50% change in the resonance frequency of the driver (this can be difficult when measuring low resonance drivers, since f_{sa} could be below 10Hz and be very difficult to measure accurately even with expensive professional test equipment, so keep f_{sa} above that frequency). This is approximately 10 grams for smaller diameter cones of 6″ or less, and up to 40 grams or more for larger cones. The cone mass M_{md} is calculated by:

$$Mmd = \frac{M_a}{(f_s/f_{sa})^2 - 1} \quad \text{in grams}$$

Then $M_{ms} = M_{md} + M_{mr}$.

B. Delta Compliance (test box) Method.

With the Delta Mass method, f_{sa} was lower in frequency than the free air resonance of the driver. This can present problems with extremely low resonance drivers if f_{sa} goes below 10Hz, due to the limitations of some test equipment at such low frequencies. The Delta Compliance method has the opposite effect and drives the resonance upward, and is easier to measure on less expensive test gear.

The test box must be sealed so all joints are air tight. The driver is mounted on the outside of the box as illustrated in *Fig. 8.11* for the sealed test box example. An air-tight seal between the box and the driver is also critically important. Closed-cell foam tape, the kind used for sealing door jambs, can be used for a gasket while holding the driver in place by hand, or screwed to the top of the test box.

The size of the box must be such that the resonance is changed from 50–100% of the free-air value. This depends on the driver V_{as}, and if you know what that is from manufacturer's specs, a box that is half of the V_{as} will be appropriate. Beyond that, a set of test boxes can be made which will cover most circumstances:

Diameter	Test Box Volume (cu. in.)
4–5″	216
6–7″	864
8″	1728
10″	2532
12″	3456
15″	4320

The total volume of the box should include that determined by the internal dimensions, plus the volume of the hole made for the driver. This volume can be multiplied by 1.02 to account for the air space in front of the cone. Any of the test setups in *Figs. 8.3, 8.4,* or *8.5* can be used to find the box resonance, f_c, with the same basic procedure to find driver free-air resonance. When this frequency, f_c, is determined to the nearest 0.1Hz, then calculate the box compliance, C_{mb}:

$$C_{mb} = \frac{V_{ab}}{1.42 \times E^5 \times S_d^2} \quad \text{in meters/newton}$$

where V_{ab} is given in cubic meters, and S_d in square meters.

M_{md} is then calculated by:

$$Mmd = \frac{C_{mb}^{-1} - Mmr[1.85(6.283f_c)^2 - (6.283f_s)^2]}{(6.283f_c)^2 - (6.283f_s)^2} \quad \text{in kg}$$

where C_{mb} is given in meters/newton, and M_{mr} in

kilograms. Then $M_{ms} = M_{md} + M_{mr}$.

8.42 CALCULATE DRIVE COMPLIANCE.

The driver suspension compliance, C_{ms}, is calculated from:

$$C_{ms} = [(6.283f_s)^2 \times M_{ms}]^{-1} \quad \text{in meters/newton}$$

where M_{ms} is given in kilograms.

8.43 MEASURING DRIVER BL.

BL is a measure of the motor strength of a driver and is equal to the product of the length of wire in the magnetic field multiplied by the magnetic field density. The measurement is given in units of tesla meters. Several methods are available for measuring BL, but the most repeatable bench top measurement is by the opposing force technique.[4]

The test setup for BL measurement is shown in *Fig. 8.12*. The procedure is to place the speaker horizontally on a stable surface, add a known mass (M_a) which will depress the cone to a lower position, then apply a DC voltage to the voice coil, increasing the voltage until the cone returns to its starting (rest) position. The amount of mass is not critical, but must be known accurately to 0.1 gram, and must be sufficient to depress the cone by at least 0.25″. Once the cone has returned to its rest position (if increasing voltage makes the cone go further downward, reverse the polarity), the current (i) at this point is noted. BL is calculated by:

$$BL = \frac{9.8(M_a)}{i} \quad \text{in tesla-meters}$$

where M_a is given in kilograms, and current, i, in amperes.

The trick to performing this measurement accurately is to precisely determine the location of the driver rest position before the weight is added, and when current is applied, to return the cone to the same position. A simple method for rigging a rest position indicator is to attach a piece of wire to the edge of the frame such that it bends over the top of the driver and contacts the cone at the junction of the cone and surround with the speaker in the rest position. A better indicator can be made from a handful of Radio Shack parts. The device, the "BL meter," is depicted in the photograph in *Fig. 8.13*. The list of parts is as follows (parts numbers are from Radio Shack catalog No. 459, 1991, and anything similar will work just as well):

"Helping Hands" Project Holder #64-2093
1.5V lamp #272-1139
1.5V battery holder #270-401
Self-Adhesive Alarm Foil #49-502
Test Probe Tip #278-705
1.5V AA battery

Assemble the BL meter by first sliding the main arm of the project holder to one side, tightening up the wingnut at the joint. Next, unscrew the probe tip from one of the test probes and unsolder the wire. Insert the bare tip in the alligator clip at the far end of the project holder arm. Place the battery in the battery holder and connect the positive red wire from the holder to the red lead on the lamp. Clip the black lead from the lamp in the other alligator clip on the project holder. Take the battery/lamp assembly and tape it to the main arm of the project holder with electrical or plastic tape, leaving the black lead from the battery holder hanging free. Position the lamp so it dangles close to the holder in plain view. Next, strip about ¾″ from the end of the remaining black battery holder lead. Cut 1″ of alarm foil and fold over about ¼″ over the bare battery holder wire. This will make contact with the wire and leave ½″ of foil with adhesive surface left over.

To use the BL meter, place the device close to the woofer. Affix the foil connected to the battery holder black lead to the surface of the cone next to the junction of the surround and cone. Carefully place the probe tip so it's just barely in contact with the foil and the lamp is illuminated. When the weight is placed on the driver, the lamp will go out. When the rest position is again reached by applying DC voltage, the lamp will illuminate, signifying the rest position has been reached.

8.50 CALCULATE DRIVER VOICE COIL INDUCTANCE, L_e.

Using the test setup in *Fig. 8.3*, measure impedance magnitude (m) in ohms at a frequency of 10kHz. The voice coil inductance in henries is given by:

$$L_e = 1.592 \times 10^{-5} (m^2 - R_E^2)^{1/2}$$

A more accurate method of calculating driver voice coil inductance is to measure the magnitude and phase at 1kHz and use the formulas for computing complex impedance in *Section 8.32* to separate the inductance/imaginary component from the resistive/real component, keeping in mind that the inductance of a speaker is a frequency dependent variable and will be different at different frequencies. If phase measurements are not available the following approximation works well:[4]

1. Measure the impedance magnitude at 1kHz (Z_x) to 0.1Ω.
2. The voice coil resistive/real component at 1kHz can be estimated by:

$$R_{vc} \approx R_{evc} \times (1 + 0.038BL) \text{ in ohms}$$

3. Total reactance is determined by:

$$X_T = [Z_x^2 - R_{vc}^2]^{1/2} \text{ in ohms}$$

4. Find the inductive reactance (X_L) by removing the reactance due to cone mass by:

$$X_L = X_T \; x \; \frac{BL^2}{6283M_{md}} \text{ in ohms}$$

where M_{md} is given in kilograms, and BL in tesla meters.

5. Voice coil inductance L_{evc} is found by:

$$L_{evc} = \frac{BL^2}{6283} \text{ in henries}$$

Although voice coil inductance is typically measured at 1kHz, both the inductance and AC resistance of a loudspeaker voice coil are frequently dependent, and change substantially with frequency.

8.60 CALCULATE AMPLIFIER SOURCE RESISTANCE, R_G.

Amplifier source resistance is one of the series resistances taken into account when calculating driver Q. The easiest method is to use the manufacturer's advertised damping factor (D), usually measured at 1kHz. Calculate R_G by:

$$R_G = \frac{R_d}{(D-1)}$$

where R_d is the rated driver impedance. If impedance loading circuits described in *Chapter 7* were used in the crossover design, this figure will be constant over most of the bandwidth of the speaker. Damping factor, however, can be substantially different at different frequencies and different drive levels. Thus, a more accurate method is to measure the damping factor of your amplifier rather than use the advertised figures. This procedure, described by Small,[5] involves making the new measurement at 50Hz.

Use a 50Hz sine wave to drive your amplifier, connect a voltmeter across the output terminals with no load, and adjust the amplifier volume and generator level until you obtain a voltage equal to E_0, where:

$$E_0 = (W \times R_d)^{1/2}$$
$$W = \text{amplifier rated output in watts}$$

E_0 for a 50W amplifier working into a 4Ω speaker load would give $E_0 = 14.14V$.

With the same amplifier and generator settings, measure E_L, the output voltage, with a load resistor equal to R_d across the output terminals (the resistor should have a voltage rating equal to the amplifier output rating). Then:

$$R_G = \frac{R_d(E_0 - E_L)}{E_L}$$

TABLE 8.4
VENTED TEST BOX SIZE

Driver in inches	Box Volume ft³	Vent in inches
8	1.0	2″ ID × 2″
10	1.5	2″ ID × 2″
12	2.5	3″ ID × 6″
15	3.5	4″ ID × 5″

8.61 CALCULATE TOTAL SERIES RESISTANCE, R_X.

In addition to the amplifier source resistance, R_G, you must measure and include in driver Q calculations the remaining series resistance between the amplifier terminals and the woofer terminals, including the speaker cable, cabinet terminals, internal wiring, and crossover resistances (such as the inductors). To do this, measure R_t (the resistance in ohms) at the speaker cable terminals (where the amplifier would connect) with the crossover and drivers in place. Then:

$$R_X = R_t - R_E$$

8.70 CALCULATE THE VOLUME OF AIR EQUAL TO DRIVER COMPLIANCE, V_{as}.

V_{as} is one of the most difficult driver parameters to measure, since its value can change appreciably with air temperature and humidity. Here are three methods of measurement and calculation: vented-box, closed-box, and the C_{ms} method.

1. *Vented-box.* Using the appropriate box dimensions and vent sizes given by *Table 8.4*, run an impedance curve and determine the frequencies f_H and f_L (*Fig. 8.4*). Also, find the box resonance (f_c) with the vent covered.

Then:
$$f_b = (f_H^2 + f_L^2 - f_c^2)^{1/2}$$
and:
$$V_{as} = \frac{(f_H^2 - f_b^2)(f_b^2 - f_L^2)\,V_b}{(f_H^2 f_L^2)}$$

This method assumes no leakage losses, which is not always the case for all drivers. It also assumes f_b to occur at the measured impedance minimum, which, again, is not always the case, especially for large voice coil inductances.

2. *Closed-box method.* Using an unlined test box (V_t) measure the driver electrical Q (Q_{ect}) and box resonance (f_c) with the driver installed as in *Fig. 8.11* (Q_e measurement is discussed in 8.80). Use a 1ft³ test box for 8″ and 10″ drivers and a 2ft³ test box for 12″ and 15″ drivers. The driver must have an air-tight seal. Then:

$$V_{as} = V_t \left[\frac{f_{ct} Q_{ect}}{f_s Q_{es}} - 1 \right]$$

A faster, but less accurate, alternative is to just measure the resonance (f_c) of the test box, then:

$$V_{as} = V_t\, 1.15 \left(\frac{f_{ct}}{f_s} \right)^2 - 1$$

Both these procedures assume zero losses, and can produce inaccurate results.

3. *Driver compliance method.* Using the driver compliance (C_{ms}) calculated above, and the radiating area of the driver in square meters (S_d) (*Table*

1.15), calculate V_{as} in cubic meters by:

$$V_{as} = 1.42 \times 10^5\,(S_d{}^2)\,(C_{ms})$$

This method has the advantage of being independent of driver leakage losses. The C_{ms} method, however, as well as the other two methods described, are subject to suspension nonlinearities, which could adversely affect the results. You can minimize these effects by making the tests at the lowest possible voltage level

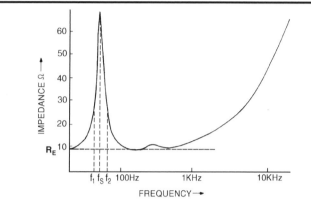

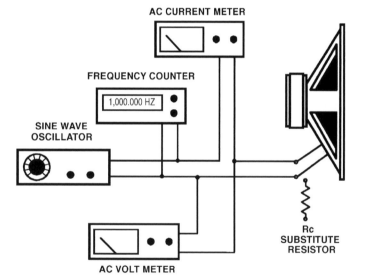

FIGURE 8.14: Test setup for measuring Q.

on your equipment. With this test, as with others in this chapter, you must take frequency measurements with a resolution of 0.1Hz, which is why a frequency counter is shown in all test setups.

In practice, methods two and three tend to produce similar results. My preference, however, is to use the Delta Compliance method to compute compliance and then use method 3.

8.80 MEASURING DRIVER "Q"—Q_{ts}, Q_{es}, AND Q_{ms}.

Driver Q may be measured in two ways. The first is the traditional one suggested by Thiele in his original paper on vented enclosures. This measurement is performed by finding the −3dB points on either slope of the resonance peak of the driver. The other is to measure driver BL and compliance and use this data to calculate Q values. Thiele's method is widely accepted in industry, but can be prone to measurement error. This is caused by nonlinear driver surrounds, and the sensitivity of the −3dB location method depending on the sharpness and shape of the resonance peak. In almost all cases the procedures described for method 2 will yield more reliable results and is highly recommended as a less error-prone alternative if you have modest equipment resources. For reference, both methods are given.

A. Method 1.
Using the test setup illustrated in *Fig. 8.14*:

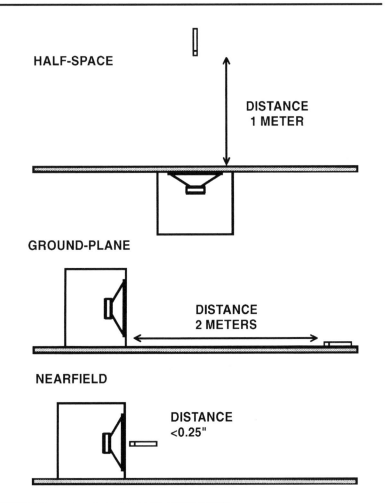

HALF-SPACE

DISTANCE
1 METER

GROUND-PLANE

DISTANCE
2 METERS

NEARFIELD

DISTANCE
<0.25"

FIGURE 8.15: Frequency response measurement domain.

1. Measure the DC resistance of the driver voice coil (R_E), preferably with an accurate bridge.
2. Select a resistor (R_c) of similar value to R_E (if $R_E = 6.5$, an 8Ω resistor is close enough).
3. Connect R_c to the test terminals, and set the generator to f_s. Take special note of the voltage at this point, since all of the following readings must be made at that exact same standard voltage. Its absolute value is not impor-

tant, so long as it remains the same in each step. Use the 100mV range if this will provide good results with your equipment. If not, increase the voltage to between 0.2–0.7V. This will probably give better results and still be within what can be considered the "small signal" range. Measure the current at the standard voltage at f_s. Label this I_c.

4. Calculate:

$$I_E = \frac{I_c R_c}{R_E}$$

5. Disconnect R_c and connect the driver. Holding the driver in midair, adjust the generator frequency for minimum current, which will be f_s. This minimum current at f_s is labeled I_0.

6. Calculate:

$$r_0 = \frac{I_E}{I_0}$$

7. Calculate:

$$I_r = (I_E I_0)^{\frac{1}{2}}$$

8. Find the frequencies f_1 and f_2, below and above f_s respectively, whose current is equal to I_r at the standard voltage (*Fig. 8.14*). Check the accuracy of measured f_s by calculating:

$$f_s = (f_1 f_2)^{\frac{1}{2}}$$

If the measured figure is less than or within 1Hz of the calculated figure, f_s can be considered a reliable measurement.

9. Calculate:
alculate:

$$Q_{es} = \frac{Q_{ms}}{(r_0 - 1)}$$

11. Calculate:

$$Q_{ts} = \frac{Q_{es} \times Q_{ms}}{Q_{es} + Q_{ms}}$$

This gives the value of Q_{ts} considering only the driver parameters. As mentioned above, you must take into account the various series resistances presented by the amplifier, connecting cables, and crossover since they will all increase the value of Q_{ts} once you install the woofer in the system. You can do this when you make your measurements by inserting a series resistor equal to the value of these resistances at the driver terminals, or by using the calculated value of Q_{es}, then:

$$Q_{es}' = Q_{es} \left(\frac{R_G + R_x + R_E}{R_E} \right)$$

$$Q_{ts}' = \frac{Q_{es}' \times Q_{ms}}{Q_{es} + Q_{ms}}$$

Once you know Q_{ts}, you may modify it for any driver to fit a particular situation, at the cost of efficiency, by:

A. Adding cone mass to increase Q_{ts}.[6]

B. Adding a small series resistor of the same power rating as the driver to increase Q_{ts}.[7]

C. Stretching porous cloth over the rear of the driver basket to decrease Q_{ts}.[8,9]

B. Method 2.

1. Calculate R_{es} by subtracting the voice coil resistance R_{evc} from the measured magnitude of the resonance peak (R) at f_s found in *Section 8.31* using the impedance measurement test procedure:[4]

$$R_{es} = R - R_{evc} \text{ in ohms}$$

2. Calculate Q_{ms} by:

$$Q_{ms} = \frac{R_{es}}{BL^2C_{ms}(6.283f_s)}$$

where BL is given in tesla meters and C_{ms} is given in meters/newton.

3. Calculate Q_{es} by:

$$Q_{es} = \frac{R_{evc}}{BL^2C_{ms}(6.283f_s)}$$

where BL is given in tesla meters and C_{ms} is given in meters/newton.

4. Q_{ts} is given by:

$$Q_{ts} = \frac{Q_{ms} \times Q_{es}}{Q_{ms} + Q_{es}}$$

There are three other readily available methods for measuring Q parameters, which utilize either computer test equipment or CAE loudspeaker software.

A. The Audio Precision System 1 has an automated Q measurement program[10] which will measure the complete parameter set and print it out in about 15 seconds. This method is essentially a computer controlled version of the Thiele method using the delta compliance method to determine V_{as}. The analyzer uses the same procedure, but at reasonably low voltages with highly accurate meter recording capability. The System 1 locates f_s by finding the zero phase point, which is usually accurate, but not always in all circumstances for all drivers. This is an excellent piece of equipment for quality controlling large quantities of drivers.

B. The DRA Labs MLSSA FFT analyzer can take a free air and delta mass or delta compliance impedance, do a scan-fit of the curve and output all the mechanical parameters. The speed is dependent on the host computer speed, but is somewhat slower than the System 1. The curve-fitting routine is more accurate than the Thiele method, and outputs all the essential driver mechanical parameters.

C. The LinearX LEAP software can import free air and delta mass or delta compliance impedance curves from the Audio Precision System 1, DRA MLSSA, Ariel SYSid, CLIO, Goldline TEF 20, and the LinearX LMS analyzers to an automated curve-fitting routine. The routine is the most elaborate and precise of the group and the results include not only the mechanical parameters, but data on the frequency dependent resistance and inductance of the upper impedance rise.

D. The LinearX LMS with the Windows 95/98 software release has an advanced parameter optimizer that uses the new and more complete driver model described in Chapter 0. This is the same T/S routine that is being used in the Windows version of LEAP (not available at the time this was published). This type of T/S calculator uses a curve-fitting algorithm in conjunction with the LEAP driver model to produce the T/S parameter set and is very accurate.

E. The last automated measurement device I want to mention is the one developed by David Clark, called DUMAX[11]. Dumax is a highly specialized device and uses a very unique and accurate methodology to derive woofer motor parameters. Mr. Clark has used the results from testing with Dumax on numerous occasions in woofer reviews he has done in both *Car Audio and Electronics* and *Car Stereo Review* magazines.

Basically, the device proceeds from the premise that parameters change dynamically as the voice coil moves through its forward and rearward range of travel. Looking at the BL curves in Chapter 0 certainly emphasizes what the problem is that the Dumax machine attempts to deal with. The device is fairly large and uses pneumatic actuators to move the driver through its entire range of motion thereby calculating the compliance and motor strength changes with different positions. Explaining the details of this machine are beyond

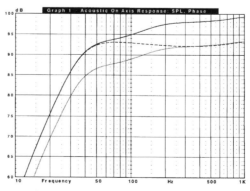

FIGURE 8.16: Three response measurements.

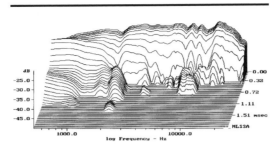

FIGURE 8.17: Cumulative spectral decay plot.

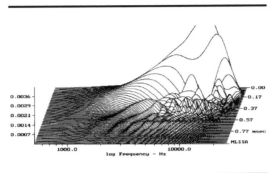

FIGURE 8.18: Wigner distribution.

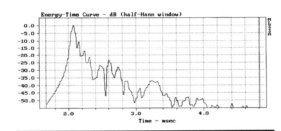

FIGURE 8.19: Energy-time curve—dB (half-Hann window).

the scope of this book, but it currently represents the state-of-the-art in dynamic parameter measurement.

8.81 FREQUENCY RESPONSE MEASUREMENT TECHNIQUES.

A. Full Bandwidth Measurements

We have three techniques for making full range, 10Hz–40kHz response measurements on loudspeakers. Each method depends upon the radiation domain of the setup. These are anechoic (also called free field, full-space, or 4π), half-space (2π), and ground-plane.

Anechoic measurements are made without any reflecting boundaries surrounding the device under test. This means nothing will reflect the sound for a full 360° in all directions, hence the term full-space or 4π. This type of measurement can only be made in either a special sound absorbing room, an anechoic chamber, such as the anechoic chamber at the National Research Council in Canada, or with the speaker suspended high off the ground in open air. Anechoic measurements can also be simulated by electronically gating the received signal from the DUT and ignoring reflections from nearby boundaries. This kind of test can be performed by FFT analyzers and gated sine wave devices. The only drawback, however, is that the shorter the gate time interval, the higher the cutoff frequency of the measurement. Getting good low end data down to 20Hz means having no reflecting boundaries for at

least 28 feet. This law of nature dictates that anechoic measurements using gated test instruments must still have a nearly free-field environment in order to get accurate data down to 20Hz. It is possible to meet such conditions indoors by making the measurements in a large warehouse area with the speakers suspended in mid-air. Getting data down to 10Hz would require no boundaries for 56 feet.

Wideband anechoic measurements are only possible in a limited number of locations, although wideband measurements produced by splicing two different domains are relatively easy. I routinely use a short 6–7′ measuring tower for gated measurements using the LinearX LMS analyzer. With the measurement microphone placed 1m from the DUT and at a 75–95″ height above the ground, reflection-free semi-anechoic measurements can be made down to 200–300Hz. These measurements can then be spliced to a 10Hz–500Hz groundplane measurement, also anechoic in nature, to produce full range 10Hz–40kHz anechoic measurements. The "splice" operation can occur anywhere between 200–400Hz and good results achieved. It is also possible to use nearfield measurements for the low-frequency part of the spliced measurement, but because of the change in radiation resistance, it is not practical to splice nearfield curves above about 100Hz.

Half-space measurements, illustrated in *Fig. 8.15,* have some of the same boundary limitations of anechoic measurements, and must be made in open air. The DUT is mounted flush to a planar surface, usually in a pit with the baffle level with the ground surface and the driver aimed at the sky. The driver is radiating into a 180° hemispheric field, hence the name half-space or 2π. Half-space measurements can be easily used for full bandwidth measurements from 10Hz–40kHz. Since the sky is an unlimited vertical boundary, the only limitation is the distance of the test position from any ground level boundaries, which should be at least 30 feet.

The third measurement domain is referred to as a ground-plane, illustrated in *Fig. 8.15.* Ground-plane measurements are set up by placing the DUT on a hard flat reflective surface, such as a cement or asphalt parking lot, with the microphone placed on the ground.[12] The measurement being made includes the sonic mirror image of the speaker, which is why the microphone is placed on the ground at the apex of the direct signal and its reflected sound image. Ground-plane measurements must also be made in open air, and the nearest boundary should also be at least 30 feet from the DUT. Given the typical parking lot situation, the latter criterion is easy to meet, making full range measurements by ground-plane relatively easy. For better results with floor standing type speakers, the speaker should be tilted forward as far as possible. The response measurements made with the ground-plane method give results similar to those obtained by anechoic measurements.

The response relationship between the three domains is shown in *Fig. 8.16*. The key to explaining the differences between the three domains is in understanding the nature of the "step" response. When sound radiates from a driver cone that has no boundary or baffle, the radiation is hemispherical up to the frequency when driver directivity starts to narrow. If the driver is mounted on a baffle, the baffle acts the same way as a reflector on a flashlight, increasing the amount of energy in a given direction. The step occurs when the wavelength of the radiated signal decreases to the approximate area of the reflecting surface, and the baffle starts to act as a sound reflector. When the baffle is made infinitely large, as is the case with the half-space measurement technique, the step occurs at all frequencies, hence the flat response shape in *Fig. 8.16*. The increase causes a doubling of the sound pressure, or 6dB of gain over the free-field measurement. When the radiating surface is limited to only the enclosure baffle, as in the case for anechoic and ground-plane measurements, the step occurs at whatever frequency is determined by the area of the baffle, as the response transforms from 4π to 2π. In the LinearX LEAP simulation in *Fig. 8.16*, the response step begins on the anechoic measurement at around 75Hz and levels out two octaves above at about 300Hz, rising at 3dB/octave. The ground-plane measurement includes two sound sources, the direct sound from the DUT and its image, so the total gain compared to the anechoic measurement is 6dB. This is why the microphone is shown at 2m in *Fig. 8.15*, since doubling the microphone distance will decrease the overall sound pressure by 6dB, and make the pressure level recorded equivalent to that of an anechoic measurement at 1m. Otherwise, the ground-plane exhibits the exact same step response as the anechoic measurement.

B. Low Frequency Measurements

Within the qualifications stated, all of the three domains, anechoic, half-space, and ground-plane, can be used for low-end measurements. Moreover, half-space measurement is probably the best of the three for determining the f_3 of a loudspeaker. There is one other technique, however, using near-field measurement,[13] which can be a quick and painless way to get low-end frequency data. Near-field measurement is simple and straightforward and is depicted in *Fig. 8.15*. The technique is basically just placing the microphone as close as possible to the driver, preferably less than 0.25″ above the dust cap of a woofer. Since the proximity to the driver swamps out any room reflections and diffraction off the baffle, the response is similar to a half-space response. Like the 2π pit measurement, near-field measurements lack the baffle step of anechoic and ground-plane measurements.

This technique works well for closed-box woofers, but requires a bit of doctoring for vented boxes. Since the low-end response of a vented enclosure is the combination of the woofer and vent output, the two must be combined. This is done by taking the two measurements separately. The vent measurement should be taken with the microphone placed in the center of the vent, flush with the baffle. Unless you can somehow isolate the woofer from the vent (try placing a soft cushion between the two if possible), the vent measurement is really only good for frequencies lower than 1.6 times the tuning frequency of the enclosure, due to crosstalk between the vent and driver. Since the port has less radiating area than the woofer, and since the near-field sound pressure is directly proportional to the area of the radiating surface, the port output will be disproportionately lower than the driver and must be adjusted. For instance, if the

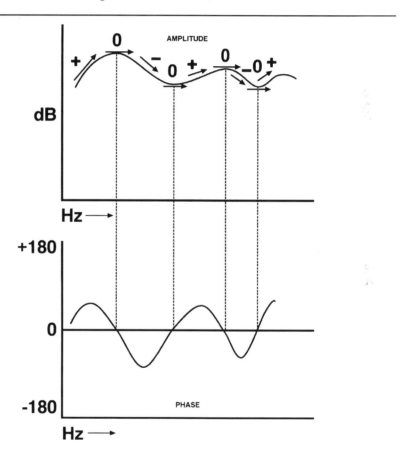

FIGURE 8.20: Relationship of frequency magnitude and phase.

port area were half that of the woofer, the port output should be scaled upward by 6dB to match the woofer.

**8.82 FREQUENCY RESPONSE
MEASUREMENT EQUIPMENT.**

Measurement equipment comes in two forms, manual and automated.

A. Manual Response Equipment

Manual measurements can be made with fairly simple equipment and the curves are generated

FIGURE 8.21: Piezo accelerometer.

by hand recording the data—point by point. The procedure is relatively slow, but with care the results can be nearly as accurate as that created by expensive analyzers. The measurement includes two processes, generating the signal source, and receiving the signal source.

1. Signal Sources

Several inexpensive signal sources are available for manual measurements, sine wave, pink noise, and warble tones (modulated sine wave). Sine wave oscillators and function generators are fairly inexpensive and can be used to drive the speaker directly, or used to drive an amplifier connected to the DUT. You can either trust the calibration on the dial of an oscillator for the frequency number, or add a frequency counter to ensure proper calibration.

Pink noise can be used to obtain frequency response data in the same manner as using a sine wave generator. This type of stimulus is somewhat immune to room modes and can be used indoors with fairly good results, although the same domain guidelines still tend to apply. Pink noise is normally generated in $\frac{1}{3}$- to $\frac{1}{10}$-octave bands and lacks the extreme detail available with sine wave measurement. Filtered pink noise generators are fairly expensive. Fortunate-ly, this test signal is readily available on a number of test records and CDs.

Warble tones are a type of modulated sine wave and, like pink noise, also tend to be somewhat immune to room modes. Warble tones are also normally generated in $\frac{1}{3}$- to $\frac{1}{10}$-octave bands, the same as pink noise. Warble tone generators, however, are not as expensive (Old Colony Sound Lab has one available in kit form), but like pink noise are also available on test records and test CDs. Despite the relative immunity to room response modes, warble tone measurements will be more accurate if

carried out under the same domain circumstances discussed in *Section 8.81.*

2. Receiving the Signal

Receiving the signal for manual frequency response measurement recording can be as simple as using a sound level meter to ascertain the level. Inexpensive meters are not likely to have a totally flat response. If it is possible, have the meter calibrated against some reasonably reliable source, then use the calibration curve to correct the readings. An alternative is to use a microphone connected directly to a voltmeter calibrated in dB. This will be only a relative measurement and not representative of actual SPL, but is adequate for many design situations (such as crossover design work). As with the SLM, calibrating against a known reliable response measuring instrument is suggested.

B. Automated Response Measuring Equipment

The chart recorder, such as instruments once made by Bruel and Kjaer and Neutrik, are pretty much a thing of the past. Virtually all audio analyzers intended for acoustic measurements are either computer based or computer interfaced. These fall into two primary categories, step sine wave and FFT type analyzers. Some analyzers will do one or the other type of analysis methodology, some do both.

Examples of computer based analyzers that can perform stepped (swept) sine wave analysis are:

1. Audiomatica CLIO[14], CLIO Lite[15]
2. Audio Precision System One and Two[16]
3. Liberty Instruments LAUD[17]
4. LinearX LMS[18]

For FFT analysis using some type of noise stimulus, such as MLS, examples are:

1. Audiomatica CLIO
2. Audio Precision System One and Two with DSP option
3. Ariel SYSid[19]
4. Goldline TEF 20[20]
5. Liberty Instruments LAUD[21]
6. DRA Labs MLSSA[22]

Each measurement type has its own advantages and all of these instruments can generate files that can be imported into CAD design programs such as LEAP, LSP CAD, Sound Easy, or Speak. The analog stepped sine wave types such as the AP System One or the LinearX LMS analyzer produce log scale sweeps with an equal number of points for each decade (3.2 octaves) of measurement. Stepped sine wave analyzers can be either ungated or gated. Ungated types are good for anechoic chambers, groundplane, and half-space measuring domains, but cannot remove nearby reflections. Gated type sine wave analyzers, like the LinearX LMS, use the computer CPU timing to selectively eliminate unwanted reflections producing the same "semi" anechoic result as applying a "window" to an FFT impulse waveform.

FIGURE 8.22: Test setup for measuring voice coil temperature.

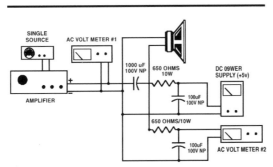

FFT (fast Fourier transform) analyzers use various noise stimulus types to produce a time impulse. This includes the standard noise impulse[23] (pink and white noise), the chirp (FM modulated sine wave) and MLS[24] (maximum length sequence). The FFT measurement sequence begins when the speaker is excited by the stimulus, the microphone receives the waveform and the resulting impulse is registered by the analyzer. The impulse, which is time domain data, is then windowed (gated) mathematically and an FFT applied to produce a frequency response curve (again remember that time and frequency are inversely related to each other—$t = 1/f$ and $f = 1/t$). The data points are recorded on a linear scale as opposed to a log scale as with stepped sine wave measurement. This tends to provide a great deal of detail at upper frequencies with very sparse detail at lower frequencies, at least at the current state-of-the-art. The most interesting advantage of any of the FFT analyzers is the ability to do time-domain measurement that allows the user to look at the energy spectrum and observe time decay. Examples of time domain type plots are given in *Fig. 8.17*, a cumulative spectral decay plot[25] of a magnitude response, *Fig. 8.18*, a Wigner Distribution,[26,27] and *Fig. 8.19*, an ETC[28] (Energy-Time Curve), all created using the DRA Labs MLSSA analyzer.

Both types of analyzers have their strengths and weaknesses. In my own work as a consultant designing products for a variety of home and car audio loudspeaker manufacturers, I prefer the gated sine wave instruments for loudspeaker system development, although both types will work for this. However, I do believe that log based analysis is the best method for system design—measuring drivers and designing crossovers. For transducer development, it is very hard to beat FFT analyzers. Their unique ability to look into time and view the nature of resonant decay is highly valuable when making decisions in the design of cone and dome drivers, as well as bracing and damping in enclosures.

8.83 MEASURING PHASE.

There are no manual methods for measuring phase. Phase measurements require a dedicated phase meter, one of the dual-channel analyzers described in *Section 8.81*, or a computer program capable of accurately deriving phase from a magnitude response measurement (LinearX's LEAP and LMS are the only programs capable of accurately doing this at the time of publication). Phase is, however, generally not well understood, and the following explanation may be helpful.

Phase of any magnitude waveform is a function of the slope of the waveform. If there is no more phase shift than that dictated by the slope changes in the magnitude, the device is said to be "minimum phase." As a general rule, loudspeakers are minimum phase devices, so phase can indeed be derived from the slope of the waveform magnitude. This is true for a loud-

speaker impedance curve and for the frequency magnitude curve. *Figure 8.20* illustrates the relationship between magnitude and phase. When the slope of the magnitude curve is flat and horizontal, the phase angle is zero degrees. When the slope is positive, phase is positive, but will return to the zero crossing by the time it reaches the flat area. Likewise, when the magnitude slope is negative, the phase angle is negative, but will return again to zero phase angle as the magnitude curve approaches another flat area. ["A Digital Phase Meter" by R. Luccassen with full construction details appeared in *Elektor Electronics USA*, June 1991, p. 32.–Ed]

8.84 MEASUREMENT MICROPHONES.

In order for any instrument to perform accurate frequency response measurements, the first priority is a microphone with a reliably flat frequency response. The following is a list of devices, descending in price, which are acceptable as measurement microphones. Microphones from ACO, B&K, Larson Davis, Rion Industries and the M-55 from Earthworks all cost in excess of $1000 to $4000 for a complete mike capsule, preamp body and power supply. The remaining microphones are priced at $100 to $1000, but have more than sufficient precision to do quality design work.

Precision Microphone Listing
1. ACO Pacific 7012[29]
2. Bruel & Kjaer Falcon Series, Models 4189/90/91 ½ free-field[30]
3. Earthworks M-55 ¼″[30]
4. Earthworks M-30 ¼″[30]
5. Josephson Engineering C550H[30]
6. Larson Davis Model 2540 ½″[31]
7. Larson Davis Model 2520 ¼″
8. LinearX M-31 ¼″
9. LinearX M-51 ½″
10. LinearX M-52 ½″
11. Mitey Mike II ¼″[32]
12. Neutrik 3382 ¼″
13. Neutrik 3384 ½″
14. Rion Industries Model UC-31[33]

In the groups, two microphones stand out as particularly good "buys." In the first group of professional precision microphones, the ACO Pacific represents an excellent value. It performs well in comparison to other precision microphones and is priced well under the competition. Among the group of less expensive microphones, the Mitey Mike designed by Joe D'Appolito and sold by Old Colony Sound Lab is as good as it gets for the price.

8.85 MECHANICAL VIBRATION ENCLOSURE MEASUREMENTS.

Measuring the vibration in the walls of loudspeaker enclosures is generally done with transducers known as accelerometers. Calibrated accelerometers are generally expensive devices costing $200–$500 or more. Accelerometers are usually made with piezo-ceramic elements, mounted so they directly couple the element to the mechanical vibration of the object they are attached to. Measurements are made with the accelerometer attached to the enclosure wall, and a sine wave or impulse type stimulus applied to the DUT. The output is directed to either a recording device or a voltmeter. The output from multiple accelerometer readings can be combined with model analysis software to create a dynamic model of the cabinet structure.[34]

There are a few less expensive alternatives, however. One is a PVDF accelerometer (Polyvinylidene) which is low cost (about $35) and has less resonance problems than the more expensive types (due to the nature of the PVDF material).[35]

Although not calibrated in decibels like more expensive units, the device is capable of making relative measurements to determine the effectiveness of bracing and wall damping materials. The unit can be used with a simple sine wave oscillator and a voltmeter, or connected to any of the automated type analyzers described in *Section 8.82*.

An even more inexpensive accelerometer can be fabricated for about $15 in parts. The device, illustrated in the photograph in *Fig. 8.21*, uses a piezo-ceramic element from a Motorola piezo-electric tweeter.[36] Disassemble the tweeter, and unsolder the element leads from the tweeter terminals. Separate the piezo element, which is a flat disk about the size of a nickel, from its paper cone. Next, cut a 0.75″ length piece of ⅜″ wood dowel and bond the dowel to the flat center of the piezo element with epoxy. When dry, the "poorman's" accelerometer can be attached to the enclosure wall with hot candle wax, beeswax, or temporarily glued to the enclosure with white glue (assuming an unfinished enclosure, of course) and used in the same manner as the PVDF accelerometer. The drive level from both types of accelerometers is quite low and may require additional amplification (20dB).

8.86 MEASURING VOICE COIL TEMPERATURE OVER TIME.

As noted in previous chapters, the dynamic performance of a loudspeaker is greatly affected by changes in voice coil temperature, which in turn cause large-scale increases in voice coil resistance. Fluctuating changes in voice coil resistance not only alter the low frequency performance and damping of a driver, but have a generally deleterious effect on the functioning of crossover circuits which are designed around nominal operating resistance. Ferrofluidics Corporation, whose products address the problem of dissipating voice coil heat, has created an indirect test to determine exact changes in temperature as a voice coil begins to heat. The test setup for this procedure is shown in *Fig. 8.22*.

The signal source can either be discrete sinewave or band-limited pink noise. Band-limited pink noise is preferred and is usually the type of stimulus that driver manufacturers use when power testing a speaker. If you choose to use a sinewave, the frequency should be just above the driver resonance, located in the vicinity of the impedance minimum of the driver. For pink noise, a variable frequency electronic high pass crossover can be used to "band limit" the pink noise signal. The high pass should generally be located about one octave above the driver resonance frequency, which for woofers would be about 100Hz, perhaps 200Hz for midrange drivers, and in the vicinity of 2kHz for tweeters.

Output voltage for the signal source (set using voltmeter 1) will vary depending on the suspected power handling of the driver under test. For woofers, somewhere between 10V to 20V is appropriate, 20V being reserved for larger voice coils that are specified as being high power handling. For tweeters and small voice coil midrange speakers, 6–7V is practical. Empirically, you will know if the voltage is too high when the driver voice coil burns out!

This test is accomplished by measuring the voltage across the capacitor at voltmeter 2, and substituting the value in this equation:

$$T_1 = 1 \left[\frac{V_S - V(T_O)}{a \, V_S - V(T_1)} \times \frac{V(T_1) - 1]}{V(T_O)} \right] + T_O$$

Where:

a = thermal coefficient of resistance for the voice coil wire. For copper wire use 0.00385; for aluminum wire use 0.00401.

T_O = the nominal initial voice coil temperature (normally use 25°C as the ambient temperature).

T_1 = the elevated temperature being measured.

$V(T_O)$ = voltage measured at voltmeter 2 at the beginning of the test (ambient voltage).

$V(T_1)$ = voltage measured at voltmeter 2 during the procedure (normally is accomplished at specified time intervals such as 10 minutes, 30 minutes, 60 minutes, etc.).

The procedure is basically simple. The voltage is measured across voltmeter 2 at the beginning of the test, and the value substituted into the temperature equation to calculate a starting value for T_1. At specific time intervals, the voltage across voltmeter 2 is again taken and the value substituted into the temperature equation. When all successive measurements are complete, a graph can be created with a temperature scale for the vertical axis and a time scale for the horizontal axis which shows the temperature rise with time for the driver being tested. This test could easily be used, for example, to calculate the difference in driver protection afforded by different crossover frequencies and slopes for a given driver.

8.90 USEFUL CONVERSION FACTORS.

Linear Measurement

1 mm	=	0.03937″
1 cm	=	0.3937″
1 m	=	39.37″
1 m	=	3.2808′
1 ″	=	25.4mm
1 ″	=	2.54cm

Square Measurement (area)

1 cm^2	=	0.155 in^2
1 in^2	=	6.452 cm^2
1 ft^2	=	929.0341 cm^2
1 m^2	=	10.76307 ft^2

Volume Measurement

1 mm^3	=	6.1×10^{-5} in^3
1 liter	=	0.0353 ft^3
1 ft^3	=	28.317 $liters^3$
1 ft^3	=	0.02831 m^3
1 ft^3	=	1728 in^3
1 m^3	=	35.314 ft^3

Mathematic Conventions

$$(X)^{-1} = \frac{1}{X}$$

$$(X)^{1/2} = \sqrt{X}$$

REFERENCES

1. N. Crowhurst, "Audio Measurements Course, Part 15," *Audio*, August 1976.

2. V. Dickason, "Industry News and Developments—LinearX VI Box," *Voice Coil*, Volume 10, Issue 7, May 1997.

3. V. Dickason, "How to Plot and Understand Complex Impedance," *Speaker Builder* 2/88, p. 15.

4. LEAP (Loudspeaker Enclosure Analysis Program), Version 3.1, Operating Manual, by LinearX Systems.

5. R. Small, "Direct Radiator Loudspeaker System Analysis," *JAES*, June 1972.

6. J. N. White, "Loudspeaker Athletics," *JAES*, November 1979.

7. H. J. J. Hoge, "Switched on Bass," *Audio*, August 1976.

8. R. Small, "Vented-Box Loudspeaker Systems," *JAES*, June–Oct. 1973.

9. J. Graver, "Acoustic Resistance Damping for Loudspeakers," *Audio*, March 1965.

10. R. C. Cabot, "Automated Measurement of Loudspeaker Small Signal Parameters," 81st AES Convention, preprint no. 2402.

11. D. Clark, "Precision Measurement of Loudspeaker Parameters," *JAES* Volume 45, No. 3, March 1997.

12. M. R. Gander, "Ground-Plane Acoustic Measurement of Loudspeaker Systems," *JAES*, October 1982.

13. D. B. Keele, Jr., "Low-Frequency Loudspeaker Assessment by Nearfield Sound Pressure Measurement," *JAES*, June 1989.

14. J. C. Gaetner, "CLIO Test System," *Speaker Builder*, Volume 16, Number 5, July 1995.

15. D. Pierce, "Product Review: CLIOLite," *Speaker Builder*, Volume 19, Number 3, May 1998.

16. V. Dickason, "New Software," *Voice Coil*, Volume 12, Issue 4, February 1999.

17. V. Dickason, "Liberty Instruments Releases," *Voice Coil*, Volume 11, Issue 10, August 1998.

18. V. Dickason, "New Update for LMS," *Voice Coil*, Volume 6, Number 4, February 1993.

19. V. Dickason, "SYSid Analyzer Update," *Voice Coil*, Volume 11, Issue 7, May 1998.

20. V. Dickason, "Windows for TEF 20," *Voice Coil*, Volume 11, Issue 8, June 1998.

21. V. Dickason, "Liberty Audiosuite Version 2.2," *Voice Coil*, Volume 10, Issue 10, August 1997.

22. V. Dickason, "Acoustic Analyzer News," *Voice Coil*, Volume 11, Issue 3, January 1998.

23. L. R. Fincham, "Refinements in the Impulse Testing of Loudspeakers," *JAES*, March 1985.

24. Rife and Vanderkooy, "Transfer-Function Measurement with Maximum-Length Sequences," *JAES*, June 1989.

25. Lipshitz, Scott, and Vanderkooy, "Increasing the Audio Measurement Capability of FFT Analyzers by Microcomputer Postprocessing," *JAES*, September 1985.

26. Janse and Kaizer, "Time-Frequency Distributions of Loudspeakers: The Application of the Wigner Distribution," *JAES*, April 1983.

27. Verschuur, Kaizer, Druyvesteyn, and de Vries, "Wigner Distribution of Loudspeaker Responses in a Living Room," *JAES*, April 1988.

28. Vanderkooy and Lipshitz, "Uses and Abuses of the Energy-Time Curve," *JAES*, November 1990.

29. *Voice Coil*, May 1989.

30. R. Honeycutt, "Test Microphone," *Voice Coil*, Volume 11, Issue 5, March 1998.

31. *Voice Coil*, June 1990.

32. D. Queen, "Product Review: Mitey Mike II," *Speaker Builder*, Volume 20, Number 5, August 1999.

33. *Voice Coil*, November 1989.

34. Hoffman, Matthiessen, and Veirgang, "Measurement of Operating Modes on a Loudspeaker Cabinet," presented at the 87th AES Convention, preprint no. 2848.

35. *Voice Coil*, February 1991.

36. Patent pending by Genesis Technology Inc.

CHAPTER NINE

CAD SOFTWARE FOR LOUDSPEAKER DESIGN AND LOUDSPEAKER ROOM INTERFACING

When I first started updating revisions for this book, computer software for loudspeaker design was only just becoming widely available. Since that time, numerous products have been introduced to this highly specialized market. The list of loudspeaker software is now quite extensive, and not only are there more programs, but programs with more sophisticated modeling and better quality programming. Along with substantial advances in loudspeaker engineering software for box and crossover design, a number of programs have also been introduced for making your speaker work better in the room it was designed for. Studies by Dr. Floyd Toole and his associates at Harman International and others have shown that room placement and the various other acoustic aspects of a room drastically affect the sound quality of a loudspeaker. These effects are so strong that the differences in room placement cause greater perceptual differences in loudspeaker sound quality than the sometimes large differences between entirely different speakers[1,2]. The engineers at Harman International (JBL, Infinity, Harman Kardon), under Dr. Toole's direction, knowing how critically important room placement is to speaker evaluation (yes, all those years you spent listening to rows of loudspeakers in hi-fi store showrooms trying to "pick" the best sounding speaker was mostly a waste of time), built a rather complicated and expensive machine that physically moves speakers rapidly into and out of the exact same listening position so that meaningful A/B listening can be done[3]. For this reason, this chapter now contains three sections, one for loudspeaker design software, and two new sections, one for room design software and another for room measurement software that is designed to be used with computer sound cards. Please note the prices quoted on these products were current as of July 2005. The following is an alphabetical listing of what is currently available at the time this revision was produced.

9.1 LOUDSPEAKER DESIGN SOFTWARE.

Active Filter Workshop—by Frank Ostrander. AFW is a Windows based suite of active filter-design utilities. This includes different circuit topographies like Sallen-Key filters, State-Variable, shelving, parametric EQ, and all-pass delay filters. This software is available from Old Colony Sound Laboratory (PO Box 876, Peterborough, NH 03458, 603-924-9464, Fax 603-924-9467, custserv@audioXpress.com) and is priced at $79.95.

AkAbak v. 2.1—by Jorg W. Panzer. This Windows based software is one of the more sophisticated modelers on the market and has an open architecture that allows the user to experiment with a wide variety of box configurations, previously used designs as well as ideas not yet tried[4]. In its fifth year, the current version is 2.1[5,6]. This program does both enclosure and network design and uses a combination of menus and scripts to describe the task being undertaken. Features include the ability to simulate both passive and active networks (models both transistors and op amps), produces directivity output as a polar or Cartesian diagram, performs non-linear analysis of voice coil resistance change due to heating and compliance non-linearity, describes driver mounting locations in an XYZ axis for predicting off-axis performance, as well as modeling loudspeaker cabinet diffraction. In the hands of a professional engineer, this is very powerful software. However, for the new and uninitiated it's probably a bit intimidating. Current price for AkAbak, distributed in the U.S. by Bang-Campbell Associates, is $700. For more, visit the website at http://users.rcn.com/rhcamp/akinfo.htm.

Bass Box 6 Pro—by Harris Technologies. Various versions of Bass Box have been around since the mid '90s and is now in its v. 6.0 Win95/98/NT 32-bit format[7,8,9]. This has always been a great low-cost box design program for loudspeaker enthusiasts. The new 32-bit Bass Box 6 Pro version includes the usual box design features from the previous version, but also contains simulation of vent pipe resonances,

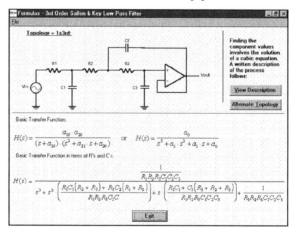

ACTIVE FILTER WORKSHOP

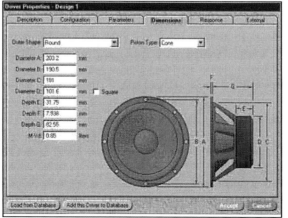

BASS BOX 6 PRO

the ability to add the effects of passive and active networks to all box-design types (such as series inductance added to bandpass enclosures to attenuate port anomalies or second-order high-pass active filters to boost the low-end of vented enclosures), and diffraction modeling on some cabinet types. Included is a 1000-driver database with search and edit facilities. For more information, you can visit the Harris Technologies website at www.ht-audio. com. The program is priced at $129 and is available from Old Colony Sound Laboratory.

Bass Horn Design—by A. L. Senson. This is basically a tractrix horn design DOS type program that calculates dimensions for a catenoid, exponential or hyperbolic bass horn and prints out in less than a minute. Not only does it provide a horn's general data, such as mouth and throat areas and back-chamber volume, but it will also give you the dimensions to design your enclosure. Bass Horn De-

sign is priced at $19.95 and is available from Old Colony Sound Laboratory.

CALSOD v. 3.10 Professional—by AudioSoft. CALSOD (Computer Aided Loudspeaker Design)[10] was written by Australian engineer Witold Waldman and was originally described in the *JAES* September 1988 article titled "Simulation and Optimization of Multiway Loudspeaker Systems Using a Personal Computer" written by the program's author. This is a DOS based full-featured loudspeaker simulation program that does both enclosure simulations (sealed, vented, vented and sealed bandpass, passive radiator and assisted vented enclosures) and crossover optimization. CALSOD can work from manually modeled data or will import SPL and impedance data from AP System 1 or 2, DRA MLSSA, IMP, active as well as passive filters, impedance optimization for conjugate networks, optimizing T/S parameter calculator, simulation of room gain, SPL optimization of up to five points on and off-axis, and XYZ specification of driver location. Price of the full version 3.10 is $269 from Old Colony Sound Laboratory. Also available from Old Colony is an abbreviated version, CALSOD v. 1.4, priced at $69.95 (will not import analyzer files).

Easy Loudspeaker Design Software Suite—by Marc Bacon. This software trilogy consists of three separate Win95 based programs, Easy Speak, EasyRoom and EasyTest, all written for the popular Microsoft Excel spreadsheet software (not included with the program). EasySpeak performs both box design and crossover calculations (crossover calculations are based on resistive termination formulas). EasyRoom predicts the acoustic behavior of your listening room and gives information for developing absorbers and diffusers to help correct unwanted behavior. EasyTest follows Joe D'Appolito's book, *Testing Loudspeakers*, to assist in performing various loudspeaker test procedures described in Mr. D'Appolito's book. Priced at $129.95, the software can be purchased from Old Colony Sound Laboratory.

FilterShop 3—by LinearX. FilterShop[11,12] provides the user a substantial collection of features and capabilities that enable extremely complex filter designs to be constructed with high accuracy and relative ease and speed. A full AC circuit simulator is included with advanced components specialized for filter design work and not generally found in SPICE type simulators. FilterShop provides a highly integrated target generation system that is capable of modeling virtually any type of analog or digital filter type. The program can optimize analog 1–16th-order filters (Butterworth, Chebychev, Bessel, Legendre, linear phase, transitional 6dB, transitional 12dB, transitional 3dB, synchronous, Gaussian, MCP Butterworth and MCP Chebychev all available as either low-pass, high-pass, all-pass, bandpass, or bandreject transformations) and perform digital filter synthesis for FIR and IIR filter types. FilterShop features over 500 predefined circuit templates with which to build filter designs (analog active, analog passive,

CALSOD

FILTERSHOP

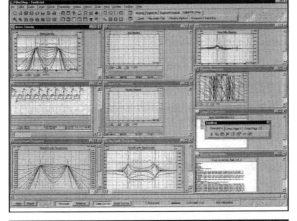

FILTER WORKSHOP

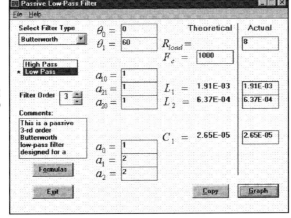

and digital filters) once a target is established. This includes such formats as RLC/RDC ladder filters, and switch capacitor filter design. FilterShop also has an extensive circuit editing facility that allows the user to quickly create any new or unusual type of filter topography, passive or active analog, or digital. Although not the real intended purpose of this powerful software package, you can even design and optimize passive loudspeaker crossovers with Filter-Shop. The recommended computer requirements for FilterShop are Windows NT4 or Windows 2000, 200MB hard drive space, 64MB of RAM, Pentium II 350 or faster, 1024 × 768 video resolution, and Adobe Type Manager. Priced at $1495 and available from LinearX direct (visit the LinearX website at www.linearx.com).

Filter Workshop—by Frank Ostrander. This program combines a useful set of passive network design tools with an instructional resource for network design. Calculations include design of attenuation networks (L-pads), high- and low-pass filters, shelving networks (both high- and low-pass), band-reject filters, inductor winding specifications (core diameter, gauge, turns, etc.) and impedance correction (conjugate) networks. Although filter values are determined using resistive termination as opposed to complex driver SPL and impedance load functions, this is still a very useful piece of Windows based software. Priced at $79.95, the software is available from Old Colony Sound Laboratory (www. audioXpress.com).

FINEBox v. 2.2—by Loudsoft. FINEBox v. 2.0[13,14,15] is a nonlinear dynamic box design program. FINE-Box is able to simulate interPort, bandpass, bass reflex, and sealed box formats while modeling compression and temperature of the motor and voice coil. Using a very intuitive interface, the user can overview the high power performance of a project and quickly optimize the box design including ports. Features include a 3D Display with "Glass" layer Time Response selection capability, a real-time "Volume Control" slider input for Vrms, Wrms versus distance, the ability to import all nonlinear T/S parameters and thermal data from FINEMotor, calculation of dynamic excursion, vent speed, impedance responses, calculation of power compression at any power level and time, an advanced thermal model that predicts heating of voice coil and motor plus the ability to view cone displacement, reflex, and interPort speeds at any power level. Priced at $900 and available worldwide from a variety of resellers. Visit the Loudsoft website for details at www. loudsoft.com.

FINECone v. 2.0—by Loudsoft. FINECone[16,17] is an acoustic finite element dome/cone simulation program. This software can rapidly calculate the frequency response for a new cone or dome profile or analyze problems with an existing driver. Features include automatic airload calculation, the ability to import FINEMotor parameter files, a frequency response overplot, a library of DXF (AutoCAD) and FINECone cone and dome models, plus the ability to export response curves in .txt format. Priced

at $3600 and available worldwide from a variety of resellers. Visit the Loudsoft website for details at www.loudsoft.com.

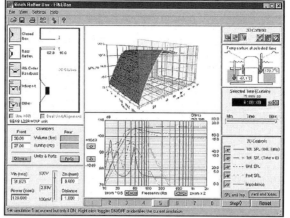

FINEBox

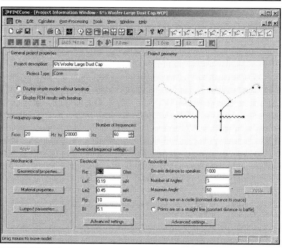

FINECone

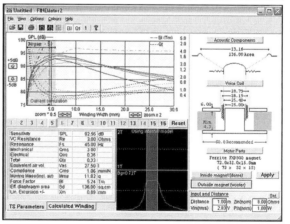

FINEMotor

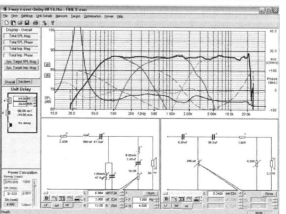

FINEXover

FINEMotor v. 3.0—by Loudsoft. FINEMotor[18,19] is a magnetic system and voice coil design program that will simulate driver SPL and T/S parameters including Xmax and wire diameter from a list of cone/dome part specifications. Features include a selection of round or edge wound copper, aluminum and CCAW voice coil wire, the ability to use dual voice coils wired in parallel, automatic Qms and Rms calculation, simulation of motor shielding, automatic magnetic compensation of air gap for different wire sizes or the number of wire layers, and the ability to specify ferrite, neodymium, or user-specified magnetic materials. Priced at $1800 and available worldwide from a variety of resellers. Visit the Loudsoft website for details at www. loudsoft.com.

FINEX-over v. 3.0—by Loudsoft. FINEX-over[20] is a full-featured crossover optimization program with some unique features. The two best features are the automatic minimum impedance limit that can be set by the user in the optimizer control menu, and the ability to iterate a component value in the network schematic using a mouse wheel to scroll through various component values while the individual driver response and the summation response change in real-time. Other features include automatic power calculation of each network component, an import routine for LMS, MLSSA, Praxis, SoundCheck and other analyzers, variable targets and network slopes, the ability to use asymmetric LP and HP filters slopes and combine them with different acoustic slopes, plus step response and time/distance compensation. Priced at $900 and available worldwide from a variety of resellers. Visit the Loudsoft website for details at www.loudsoft.com.

LEAP 5.0—by LinearX. LEAP v. 5.0[21,22,23,24,25,26], an industry standard for professionals and amateur designers alike in its DOS 4.0 version (originally released in 1985), has undergone an enormous transformation in the Windows version released in 2002/2003. The new software is now divided into two separate programs, LEAP CrossoverShop and LEAP EnclosureShop. Both programs operate in XP, Win 9X, ME or Win2000.

CrossoverShop will optimize loudspeaker passive networks, analog active networks, and FIR and IIR digital network designs, do mixed domain design (analog and digital simultaneously), optimize SPL, group delay or impedance, has a graphical schematic type component entry, a fully automated crossover design wizard, plus thermal/MonteCarlo/sensitivity circuit analysis. Graphic output includes the schematic diagram (analog active or passive and digital), SPL, voltage (network transfer function), impedance, group delay, transient (voltage vs. time), polar plots (horizontal and vertical), and ratio. CrossoverShop comes with a 462-page manual and a 174-page applications manual.

EnclosureShop has an incredible diffraction analysis engine that is like having an anechoic chamber in your computer plus a new 53 parameter transducer model that does accurate large signal nonlinear analysis. Other features include arbitrary structural enclosure analysis, far field, nearfield, and pressure analysis, 360° horizontal and vertical polar plot simulation, both finite and infinite volume domains (you can easily simulate small rooms

LEAP CROSSOVER SHOP

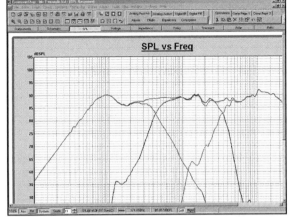

LEAP CROSSOVER SHOP

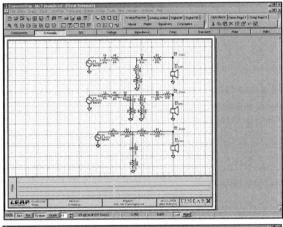

LEAP ENCLOSURE SHOP

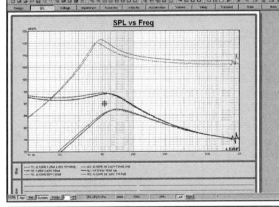

LEAP ENCLOSURE SHOP

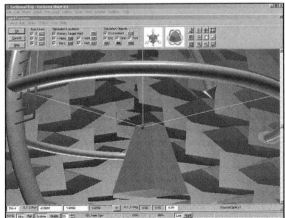

or automobile environments), a nonlinear acoustic network simulator, Quick Design and Reverse design speaker tools, will perform analysis in full, half, quarter, or eighth space measurement domains, plus OpenGL 3D graphics (EnclosureShop was used extensively in the 7th Edition of this book). EnclosureShop comes with a 576-page manual and a 178-page applications manual.

Both CrossoverShop and EnclosureShop have the same extensive set of processing tools. This includes unitary math operations (magnitude offset, phase offset, delay offset, exponentiation, curve smoothing, frequency translation, multiply by jω, divide by jω, convert to real, convert to imaginary, rectangular to polar, polar to rectangular and analytic), binary math operations (multiply, divide, add, and subtract curves), minimum phase transform, group delay transform, delay phase transform, forward Fourier transform (convert an impulse response to a frequency response), inverse Fourier transfer (converts a frequency response to an impulse response), tail correction (necessary for getting accurate phase from the minimum phase utility), curve averaging (allows you to compute averages in four different ways for any group of responses), polar converter, data transfer (for producing SPL/Z graphs), data splice (for making full range measurements from nearfield or groundplane spliced to gated upper frequency measurements), and data realign (this expands or decreases a data entry frequency range).

LEAP 5.0 uses a USB key and recommended system requirements include Win2K or XP, 300MB hard drive space, 256MB RAM, a Pentium III 1GHz or better, 1024 × 768 24/32 color video, and Nvidia OpenGL version 1.2 drivers. The complete LEAP 5.0 package is priced at $1495, while CrossoverShop and EnclosureShop purchased separately are $795 each, and the complete upgrade from LEAP 4.0 is $890. For more, go to the LinearX website at www.linearx.com.

Professional Loudspeaker Design Powersheet—by Marc Bacon[27]. This is another spreadsheet based calculator program written by Marc Bacon for DOS based Lotus 123 or QuattroPro (spreadsheet is not included with this software and will not run with Windows versions). The software performs box design calculations (sealed, vented, passive radiator, bandpass, and transmission line), crossover filter calculation, and also aids in the driver parameter measurement. An unprotected source code allows the user to customize and build upon individual spreadsheets for his own use. Individual programs are accessed through a user-friendly menu tree, and context-sensitive HELP and introductory README.1ST files are also included. The price for this program is $69.95 and is available from Old Colony Sound Laboratory.

LspCAD v. 6.0—by IJ Data. IJData has been producing loudspeaker simulation software since 1991. During that time, LspCAD[28,29,30] (Loudspeaker Computer Aided Design), the primary offering from IJData, has steadily improved and incorpor-

ed numerous features far beyond the scope of the original version. Purchased and used by notables such as B&W, Peerless, Cambridge SoundWorks, Audio Pro, BOSE, TAG McLaren, Labtec, Adire Au-

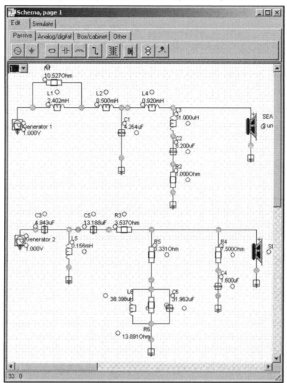

LspCAD

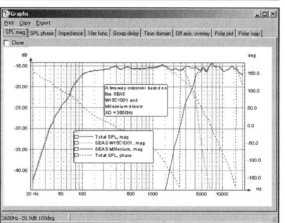

LspCAD

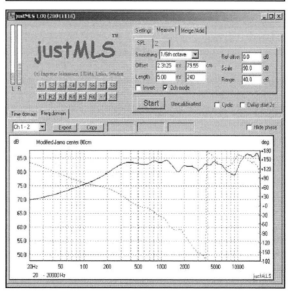

LspCAD

dio, Karl-Heinz Fink, Mission, Joseph D'Appolito, Logitech, and others, LspCAD 6 is the latest incarnation of this successful software[19].

One of the more interesting features of LspCAD v. 6.0 is its ability to model both box design and crossover design at the same time. This is basically accomplished from the schematic dialog screen ("Schema"). The dialog box allows a user to pick both box design elements *and* passive crossover components as well as analog and digital active components. Graphic output from designs includes SPL, SPL phase, impedance, transfer function, group delay, time domain, off-axis overlays, polar plots (both vertical and horizontal), and a color gradient polar map.

The Optimizer for LspCAD also has some interesting features. Tabs allow adjustment of optimizer parameters as well as range and target functions. Target functions include the selection of LP and HP functions as well as EQ, minimum impedance (very useful), and crossover frequency. The optimizer can optimize both crossover networks and box volumes.

LspCAD v 6.0 has also added an auralizer, similar in concept to the one used with the Klippel analyzer. The auralizer for LspCAD allows you to optimize crossover transfer functions and then play back the simulation through your soundcard, allowing comparative listening for different crossover slopes or topographies.

Other features include a built-in MLS soundcard analyzer (Just MLS) for making SPL and impedance measurements (up to MLS length of 32k and sampling to 96kHz) and a power compression function that allows you to compare iterative voltage analysis. Analysis for LspCAD has a number of dynamic functions including thermal analysis and nonlinear Bl and compliance functions.

LspCAD is available in two versions, Standard and Professional. The standard version does everything the Pro version does with significant exceptions. The standard version will not optimize impedance, transfer function, use minimum impedance settings, set target crossover points, optimize box volume, optimize EQ (notch filters), optimize phase response, or optimize imported targets. Also, the standard version does not contain the production tolerance analysis, dynamic thermal modeling, power dissipation in resistors, nonlinear Bl or compliance modeling, polar plots or maps, the predicted power compression function, or the snapshot function. LspCAD can be purchased on the Internet at www.ijdata.com. Pricing for the LspCAD Standard is US $200 (upgrade $150), and US $980 for LspCAD Pro version (upgrade $485). For more on LspCAD 6, visit IJData's website at www.ijdata.com.

SoundEasy v. 10—by Bodzio Software. Based in Melbourne, Australia, Bodzio Software has been producing CAD software for the loudspeaker industry since 1990, which is a long time in the simulation business. SoundEasy, previously featured in *Voice Coil* over the years[31,32,33], is now in Version 10[34] (the 12th release in a 32-bit format for this software). SoundEasy, like LspCAD, is not only a full-featured enclosure and crossover design program, but also includes a very powerful soundcard analyzer.

SoundEasy has ten primary pull-down menus. These include File, Enclosure Tools, Enclosure Calculators, Import/Export Data, Enclosure Details, Crossover Design, Crossover Tools, System Tools, Room/Car Acoustics, and EasyLab, the built-in soundcard analyzer for SoundEasy. Enclosure Tools has four sub menus: Driver Editor, Enclosure Design, Enclosure Optimization, and Edit Driver Notes. Driver Editor includes T/S parameters, SPL data and impedance data, plus the ability to calculate phase curves using a Hilbert Transform in conjunction with a tail correction system.

Box design is performed in the Enclosure Design module. You can look at and generate a total of 19 curves (SPL, phase, impedance, cone excursion, group delay, cone velocity, output power, back EMF, vent velocity, vent excursion, vent SPL, box pressure, and so on) for 13 different enclosure types. Users can choose from sealed box, vented, passive

SOUNDEASY

SOUNDEASY

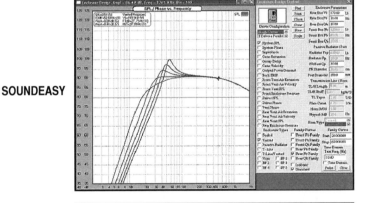

SOUNDEASY

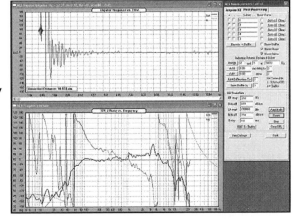

radiator, transmission line, vented transmission line, horn loaded, five bandpass types, two types of dipole configurations with either isobaric (compound) loading or single driver loading. Using the Enclosure Optimization menu, you can designate a driver parameter set and then optimize for three box formats: sealed, vented, or bandpass.

The Enclosure Calculators menu opens up six box design utilities plus a scale resolution and graph color menus. Box design utilities include a vent/passive radiator dimension calculator, a power compression analyzer, an enclosure dimension calculator for various shapes (rectangle, bandpass, slanted front baffle, pyramid, cylinder, and horns), an enclosure diffraction calculator, a nonlinear motor calculator function for $Bl(x)$ and $Cas(x)$ estimation, and a THD estimation calculator that operates from 1Hz to 2kHz.

Crossover design begins with a CAD schematic screen. SoundEasy can develop both active and passive analogy networks as well as simulate digital networks and digital equalization via your soundcard and other devices. Analog passive and active networks can be designed by either manual iteration using the Frequency/Time Domain menu or by using the circuit optimizer. The optimizer has target functions for both active and passive filters up to 48dB/octave with classic filter functions such as Bessel, Butterworth, Linkwitz-Riley, and all-pass type transfer functions.

The Crossover tools menu includes two functions for automatically displaying a single filter section without having to pick the individual capacitors, inductors, and resistors one at a time, or a complete 2-5 way crossover with various possible slope combinations. There are also six utilities that you can use to produce L-pads, CR conjugates (Zobel networks), series LCR conjugates (notch filters), parallel LCR conjugates (amplitude peak EQ), time delay lattice filters, plus transient perfect 2-way filter blocks. This menu is also set up to communicate directly with the Behringer DCX2496 digital crossover.

Once a design is completed, SoundEasy can also do a respectable job of showing performance of an optimized system design in a room simulation. This function is also generalized to include pressure field environments such as the interior of an automobile. The last menu item on SoundEasy is its built-in analog/FFT analyzer that will talk to the soundcard in your computer. The program will generate discrete sine wave tones, stepped sine wave sweeps, or high speed sweeps as well as perform gated stepped sine wave measurements. FFT analysis is done with the MLS analyzer section. Other functions on this menu include a spectrum analyzer, 2-channel oscilloscope, cumulative spectral decay plots, T/S parameter calculator, nonlinear parameter calculator, and RLC meter for measuring capacitors, inductors, and resistors.

SoundEasy also comes packaged with a separate box design program called BoxCAD. BoxCAD is a software tool focused on free-form enclosure analysis using acoustic impedance models and electrical impedance models and is recommended for advanced designers by Bodzio.

SoundEasy operates in the Windows 2000/XP OS platforms and requires a minimum of a Pentium 4 1.7GHz machine with at least 196Mb RAM. The EasyLab analyzer requires a full-duplex sound card (send and receive simultaneously). This program is available from a number of e-tailers, including *audioXpress* at www.audioXpress.com and is priced at $249.95. For more about Bodzio Software, visit their website at www.interdomain.net.au/~bodzio/.

SpeaD—by Red Rock Acoustics. Unlike all the other software listed in this section, SpeaD[35] is not a system simulation program (although the reverse synthesis part of the program does actually perform box design) for box design or crossover development, but a transducer-engineering program. SpeaD is a revolutionary tool that allows a speaker engineer to easily predict the Thiele/Small parameters for any speaker by simply inputting descriptions of its physical parts. *Reverse* SpeaD models the required T/S parameters to achieve desired Box/Speaker system performance. Together they form a suite of tools that should reduce design times and sample iterations dramatically.

In essence SpeaD is a set of integrated tools for voice coil, magnetics, mass/compliance and combined parts modeling. Each "tool" includes its own part-specific modeling and database. For example, the coil designer predicts the dimensions, weight,

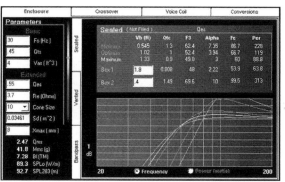

SpeaD

and DCR for virtually any coil configuration. A built-in optimizer will even search out the ideal combination wire size, DCR and winding height, given any two of the three parameters. When a design is complete, the data is passed on to the integrated parts modeler. Included in the data is the ideal front plate ID based on thermal expansion and a user-defined minimum clearance.

SpeaD's magnetics designer takes dimensions for the front plate, backplate/pole, and magnet to create a sophisticated model of the magnetic circuit. Material saturation, short-circuiting, non-ideal parts dimensions, extended poles, and many other real-world aspects are all included in the models. The resulting gap B including an accurate representation of fringe fields is sent to the integrated modeler. The percentage saturation for each of the metal parts is available via a drop-down box.

The mass/compliance modeler takes descriptions of the cone/edge, spider, and miscellaneous weights and completes the physical description of the speaker. The modeler uses universally available parameters such as cone Fo and spider deflection to create the compliance model. Once a design is complete any part can be changed to instantly see the effect on the overall parameters. Changing the coil winding height, magnet size or spider deflection takes seconds instead of days and weeks of building samples. The initial predictions that are achieved with Spea*D* are typically within 15% of the actual measured parameters on the finished speaker (if you have good parts descriptions). Generally, the more accurately the spider compliance and cone Fo are measured, the closer the predictions.

Other features for Spea*D* 2.0 (scheduled for release late 2005/early 2006) include:

- Flat voice coil wire utility for any oval wire ratio and stretch percentage
- Voice coil configurations available are single winding, 2–8 layer winding, 2 and 4 layer bifilar, edge and flat wound wire available for all types
- Thermal heating model that includes the coil, motor mass, and simple convection cooling
- Imported B curves with non-linear Bl curve conversion
- Imported spider compliance curves
- Simulated non-linear Bl (X) and Cms (X) curves
- T-pole C-yoke and MMAG motor simulations
- Motor optimizer that gives optimum dimensions for any gap height and B
- Non-linear motor model using magnet and steel BH curves
- Design tools for cone and dustcap Mmd, surround Mmd and excursion limit, and spider excursion limit
- Complete parts databases with drawing generation
- PDF output support
- Box response simulations for infinite baffle, vented, and bandpass enclosures (includes non-linear data)
- Spea*D* FEA 1.0 is an optional and integrated magnetic FEA utility with AutoCad dxf motor drawing import capability

Reverse Spea*D*[36] fills the gap between CAD box design programs and idealized system design. Instead of predicting the performance of a known speaker in a particular box, it allows you to model the ideal speaker to achieve desired performance in any box.

For example, input a desired f_3, box volume, and Q_{TC} for a sealed box. Then describe some of the basics about the speaker, such as size, DCR and mass. *Reverse* Spea*D* predicts the rest of the parameters required to hit the target. *Reverse* Spea*D* features models for sealed, vented (four types), and single reflex bandpass boxes. Graphical predictions for frequency response, impedance, and displacement power are shown for each design.

Spea*D* and *Reverse* Spea*D* are sold as a package for $2030 USD. Spea*D* is sold separately for $1730 and *Reverse* Spea*D* for $530. The Spea*D* FEA 1.0 option for Spea*D* is priced at $500. For more information, visit the Red Rock Acoustics website at www.redrockacoustics.com.

Speak v. 2.5.112—by Gedlee Associates. Speak v. 2.5.112 is the latest version of the full-functioning speaker simulation software written by noted industry engineering authority, Dr. Earl Geddes. Aimed at the knowledgeable designer of loudspeakers, Speak_32 is a highly sophisticated software package with a straightforward user interface. The program functions using four different databases titled Project, Driver, Enclosure, and Crossover. The Project database holds the specified driver parameters, enclosure details, and any crossover information, active or passive. Speak's driver database contains not only a T/S description by nonlinear parameters for Bl and compliance but also thermal coefficients for doing dynamic analysis. Like most simulation programs, Speak will model sealed, vented, passive radiator, and bandpass. However, unique to Speak is the ability to model low-frequency horns. The user can choose between conical, oblate spheroidal, exponential, or square conical. Also unique to Speak

SPEAK

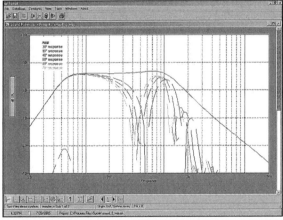

SPEAK

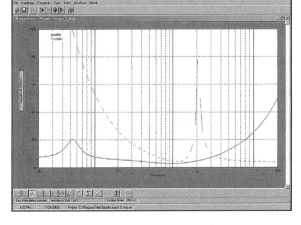

SPEAK

is the ability to model Dr. Geddes' "acoustic lever," which is a patented high-efficiency type of bandpass enclosure. Purchasing Speak licenses the user to build an acoustic lever for his own use.

Crossover design can be either passive or active. Active filter choices are Bessel, Paynter, Butterworth, 1–3dB ripple Chebychev, and Linkwitz-Riley along with parametric EQ filter options. The layout for passive crossovers has an open architecture and is based on a fourth-order topography.

The program can display up to seven curves simultaneously. Plots available in the various analysis modes for Speak include SPL, cone displacement, impedance, voltage, current, power response, and harmonic distortion.

Speak will run in Windows XP, 2000, or NT and is priced at $299.95. For more information, visit the Gedlee website at www.gedlee.com.

TLwrx v. 3.0—by Perception Inc. TLwrx accurately predicts the behavior of transmission speakers at the design stage. Including basic performance relationships similar to the Thiele/Small analysis of vented boxes, TLwrx Version 3.0 provides designers with valuable information about expected system response prior to construction. The software is based on extensive research by G. L. Augspurger, longtime technical manager for JBL's professional division and designer of over 100 custom installations in recording studios worldwide. Information gained from this work is the basis behind Joe D'Appolito's well-known design for the Thor speakers. Included on the CD are TLwrx Version 3.0 software, extensive notes, an alignment chart, and Augspurger's three articles describing his research which appeared in *Speaker Builder* magazine. Priced at $129.00 and available from *audioXpress* at www.audioxpress.com.

WinSpeakerz v. 2.50/MacSpeakerz v. 3.5—by True Audio. True Audio produces a box design program that has been coded for two different platforms, Windows 95/98/NT and the Macintosh. First released in 1989, MacSpeakerz[37] was one of the first speaker simulation software packages, and about the only choice for the Apple platform, and later the Macintosh computers. The new versions of this software have been completely rewritten, while at the same time retaining the user interface familiar to loudspeaker designers who have previously used the product. The latest release adds several new modeling options including the ability to simulate the response of the speaker in an auto cabin environment and the ability to show the bass reduction effects due to diffraction loss (full space as well as half space responses).

Many other aspects of the user interface have been fine-tuned based on user feedback. The program also includes a library of over 1000 loudspeaker drivers from various manufacturers. The user can choose any of these drivers and work with any of the 18 different enclosure types, including third-order closed and fourth, fifth, and six-order bandpass box types in addition to standard second-order closed and fourth-order vented type enclosures (six differ-

ent box design types and compound [isobarik] versions of each).

MacSpeakerz will operate on any Power PC Macintosh or any 68k Mac with a 68020 processor or later, with 35MB of free hard drive space and Macintosh OS 8.1 or later. WinSpeakerz[38] will work well on any Pentium type PC and has installation

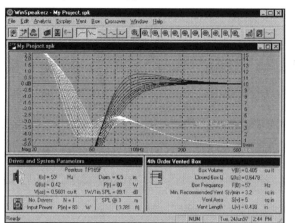

WINSPEAKERZ

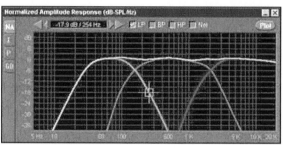

X-OVER PRO

software that will detect the OS being used and load the appropriate files for Win9X, ME, NT, or 2000. The price for WinSpeakerz or MacSpeakerz is $79. Available from True Audio at www.trueaudio.com.

X-Over 3 Pro—by Harris Technologies[8]. X-Over 3 Pro is another design program for calculating network values using resistive termination, but it is also capable of a whole lot more. While this program will not optimize filters, it will allow you to import real SPL and impedance curves from CLIO, IMP, LMS, Smaart Pro, MLSSA, and TEF-20. Using this data X-Over 3 becomes a great computer-based cut-and-try program. Needless to say, using resistive termination formulas to get values for crossovers is barely even a starting place, but the ability that this program has for iterating network values and using real SPL data makes it very useful. X-Over 3 Pro is also useful for coming up with conjugate circuits (CR and LCR), L-pads, and various EQ type circuits. Priced at $99, the software is available from Old Colony Sound Laboratory. For more, visit the Harris Technologies website at www.ht-audio.com.

9.2 ROOM DESIGN SOFTWARE.

AcousticX—by Pilchner-Schoustal, Inc. AcousticX[39] is specifically aimed at small room design as opposed to acoustic programs that will perform analysis of large and medium sized venues, such as EASE from Renkus-Heinz. Given its application to small acoustic spaces, AcousticX should be a valuable tool for home theater and recording studio installation.

The program is divided into four working modules: Model Response, Speaker Boundary Interference Response, Ray Tracing, and Reverb Time. In concert these modules guide the user to making decisions on room layout, dimensions, and treatments. Each module allows the user to enter specific information or allow the software to focus on a range of possibilities. Each of the four modules will allow the acoustic information to be displayed in three ways: room view (3-D room presentation), chart view, or data view.

The Model Response module examines the effect of room size and the ratio of dimensions on the distribution of resonant frequencies. As the user enters room dimensions, AcousticX reports on the degree to which the room fits a prescribed criterion. The module will also allow the user to fix one dimension and vary other dimensions. This information can be displayed as planes on the three axes of the room and can be used to locate the most effective placement of bass traps.

Speaker Boundary Interference Response module explores the interaction of low-frequency radiation and the speakers' relative location to room boundaries. The software will chart the results and automatically display the listening position, size and placement of absorption material, and the size of a valid listening area. If desired, AcousticX will find the best loudspeaker placement based upon the lowest interference level.

The Ray Trace module takes information about room size, speaker type and placement, room absorption size and placement and traces reflection paths with the space. Trace precision can be set to less than 1° increments. Each trace can be followed for any number of reflections with direct energy shown in red, first reflections shown in blue, second reflections in light blue, and later reflections shown in green. This module also displays polar energy on the horizontal plane. Once processed, the room view display will show the direction and magnitude of the energy arriving at the listening position.

Last, the reverb module allows the user to compare and select absorptive treatments by graphically displaying their coefficients and double-clicking to add the material to a room surface. The reverb time is automatically calculated and graphically displayed against a set of preferred criteria.

The program also includes an Acoustic Calculator which performs a variety of common acoustic equations and operations such as unit conversion, level addition, inverse square law, frequency/wavelength/period/velocity calculations, and comb filter calculation. Priced at $399, AcousticX is available directly from Pilchner Schoustal, Inc. You can visit the company website at www.pilchner-schoustal.com.

CARA v. 2.2 Plus—ELAC Technishe GMBH. CARA is acoustic room design software that will help with room design by letting you produce a floor plan up to 100m × 100m with added features for ceiling angles, pillars, and so on. Databases for room materials and loudspeakers are enhanced by editing capabilities allowing you to add your own custom elements. Numerous calculation features evaluate a variety of room effects. Graphs and diagrams including 3D picture sequences will help locate the optimum position of loudspeakers in the described space. New features of the 2.2 release include a New Room Design wizard which offers a number of predefined floor plan templates and allows all dimensions to be entered in non-metric units (inches and feet). The Loudspeaker Editor now allows 1-5 way loudspeakers. Support for Surround 6.1, 7.1, and 8.1 is now available. The program runs in Windows 95/98/Me/NT 4.0/2000, and is priced at $74.95 and available from *audioXpress* at www.audioXpress.com.

Modes For Your Abodes—by Joseph Saluzzi. Modes for your Abodes[40] is available in Win95, Win3.1, and DOS versions. This program is menu driven and allows the user to input room dimensions. From this, the software will calculate and display axial,

ACOUSTICX

CARA

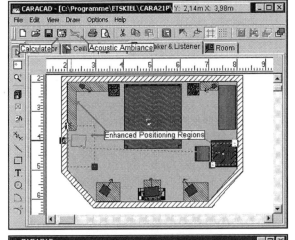

CARA

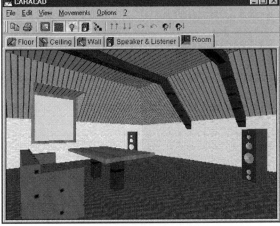

tangential, and oblique acoustic modes as well as predicting axial coincidences. Price is $25 in DOS, and $49.95 in Win95 or Win3.1. Available from Old Colony Sound Laboratory.

Room Optimizer—by RPG Inc. Room Optimizer[25] is a Windows 95/98 based program that automatically and simultaneously optimizes both the room modes and the speaker boundary interference response to determine optimal speaker placement and listening position. The program functions by first locating a random set of listener and loudspeaker placement locations and evaluates these by calculating the energy impulse response via an image model. Following this process, two FFTs are performed on the impulse response to reflect transient and long-term aspect of the way music is perceived. A short-term FFT of the low-order reflections determines the speaker boundary interference response (SBIR) and a long-term FFT of the entire windowed impulse calculates the room model response. A weighted sum of the standard deviation of each FFT response of a defined frequency range is then compared. If the result is below a given error tolerance, the process continues, doing successive iterations until a solution set is determined.

Room Optimizer is quite comprehensive and can suggest such things as specific height of speaker stands, optimum height for listener seating, and placement of acoustic treatments. Any type, number and combination of monopole, dipole, bipole, multipole loudspeaker configurations can be modeled with this software as well as a wide variety of surround placement configurations. Room Optimizer is priced at $99. For more information, visit the RPG website at www.rpginc.com.

9.3 SOUND CARD ROOM ANALYZERS.

ETF v. 5.0—by Acoustisoft. *Voice Coil* featured two previous versions of ETF (Energy Time Frequency) room design and analyzer software in the December 1998 (V 4.0) and October 1997 (V 3.0) issues. Acoustisoft's current version is ETF 5.0[41]. New features include a "maximum length sequence" (MLS) test stimulus, two channel operation, pseudo real-time analysis, increased resolution of room resonance, impulse response measurements, more clear identification of room resonances, the ability to post process measurement files, fractional octave displays (1/3, 1/6, 1/12) plus phase and delay measurements.

ETF 5.0 runs on Windows 95/98/2000/NT 4.0 OS and will work with any soundcard including the new 48kHz PCI soundcards. Unlike the previous version, ETF 5.0 does not require or include a test CD or have 3D graphics. The MLS stimulus is software generated and is not user controllable and contains a 262,143 point sequence. FFT size varies with the gate (window) time selected. The setup is fairly basic with the microphone going into one channel of a soundcard and the output connected to a measurement amplifier and DUT.

Included in the software is a Device Design section that aids in the design of Helmholtz-type bass

traps and rear wall diffusers. ETF 5.0 also includes a well-written manual with an excellent tutorial on room acoustics. The deliver price of ETF 5.0 is $150. Acoustisoft also has available a ¼" calibrated microphone and dual channel preamp (same type as supplied by Liberty Instruments) for $325. For more information on this program, visit the Acoustisoft website at www.acoustisoft.com.

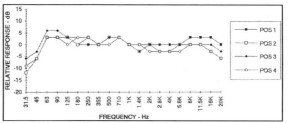

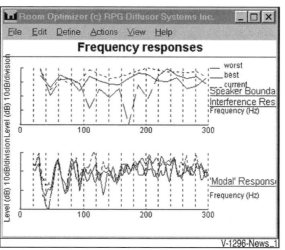

MODES FOR YOUR ABODES

ROOM OPTIMIZER

SIA SmaartLive 5.0—by SIA. SmaartLive 5.0 in its current incarnation is a 32-bit Windows 9X, XP, ME 2000 program that uses the A/D converters of your sound card to perform FFT-based real-time spectrum analysis with real-time transfer function capability, including a built-in delay locator. The real-time module has a plug-in architecture for external MIDI/series/parallel port controlled equalizers (equalizer manufacturers supporting this implementation include BSS, Shure Brothers, Ashley,

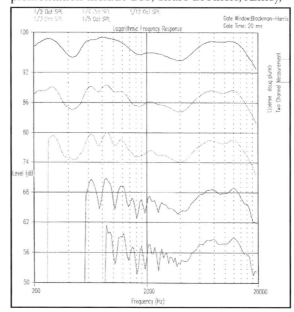

ETF

Level Control Systems, T.C. Electronics, Rane and Klark Teknik), a built-in signal generator for use with duplex soundcards (pink noise and sine wave), and FFT sizes up to 16k points.

Smaart Pro works pretty much like any other two-channel FFT. Once you enter the Analysis Mode and the signal is applied to the speaker and recorded, you window out the reflections and perform an FFT to achieve the in-room response. Smaart Pro can also function as a 1/3 to 1/24 octave RTA.

The current price for SmaartLive v. 5.0 is $695. SIA also offers various upgrades for the different versions from different sellers. For more, visit the SIA website at www.siasoft.com.

SIA SmaartLive

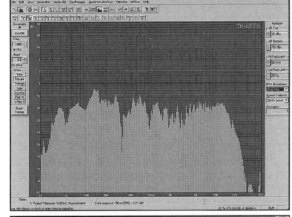

TrueRTA

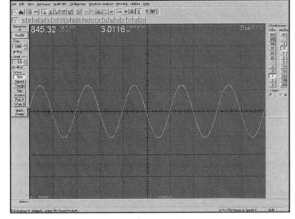

TrueRTA

WinAIRR v. 4.0—by Julian J. Bunn. Available in both DOS and Win 95/98/NT versions, WinAIRR (Anechoic and In-Room Response) works with any full-duplex soundcard such as the Turtle Beach Fiji

or Pinnacle cards. Since the software uses a duplex card, it generates its own noise stimulus, either a pulse, MLS, white noise, sine sweep, or square wave. WinAIRR operates as a dual channel FFT that can display SPL, phase, and time-decay waterfall plots. Priced at $49.95 for either version, WinAIRR is available from Old Colony Sound Laboratory.

TrueRTA v. 3.2—by TrueAudio. TrueRTA is a software measurement tool developed by TrueAudio, the software company that developed the WinSpeakerz/MacSpeakerz loudspeaker design software. TrueRTA is a software-based audio analyzer for testing and evaluating audio systems using any PC with basic sound card capability. The instruments found in TrueRTA include a low distortion signal generator, a digital level meter, a crest factor meter, a dual trace oscilloscope, and a high-resolution real-time audio spectrum analyzer that will display from 1/1 octave to 1/24 octave resolution. This instrument is ideal for installing home theater systems.

New version 3.2 has added several new waveforms to its low-distortion generator including square, triangle, sawtooth, and impulse waveforms. Besides the new waveform generator changes, v. 3.2 also has a new dialog window for adjusting the square-wave duty cycle, added a peak dB number display above dB RMS, upgraded the metering display from 4 to 5 digits, and revised and updated the Help Topics menu content.

By producing these test instruments in software and by calibrating signal input and output capability of your PC's sound card, you are able to achieve a level of performance that could only be replaced by a fairly high quality discrete hardware RTA. TrueRTA is priced at $99.95 for the full version that does up to 1/24 octave resolution analysis. For more information on TrueRTA, visit the TrueAudio website at www.trueaudio.com.

A Closing Note: If you could utilize all the information in this book, and even implement the facilities of some of the outstanding software described above, you will still not necessarily be equipped with what it takes to produce truly superior loudspeakers. Building a well-damped, and correctly designed enclosure is critically important, as is the appropriate design of the crossover network, but it remains that much of what you consider as the "sound" of a loudspeaker has to do with the choice of materials, magnet geometries, and adhesives, things out of the control of most of us doing loudspeaker design work. If you cannot design your own drivers, then the selection of drivers and their eclectic combinations becomes strategically important.

Performing a critical analysis of the drivers you select and identifying and correcting their resonant flaws by modification can be paramount in producing a great speaker. The art of loudspeaker design is knowing how to skillfully combine different driver types and materials to arrive at some sort of "timbre soup" to which others will listen and judge as "musical."

If you wish to better understand the timbre bias which you face in designing loudspeakers, listen to track five of *Stereophile*'s test CD1($6.95 from *Stereophile* at www.stereophile. com). This short demonstration presents the human voice of J. Gordon Holt, the magazine's founder, speaking through a wide variety of typical microphones often used in the recording process. No two of them sounds exactly the same. Each produces an entirely different spectral content from an identical source. Of course, it comes immediately to mind that this would make judging decisions about the timbre bias of a new loudspeaker design exceedingly difficult, which indeed it is.

The loudspeaker industry is still light-years away from being able to replicate original acoustic events in a scientifically controlled and measured manner. In fact, replicating original events probably isn't, in my view, a reasonable goal. Perhaps all the medium will ever be capable of is merely creating a good illusion of the original event. Nonetheless, the best illusions have yet to be produced, and designing and listening to music through loudspeakers you have designed and built yourself is still an exciting and absorbing pursuit.

REFERENCES

1. Soren Bech, "Perception of Timbre of Reproduced Sound in Small Rooms: Influence of Room and Loudspeaker Position," The Perception of Reproduced Sound, Proceedings of the AES 12th International Conference (Copenhagen, June 1993).

2. Olive, Schuck, Sally, and Bonneville, "The Effects of Loudspeaker Placement on Listener Placement Ratings," *JAES* Vol. 42, Number 9, September 1994.

3. Vance Dickason, "Harman's Moving Speakers," *Voice Coil*, August 1999.

4. Jorg W. Panzer, "Multiple Driver Modeling with a Modern Lumped Element Simulation Program," 102nd AES Convention, March 1997, preprint No. 4441.

5. Marshall Leach, Jr., "Software Review: AkAbak," *Voice Coil*, August 1995.

6. Vance Dickason, "New Version of AkAbak," *Voice Coil*, March 1998.

7. Vance Dickason, "BassBox 5.1 by Harris Technologies," *Voice Coil*, May 1996.

8. Vance Dickason, "Bass Box 6 Pro/X-Over 3 Pro," *Voice Coil*, August 1998.

9. Vance Dickason, "BassBox Lite," *Voice Coil*, April 2000.

10. Vance Dickason, "CALSOD 3.0 Update," *Voice Coil*, November 1995.

11. Vance Dickason, "DSP/Analog Filter Design Software," *Voice Coil*, July 1999.

12. Vance Dickason, "Analog and Digital Filter Design," *Voice Coil*, October 1999.

13. Vance Dickason, "New Box Design Program from Loudsoft," *Voice Coil*, January 2003.

14. Vance Dickason, "FINEBox 1.0 from Loudsoft," *Voice Coil*, September 2003.

15. Vance Dickason, "FINEBox 2.0," *Voice Coil*, September 2004.

16. Vance Dickason, "Another Look at FINECone," *Voice Coil*, June 2001.

17. Vance Dickason, "FINECone Version 1.01," *Voice Coil*, February 2002.

18. Vance Dickason, "WinMotor Second Edition," *Voice Coil*, June 2002.

19. Vance Dickason, "New CAD Software Loudsoft," *Voice Coil*, September 2002.

20. Vance Dickason, "User Report: Loudsoft FBX," *Voice Coil*, October 2004.

21. Vance Dickason, "LEAP 4.0," *Voice Coil*, February, March 1991.

22. Vance Dickason, "LEAP 5.0 for Windows—First Look Part 1," *Voice Coil*, January 2002.

23. Vance Dickason, "LEAP 5.0 CD, Part II," *Voice Coil*, August 2002.

24. Vance Dickason, "LinearX LEAP 5.0—A Quantum LEAP," *Voice Coil*, June 2003.

25. Vance Dickason, "LinearX LEAP 5 Enclosure Shop," *Voice Coil*, July 2003.

26. Vance Dickason, "LinearX LEAP 5 Enclosure Shop—Part III," *Voice Coil*, August 2003.

27. Marc Bacon, "The Danielle," *Speaker Builder*, July 1992.

28. Vance Dickason, "LspCAD Loudspeaker Design Software V. 4.0," *Voice Coil*, May 2000.

29. Vance Dickason, "LspCAD 6 Demo Released," *Voice Coil*, December 2004.

30. Vance Dickason, "LspCAD 6 from IJData," *Voice Coil*, June 2005.

31. Vance Dickason, "SoundEasy Version 3.01," *Voice Coil*, April 1998.

32. Vance Dickason, "Box Cad 1.10," *Voice Coil*, February 1997.

33. Bohdan Raczynski, "The Auto Passenger Compartment as a Listening Room," *Speaker Builder*, January 2000.

34. Vance Dickason, "SoundEasy Version 10," *Voice Coil*, August 2005.

35. Vance Dickason, "SpeaD Hits the Streets," *Voice Coil*, January 2000.

36. Vance Dickason, "Reverse SpeaD Released," *Voice Coil*, June 2001.

37. Vance Dickason, "MacSpeakerz Version 3.5," *Voice Coil*, September 1999.

38. Vance Dickason, "WinSpeakerz for Windows 95," *Voice Coil*, August 1997.

39. Vance Dickason, "CAE/CAD Software News: AcousticX," *Voice Coil*, February 1998.

40. Joseph Saluzzi, "What Makes Your Room Hi-Fi? (three part article)," *Speaker Builder*, December 1992, January 1993, March 1993.

41. Vance Dickason, "ETF 5.0," *Voice Coil*, December 1999.

HOME THEATER LOUDSPEAKERS

10.1 HOME THEATER vs. HI-FI.

Loudspeakers differ based on their applications, not only in terms of the transducer specifications but also in the overall requirements of system designs. Pro sound speakers are vastly different from those designed for home listening, which in turn differ from car audio speakers. Over the last several years a new category of home listening loudspeaker has been created for the reproduction of movie soundtracks, called home theater.

An entirely new range of special requirements for home theater has caused the development of specific design criteria. Whether or not home theater speakers should be designed any differently from two-channel speakers raises a somewhat controversial issue on which no firm consensus exists among industry practitioners. On the surface, you would tend to assume that, since movies are almost always choreographed with music, there would be no special design considerations for a home theater speaker as opposed to normal stereo speakers. It seems obvious, but both media require the reproduction of music and speech (if singing can be considered a form of lyrical speech). A closer look at the film sound mixing process, however, yields some interesting contrasts between these two specialized types of sound reproduction.

Music recorded for CDs is for the most part mixed in relatively small studios (compared to a film studio). The mix console and engineer are perhaps less than 10–15' from the main monitors and only 3–6' from the console monitors, therefore the music is mixed near field (close field). The physical layout of a movie dubbing stage is much different. Motion picture soundtracks are mixed in a small theater or soundstage, where the console is perhaps 30–35' (80' in the case of the Disney studios) from the viewing screen and speakers. Mixed in a large acoustic space, the final product sounds great in a full-sized theater equipped with the usual complement of large, equalized horn-loaded speakers (equalized with the SMPTE A202M/ISO 2969 "curve X" or "house curve," which compensates for theater acoustics with a 3dB/octave rolloff above 2kHz[1]). When that same soundtrack mix is dubbed to a VHS VCR tape or laserdisc and played back through typical near-field home stereo equipment, however, it invariably sounds very "bright" in comparison to music recordings, especially on special effects.

Unfortunately, film soundtracks are not generally remixed for home playback (although there is currently a new trend for DVDs to be remixed for a flat response), and this situation must be dealt with at the playback end of the process. It was this observation which led Tom Holman to begin work on the THX (Tomlinson Holman eXperiment) home specification.[2] His answer was to include some upper-frequency equalization in the sound processor (using the same Dolby Pro Logic chip as non-THX equipment), which began gradually

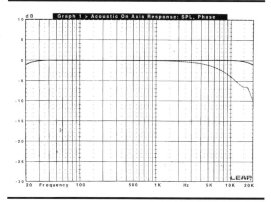

FIGURE 10.1: Response curves for Dolby Pro Logic (solid) and THX (dot) surround-sound processor front/center wide mode.

rolling off the response of the front speakers at around 3kHz, making the response about 1.5dB down at 10kHz and 6dB down at 20kHz (*Fig. 10.1*).

Obviously, this is more of an EQ problem than a loudspeaker design problem, but it does establish a perspective on the differences between speakers intended for music and those for video. It also implies that the spectral balance of video speakers is even more critical than normal two-channel speakers and that, at a minimum, the former should have a flat response. This would suggest that an upward-tilted spectral balance (biased toward the high-frequency end of the spectrum) should be specifically avoided in video playback speakers; a somewhat downward-tilted spectral balance (often favored by two-channel audiophiles) is a reasonable response goal which complements both types of reproduction. (It also implies that the use of a 1/3-octave equalizer on non-THX systems is a good option.)

Considering all of this, does it mean that speakers that are meant to reproduce music are to be designed differently than home theater speakers and vice versa? The answer is no, definitely not. Requirements for accurate timbre are generally reflected in a flat frequency response both on and off-axis and is not different no matter how many channels are in a system[3]. As will be seen, speakers intended for home theater center channels and surround channels do represent different configurations, but again, have the same overall sonic requirements as any previously considered necessary for two-channel only speakers.

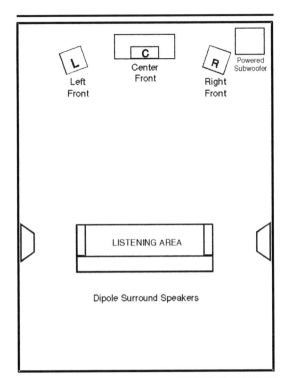

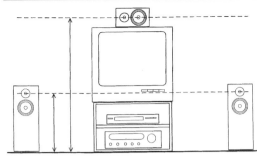

FIGURE 10.4: Improper placement of left/center/right front speakers.

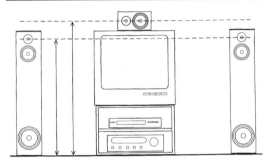

FIGURE 10.5: Compromise placement of left/center/right front speakers.

FIGURE 10.2: Typical loud-speaker placement in a home theater system.

FIGURE 10.3: Proper place-ment of left/center/right front speakers.

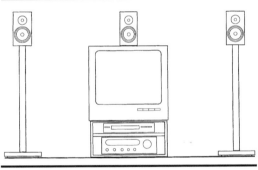

There are two issues of contention among speaker designers and system practitioners which have come up with home theater that did not exist with 2-channel audio-speaker directivity and monopole versus dipole surround speakers. Enhanced vertical directivity and the use of dipole surround speakers are both enhancements that are part of the Lucasfilm THX Ultra and Select home theater certification standards. Controlled-directivity speakers, long used in large acoustic spaces to improve intelligibility (by limiting reflections) and provide uniform coverage, have never been popular for music listening in small acoustic spaces such as the home or music studio. However, experiments designed to explicate the difference between controlled-directivity speakers and wide-dispersion types used in movie theaters provided some of the inspiration for Tom Holman in the development of the THX motion picture theater speakers, and then later in establishing the criteria for THX home theater units.

While no real substantive conflict exists between Dolby Labs' (Dolby licenses all of the processing chips used for home theater systems, THX or not) recommendations for home theater loudspeakers[4] and those licensed by Lucasfilm for THX home systems, there are some marked differences. What follows are general guidelines which will be helpful in designing various home theater loudspeaker elements.

10.2 OVERVIEW OF HOME THEATER LOUDSPEAKER SYSTEMS.

All home theater loudspeaker systems are composed of four basic elements: left/right front speakers, a center front channel located between the left and right speakers, left/right rear (surround-sound) speakers, and a subwoofer, as illustrated in *Fig. 10.2*. While a six-speaker system is the typical configuration, there are a number of variations. These include eliminating the subwoofer and using two full-range left/right front speakers, or using two subwoofers instead of one, for a total of seven speakers per system. Either way, the six-speaker system is derived from four channels using the Dolby Pro Logic processor (left, right, center, and surround) and 5.1 channels for the Dolby Digital format (left/right/center front, left/right rear, and a subwoofer).

Regardless of configuration it is generally agreed that, at a minimum, the left, right, and center speakers should have matching timbres (tonality). To meet this requirement all three front speakers should use identical woofers, midranges, and tweeters, as well as the same crossover elements, with the system response matched as closely as possible. Ideally, the surround channels should incorporate the same woofers, mids, and tweeters as the front speakers, although this requirement is not as stringent as the need for matching timbre in the front speakers. As the industry converts to the new 5.1-channel, Dolby Digital discrete surround standard, the requisite for matched driver timbres in the rear channels will be greater given the surround standard's full-range aspect.

10.3 LEFT/RIGHT FRONT SPEAKERS.

One of the most important parameters for the left/right/center front speakers in a home theater system is height. *Figure 10.3* shows the ideal arrangement of the speakers in relation to the viewing screen: all three speakers are aligned on the same horizontal plane. This establishes a firm image height when special audio effects are panned from one side to the other, such as a speeding car moving across the screen. When a large discontinuity occurs between the heights of the center front channel and the left/right front speakers, as depicted in *Fig. 10.4*, the visual image of the moving object becomes confused and somewhat unnatural due to shifts in image height. Relatively small differences of 12″ or less (illustrated in *Fig. 10.5*) are acceptable.

Height should be an important design factor for determining cabinet dimensions and driver placement for all three front speakers. Also note that a video system's left/right front speakers are not usually placed as far apart as normally recommended for two-channel sound, where a typical angle for good imaging is 60°. The greater placement angle tends to create an image that is wider than desirable for video, so 45° is generally prescribed (*Fig. 10.6*).[5,6,7] Besides the LRs being closer together than the usual 2-channel setup, the speakers should also be angled toward the listening area and the same distance as the center channel to listener. The delay caused by the minor distance difference is actually pretty trivial, but that is the typical recommendation in the industry.[8]

10.4 SHIELDING REQUIREMENTS.

Due to the nature of direct-view, CRT-type television receivers—the type found most often in home theater (front- and rear-projection TVs make up a small percentage of the total)—shielding the picture tube from the high-strength magnetic motors in loudspeakers is essential. Placing a strong magnetic field, such as a loudspeaker, in close proximity to a CRT will distort both color and image. If exposure to the field occurs over a long enough period and at a sufficiently high intensity, the CRT will become discolored even after the source is removed, and will require degaussing. The larger the CRT the more sensitive it is, with the 35″ and larger direct-view sets being the worst. Also, the new 16×9 screens are more sensitive than the normal 4×3 aspect ratio screens.[9,10]

Magnetic shielding is an absolute necessity for center front speakers, and is highly recommended for left/right front speakers. Subwoofers placed in close proximity to a direct-view CRT, or the projection devices in a rear-projection receiver (which are close to the floor), can also require some type of shielding.

Two different levels of shielding which can be incorporated into the motor system of a transducer are illustrated in *Fig. 10.7*. In the first, an additional magnet is attached to the back plate of the motor assembly, with its magnetic polarity reversed in relation to the main magnet. Often referred to as a bucking magnet, this second

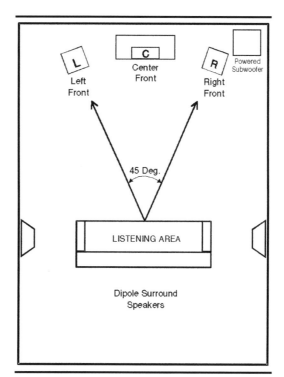

FIGURE 10.6: Placement angle from center listening position for left- and right-channel video speakers.

FIGURE 10.7: Shielding formats for woofer motors.

magnet is usually smaller in diameter and height than the primary magnet. Its exact size is determined by experimentation. This will suppress the magnetic field extending laterally from the driver, but not the field being emitted from the front and rear of the motor. Shielding of this type can be adequate for left/right front speakers and subwoofers when placed at least 1′ or 2′ from the CRT, depending on the sensitivity of the TV used. Generally, a bucking magnet will only slightly change driver sensitivity and will not radically alter the driver's T/S parameters, at least not enough to affect box design.

The second level of shielding incorporates a bucking magnet and a metal shielding "cup" which fits around the magnet structure. Developing a well-shielded driver can be fairly difficult and generally involves a substantial amount of trial and error experimentation to simultaneously achieve a target set of T/S param-eters, the response profile, and the required level of shielding.

Several important criteria determine the effectiveness of a completely shielded driver. The bucking magnet dimensions tend to be the same when used without the cup, the thickness of which is

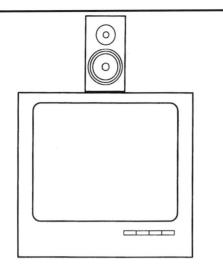

FIGURE 10.8: Vertical placement for a single-woofer center-channel speaker.

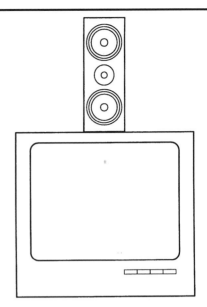

FIGURE 10.10: Vertical placement for a double-woofer center-channel speaker.

FIGURE 10.9: Horizontal placement of a single-woofer center-channel speaker.

FIGURE 10.11: Horizontal placement of a double-woofer center-channel speaker.

also important, as too thin a metal will not shield adequately. Simply placing a cup around the assembly and gluing it to the bucking magnet often will not produce the desired result. Depending upon the specific driver parameters and motor strength, it is also frequently necessary for the cup's rim to be nearly touching the front plate of the motor structure. This invariably requires tooling a special front plate for the driver. When such a method is used, however, the T/S parameters are generally more than moderately affected. Consequently, achieving the same target parameters as the driver's unshielded version can be difficult and require significant juggling of voice coil and motor specifications.

Using fully shielded woofers, mids, and tweeters for center-channel speakers is always required, but this may still be insufficient to completely shield a speaker which is directly on top of a large and magnetically sensitive, direct-view TV. Because some residual magnetic field is always present even in a well-shielded driver, magnetic sum and difference areas can be generated by the interaction of these stray fields when using multiple drivers (magnetic phase anomalies). While a single driver may be suffi-

ciently shielded so as not to visually affect a screen, the close proximity of several such drivers sometimes creates a composite stray field which disturbs the picture, as with the popular woofer/tweeter/woofer format used in center channels. These field problems are further attenuated by adding one or two layers of a galvanized-steel sheet metal lining to the inside walls of the cabinet.

10.5 CENTER-CHANNEL SPEAKERS.

The center channel is perhaps the most important speaker in a home theater system, and actually delivers close to two-thirds of the total system acoustic energy.[11] A center-channel speaker not only delivers the dialogue but pins the entire acoustic presentation to the screen image. Besides matching drivers and response curves (timbre), matching acoustic polarity (the same vertical or horizontal driver placement) is also desirable for left/right/center speakers. Yet, while this is a requirement for THX systems (*Fig. 10.3*), many non-THX manufacturers

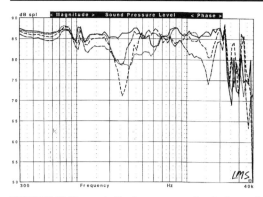

FIGURE 10.12: On- and off-axis response of a single-woofer center-channel speaker with horizontal acoustic polarity (0°= solid, 15°= dot, 30°= dash, 45°= dash/dot).

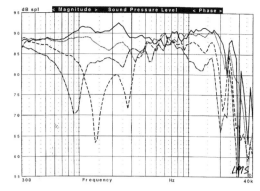

FIGURE 10.15: On- and off-axis response of a double-woofer, flat-baffle, center-channel speaker with horizontal acoustic polarity (0°= solid, 15°= dot, 30°= dash, 45°= dash/dot).

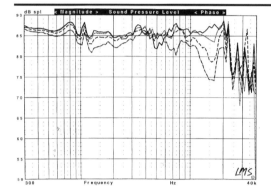

FIGURE 10.13: On- and off-axis response of a single-woofer center-channel speaker with vertical acoustic polarity (0°= solid, 15°= dot, 30°= dash, 45°= dash/dot).

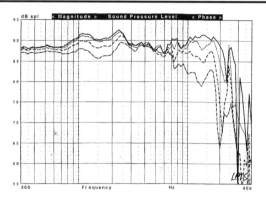

FIGURE 10.16: On- and off-axis response of a double-woofer, flat-baffle, center-channel speaker with vertical acoustic polarity (0°= solid, 15°= dot, 30°= dash, 45°= dash/dot).

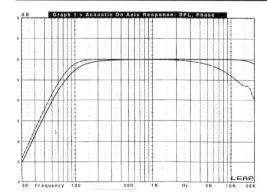

FIGURE 10.14: Response curves for Dolby Pro Logic (solid) and THX (dash/dot) surround-sound processor center normal mode.

sell systems which mix the polarity (*Fig. 10.5*). In this regard, all left/right speakers have a vertical polarity—the drivers are mounted one above the other. *Figures 10.8–10.11* illustrate two typical formats for center front speakers, single- and double-woofer, placed in both vertical and horizontal polarities on top of a TV receiver. The center front channel speaker is commonly positioned horizontally, with the drivers in opposite acoustic polarity to the left/right front speakers. This is done strictly for aesthetic reasons. Most users seem to prefer

the aspect ratio (height to width) of the TV set to be similar to the physical aspect ratio of the loudspeaker cabinet.

For single-woofer models, depicted in *Fig. 10.9*, the frequency response consequence of horizontal placement is a phase cancellation in the crossover region as you move off-axis, as illustrated in *Fig. 10.12*. Compared with the smooth off-axis response of the vertically polarized driver placement (*Fig. 10.13*), the consequence for the horizontally polarized speaker at 30° off-axis is a dip in the response at the crossover frequency of nearly −15dB (indicating the woofer and tweeter are relatively out of phase). While human hearing is not the same as single-point microphone measurement, the result is a noticeable (though not radical) response shift for those seated well off-axis of the speaker. If perfection is the goal, the obvious solution is to follow the THX specification and place the speaker as in *Fig. 10.8*, regardless of cosmetic appearance.

Both Dolby Pro Logic and Dolby Digital based processors generally offer the option of the center-channel speaker functioning either full range (20Hz–20kHz) or with an 80–100Hz high-pass filter (*Fig. 10.14*). Since many systems utilize the "center wide" output configuration, dual-woofer formats have become quite popular, mostly because their increased radiating

FIGURE 10.17: Comparison of flat-baffle and angled front baffle, double-woofer, center-channel speakers.

FIGURE 10.20: Controlled directivity two-way speaker format.

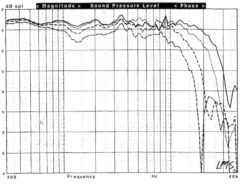

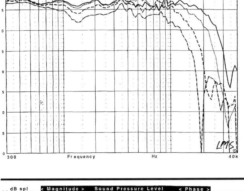

FIGURE 10.18: On- and off-axis response of a double-woofer, angled baffle, center-channel speaker with vertical acoustic polarity (0°= solid, 15°=dot, 30°= dash, 45°=dash/dot).

FIGURE 10.21: Controlled directivity three-way speaker format.

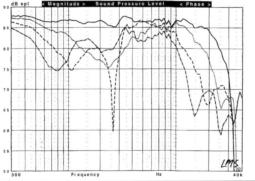

FIGURE 10.19: On- and off-axis response of a double-woofer, angled baffle, center-channel speaker with horizontal acoustic polarity (0°= solid, 15°=dot, 30°= dash, 45°=dash/dot).

physical placement will be less severe. *Figures 10.18* and *10.19* show, respectively, the vertical and horizontal response curves for a double-woofer system, with the woofers angled 20° toward the back of the cabinet. (This obviously entails complex cabinet construction, which is one reason you don't very often see this type of

area allows better handling of low-frequency transients. Unfortunately, this type of center-channel speaker's horizontal placement (*Fig. 10.11*) creates twice the off-axis frequency response problem as single-woofer models (*Fig. 10.15*). Compared with the even off-axis response curves for vertical placement shown in *Fig. 10.16* (which looks quite awkward because of the cabinet height and different aspect ratios, as depicted in *Fig. 10.10)*, the off-axis response for a horizontally placed double-woofer speaker has dips caused by cancellation between woofer and tweeter at the crossover frequency, as well as between the two woofers' off-axis response. The frequency at which this off-axis phase cancellation occurs is a function of the center-to-center distance between the woofers. With this particular speaker, the worst dip comes at 1.5kHz, 30° off-axis, with a −25dB depression.

If horizontal placement is to be used for a double-woofer format speaker, one design solution will ameliorate the problem somewhat. By mounting the woofers at a 20–30° angle toward the rear of the enclosure, as illustrated in *Fig. 10.17*, the off-axis attenuation caused by their

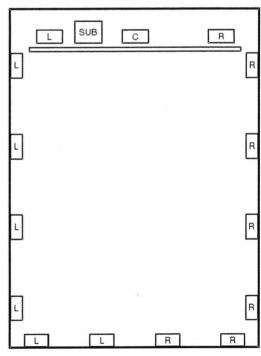

FIGURE 10.22: Surround speaker layout for a typical motion picture theater.

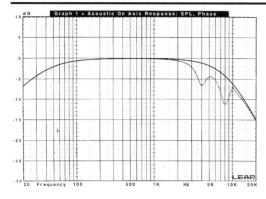

FIGURE 10.23: Response curves for Dolby Pro Logic (solid) and THX (dash/dot) surround-sound processor, rear-channel mode.

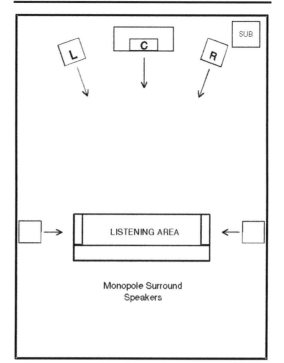

FIGURE 10.24: Suggested placement toward the listening area for surround-sound rear-channel speakers.

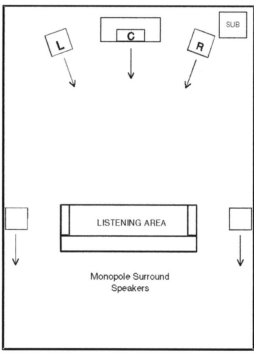

FIGURE 10.25: Alternate placement (toward the back wall) for surround-sound rear-channel speakers.

configuration.) The horizontal placement curves show the off-axis dips caused by woofer cancellation to be only −10dB, a 15dB improvement over a flat-baffle design (the sharp dip is crossover cancellation, because this model had a different driver and crossover than the speaker depicted in *Fig. 10.16*). Renkus Heinz, Inc., produces a software package, ALS-3, which can predict off-axis lobing.[12] It will model different cabinet configurations, such as the one discussed (reviewed in *Voice Coil,* July 1993).

10.6 CONTROLLED VERTICAL DIRECTIVITY.

Part of the development for both THX theater and home theater systems included experimental evidence indicating that speakers with controlled directivity, adjusted to cover the audience area but minimize sound in other directions (notable up and down in the vertical plane), produced greater dialogue clarity. This design aspect has become an integral part of the approved "standard" licensed by Lucasfilm to loudspeaker manufacturers for THX home systems. *Figures 10.20* and *10.21* illustrate two formats which produce this kind of vertical directivity.

Although double-tweeter arrangements like the one in *Fig. 10.20* have never been popular in hi-fi speakers because of the comb filter effects occurring at high frequencies, this driver layout has become a fairly common design approach for high-end loudspeakers. Companies such as B&W, Snell, and Duntech have used the three-way D'Appolito format for a number of years. The exact specification for these types of speakers is proprietary to Lucasfilm, yet enhanced vertical directivity is an unavoidable by-product of this design, especially when the midrange-to-tweeter crossover frequency is extended to the upper response region of the midrange. This format also allows a horizontal aspect ratio, so the center front speaker has the same mid- and upper-frequency acoustic polarity as the left/right front speakers. This is especially effective if the woofer-to-midrange crossover is kept as low as possible (200–400Hz). Whether or not you optimize this design for restricted vertical directivity, the driver layout still makes a lot of sense for a home theater

system. The Atlantic Technology 370 and 450 THX Ultra systems are both good examples of this type of enhanced directivity loudspeaker design.

10.7 REAR-CHANNEL SURROUND-SOUND SPEAKERS.

The setup for rear-channel surround speakers in a full-size theater (*Fig. 10.22*) is substantially different from that of home theater. Due to the size of the acoustic space, multiple speakers are located high above the listeners' heads and arrayed along the theater's side and back walls to provide coverage for the entire listening area. Delay circuits are required to prevent arrival time conflicts, which

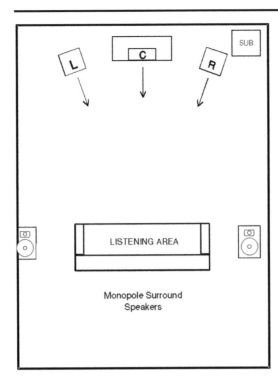

FIGURE 10.26: Alternate placement (aimed at the ceiling) for surround-sound rear-channel speakers.

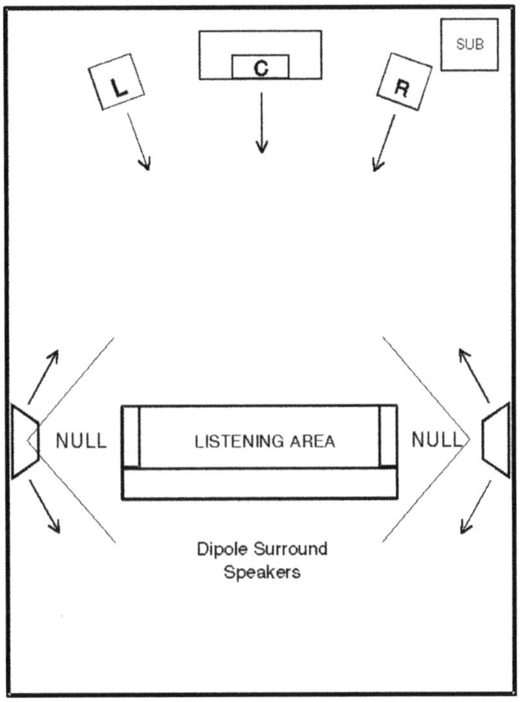

FIGURE 10.27: Placement of dipole-type, surround-sound, rear-channel speakers.

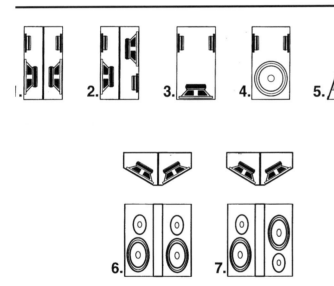

FIGURE 10.28: Different formats for dipole-type, surround-sound, rear-channel speakers.

could disturb the performance's intelligibility. It is also important to note that, unlike the Dolby Pro Logic format, the rear channels of a motion picture theater are discrete rather than a derived matrix of the front channels. This difference will obviously be less for the Dolby Digital standard, which also uses discrete channels for the rear speakers.

Whether the home system uses the Dolby Pro Logic matrix arrangement or the Dolby Digital discrete-channel standard, the objective is to provide the same acoustic effect as obtained in a full-size theater. The surround speakers should be able to produce a diffuse enveloping ambient sound field, which adds to the spaciousness of the experience while concurrently allowing for the directional location of special surround effects. This multi-purpose goal is best attained when the source of the sound cannot be localized by the listener. To this end, Dolby Pro Logic processors modify the response going to the surround channels by providing a high-frequency rolloff above 7kHz (*Fig. 10.23*). While the ear's process for localizing sound is rather complex,[13] diminishing the high frequencies above 5kHz (usually the tweeter's range) makes localization more difficult.

The THX equalization curve also improves the problem timbre shift, which can occur when sounds move from the front to surround speakers. The Dolby Pro Logic processor also rolls off the low frequencies below 100Hz, which means the response of a surround speaker is 100Hz–7kHz. The Dolby Digital standard has options for both full-range 20Hz–20kHz and 100Hz–20kHz surround channels, plus the possibility for more localized surround effects.[14] The discrete nature of the Dolby Digital rear-channel standard does not necessarily imply, however, that surrounds should become totally directional monopoles. In an Internet dialogue in April 1995, Tom Holman stated that past THX and recent Dolby Digital experiments performed at Snell Acoustics by Kevin Voecks indicate experienced and novice listeners substantially preferred the diffuse radiation null pattern of dipole surrounds over direct monopole full-range radiation aimed at the listening area. Later informal tests done by *Home Theater* magazine in April 2000[15] also indicates an overall preference for dipole speakers used as surrounds, however, this is somewhat up to personal taste. Experiments done by the author are in agreement with many in the industry in that movies with a lot of discrete surround information, such as action type films, tend to sound somewhat more "spectacular" with direct radiator surrounds, while the majority of films that use surround information for ambient "fill" certainly sound better with dipoles. One company, M&K, has developed a surround speaker they call a "Tripole" speaker, which has switchable direct and dipole speaker effects. Personally I think this is way too much trouble for most users, so I would have to agree with industry professionals[4,16] on this issue. If you are to choose one type of surround, dipolar is probably the best overall choice.

Dolby Labs recommends placing the surround speakers directly opposite the listening area, and aiming them across the area at a height of 2–3′ above the listener, as illustrated in *Fig. 10.24*. Considering the placement height of the high-frequency rolloff, this technique is certainly viable, but is probably the least preferred if a more diffuse application is possible. Variations on this monopole surround speaker placement which will lead to improved diffusion include aiming the speakers toward the rear wall or the ceiling, as shown in *Figs. 10.25* and *10.26*.

Dipolar surrounds present a whole separate category of surround-sound speaker. They are under the exclusive patent license of Lucasfilm's THX division (patent #5,222,059 and #5,109,416). As with any dipolar speaker, such as a Magnepan screen speaker, the front-to-back cancellation due to the out-of-phase radiation produces a "null" area in the response to the sides of the speaker (like a figure-eight cardioid microphone pattern). Although the original dipole surround application patent did not originate with Lucasfilm, they have refined and championed its use for home surround sound and strongly believe that the dipole provides the very best simulation of typical left/right/back wall arrays found in theaters.

For effective operation, dipole surround speakers should be located on the walls 2–3′ above and on opposite sides of the listening area, with the null firing directly into the area, as illustrated in *Fig. 10.27*. Note that the positive-going drivers (an outward cone motion with + DC applied to the speaker's + terminal) are aimed toward the front speakers.

Designing dipole surrounds is not always as easy as merely reversing the polarity of the opposite driver set. Distance between drivers, which will determine the frequency response and depth of null produced, may or may not require separate manipulation of high- and low-pass crossover sections for each driver set. A number of variations on the dipole theme are currently in production (*Fig. 10.28*). It is important that the woofers be placed in separate chambers, or complete bass cancellation will result. Some manufacturers put a 100–200Hz high-pass filter on one of

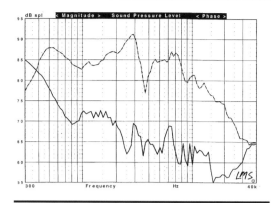

FIGURE 10.29: Frequency response curves for a dipole surround speaker (null axis=solid, on-axis with left driver pair=dash/dot).

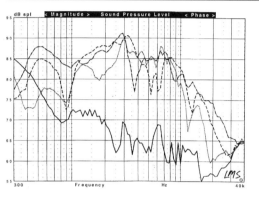

FIGURE 10.30: Frequency response curves for a dipole surround speaker (null axis=solid, 15° left of null=dot, 30° left of null= dash, 45° left of null =dash/dot).

the woofers so less acoustic cancellation occurs at low frequencies, thus giving the speaker more bass.

Figure 10.29 shows the frequency response of a dipole speaker (type 7 in *Fig. 10.28*) measured at the maximum null, which occurs at 0° at a point between the two driver pairs and on-axis (45° off-center) to one of the speaker pairs. It is readily apparent that the null is greatest at frequencies above 2kHz. Also note that this THX-licensed dipole (produced by Triad Speakers) obviously has some deliberate attenuation of frequencies above 10kHz (on-axis to a driver set), suggesting a further attempt to create a nonlocalizable sound source. This extra-high-frequency attenuation is not necessarily typical of dipoles. Many, such as the first Snell THX-licensed dipole, have frequency responses flat out to 20kHz without any additional attenuation (superfluous for Pro Logic processors but possibly relevant for AC-3s).

Figure 10.30 shows the response at 0°, 15°, 30°, and 45°off-center, indicating that the maximum null area is fairly narrow. Some dipole surround speaker manufacturers have added the ability to switch the drivers back into phase and change from dipolar to bipolar radiation. A good bipolar radiation pattern is not necessarily achieved as easily as flipping the phase of the drivers, however, and some well-designed bipole speakers currently on the market have separate high- and low-pass sections for the two driver sets to achieve an even response at extreme off-axis points. Bipoles provide more localization than dipoles, as well as a diffuse surround field. Some manufacturers have gone so far as to connect the two woofers bipolar, with only the two tweeters switchable

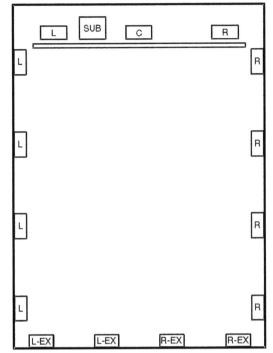

FIGURE 10.31: Surround EX speaker layout for a typical motion picture theater.

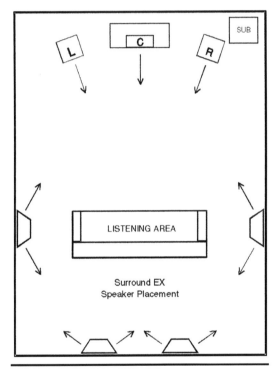

FIGURE 10.32: Surround placement for Surround EX in a home theater system.

sound effect mixed into the surrounds channels. Effects were primarily heard to come from the sides of the listener, but lacked spatial cues that could give the illusion of sounds moving overhead or from behind the listener. Surround EX in a theater (see *Fig. 10.31*) separates the left and right rear surround speakers that were previously part of the left and right wall surround arrays (see *Fig. 10.22*), and connects them to two new amplifier channels driven by a pro-logic type processor that derives the left and right rear channels from the discrete left and right surround channels[16]. You now have the same 5.1 channels of discrete information that is part of the Dolby Digital standard, but in addition the recording process includes rear surround information coded into the left and right surround channels. This is accomplished in the same manner as the center channel was decoded from the left/right channels or the original Dolby Prologic format. The result is nothing short of spectacular. At the 1999 CES in Las Vegas, Lucasfilm THX invited me to preview their new Surround EX (SEX?) demo that was being made to various industry individuals prior to the release of Star Wars Episode I—The Phantom Menace. My thought going into the demo was that the last thing the home theater-buying public needs is two more amplifiers and a room with two more speakers. After viewing and listening to a very spectacular demo that had helicopters flying over my head, a bee buzzing from the left side of the room to the rear of the room, behind my back and back to the right side of the room, my only question was "when can I get a processor for this!"

The setup in a home theater is similar to that of a movie theater (see *Fig. 10.32*). The current consensus seems to be that dipole surrounds provide the best effect and that you need two of them to really get it right.[15] So now, a complete state of the art home theater will have 7.1 channels with a front LCR, dipole left and right dipole (or direct radiator) surrounds, left and right rear dipole surrounds, and the LFE subwoofer channel. Gone are the days of two speakers for stereo, and now it's eight. And not only is this system intended for home theater, 5.1 music is beginning to make headway, and future music releases may well someday all be multichannel like home theater.

10.9 SUBWOOFERS.

When used for video, subwoofers are called upon to produce much more than low pedal tones on a pipe organ or low notes on a sampling synthesizer. Many film special effects are extremely demanding in the low-frequency range and include such sounds as explosions, earthquakes, a jet fighter squadron making a low-altitude fly-by, and even the footsteps of a Tyrannosaurus Rex (try listening to *Jurassic Park* at +105dB!). These effects are not only demanding in terms of spectral content. Film producers tend to exaggerate them in the sound mix to give their work a larger than life feel, which

from bipolar to dipolar radiation. This affords the choice of more or less localization, while also having normal bass loading for both options.

10.8 SURROUND EX—7.1 CHANNEL DOLBY DIGITAL.

Dolby/THX Surround EX is a relatively new enhancement to the Dolby Digital 5.1 standard and adds two more surround channels to the back of the theater. The rationale for the new rear surround in theaters was that the current array of left/right surround information coming from the left/right and rear walls of the theater limited the ability to pinpoint the direction of a

places even greater demands upon a subwoofer. As such, the need for long excursion and linearity are even greater than for a woofer intended only for music reproduction.

The low-frequency specification for video subwoofers is −3dB at 30Hz, which is not all that hard to get out of a moderate-sized cabinet. The real problem is simply excursion and transient capability. For this type of application, single small-diameter woofers (6.5″, 8″, and 10″) are probably best used in pairs, and any sub with 6.5″ drivers should not be expected to reach high volume levels in large rooms at any frequency below 50Hz. Maintaining high SPLs in larger rooms (3,000ft^3 or more) requires a minimum of two 12″ or 15″ drivers per cabinet.[17] Video sub manufacturers frequently use double 10″ and single 12″ and 15″ formats for their products. Obviously, this is all from the perspective of producing very loud and credible home theater performances on the order of what you experience in a real theater. This is not to say that if you really don't listen that loud, then a good single 8″–10″ woofer will certainly provide good performance.

Regarding placement, there are several good criteria for subwoofers. The easiest and probably best location for a home theater subwoofer is in a corner adjoining the LCR array. The corner location allows the woofer to couple to the maximum number of boundaries in the room and provides substantial "gain" over placement locations along a wall or away from a wall. In this location, a single sub can virtually outperform multiple subs in different non-corner locations[18].

If the corner location doesn't sound "right" and seems too robust ("boomy") even when you adjust the levels, try moving it along the wall leading toward the listening area until things begin to sound right. If this proves unsuccessful, try locating your woofer in the center of your listening position, then walk round your room until you locate a position that seems to deliver the best bottom end performance[8]. Once located, place the subwoofer in that location and listen again to the result.

REFERENCES

1. T. Holman, "New Factors in Sound for Cinema and Television," *JAES*, July/August 1991.

2. T. Holman, "Home THX: Lucasfilm's Approach to Bringing the Theater Experience Home," *Stereo Review*, April 1994.

3. K. Voecks, "Multichannel Sound: Reading the Promise & Dangers of Convergence–Kevin Voecks, Designer, Revel Loudspeaker System, Summing Up," *The Absolute Sound*, Issue 114, October 1998.

4. "Dolby Laboratories Information: Speaker Systems for Multi-Channel Audio," available from Dolby Laboratories, reprint no. S88/8272.

5. R. Dressler, "Important Considerations for Dolby Surround," *S&VC*, September 1991.

6. T. Holman, "The Center Channel," *Audio*, April 1993.

7. "A Listener's Guide to Dolby Surround," ©1994, Dolby Laboratories, reprint no. S94/10258.

8. M. Peterson, "Adventures in Loudspeaker Placement: The Agony and the Ecstasy," *Home Theater* magazine, Volume 7, No. 3, March 2000.

9. "User Friendly Shielding Plots," Vifa Newsletter, Winter 1995.

10. "Speakers for Surround Sound–Stray Fields and How to Maintain Them," Peerless International Newsletter 1/94.

11. D. Kumin, "Center Field: The Right Speaker for the Critical Center Channel," *Stereo Review*, April 1993.

12. ALS-3 Software, Renkus Heinz, Inc., 17191 Armstrong Ave., Irvine, CA 92714.

13. F. Alton Everest, *The Master Handbook of Acoustics*, 3rd ed., Tab Books, 1994.

14. R. Dressler, "The (Near) Future of Multichannel Sound," © 1994, Dolby Laboratories, reprint no. S94/10009, p. 51.

15. M. Wood, "Surround-Speaker-Configuration Wars," *Home Theater*, Volume 7, No. 4, April 2000.

16. P. Sun, "THX Surround EX–The Third Dimension in Surround Sound for Home Theater," *Widescreen Review*, Issue 35, Nov/Dec 1999.

17. R. A. Greiner, "Lowdown on Subwoofers," *Audio*, August 1993.

18. T. Nousaine, "A Tale of Two Rooms," *Stereo Review*, January 1999.

CAR AUDIO
LOUDSPEAKERS

11.1 CAR AUDIO LOUDSPEAKERS vs. HOME LOUDSPEAKERS.

With any loudspeaker category, be it home theater, two-channel, pro sound, or car audio, the application dictates the nature of the design. Car audio installations present a unique challenge and place the speaker builder in an almost acoustically "hostile" environment. Designing speakers for cars can literally be described as fitting small boxes inside slightly larger boxes!

The most frequently asked questions I have encountered over the last several years, when lecturing before groups of car audio installers, pertain to the relevancy of Thiele/Small parameters in the design of woofer enclosures, and how to create the kind of center-channel imaging found in high-quality home systems. In this chapter I will address some of the primary issues involved when installing hi-fi loudspeakers in automobiles.

11.2 CLOSED-FIELD vs. FREE-FIELD ACOUSTIC SPACE.

When a loudspeaker is in a large, open acoustic space, such as suspended from a cable 50′ in mid air, there are no boundaries to cause reflections. The frequency response measurement will represent only the drivers and the enclosure. This acoustic situation defines the term "free-field," or anechoic environment, as discussed in *Chapter 8, Section 8.81.*

Any enclosed area where sound energy is propagated is an acoustic space, which can be loosely divided into two categories. A large acoustic space includes a coliseum or auditorium, while a small acoustic space can range from a music recording studio or home listening room to the extreme of the room size spectrum, the passenger compartment of a 348 Ferrari. All are examples of closed fields. As an enclosed room size decreases, and the dimensions of the space diminish in comparison to the sound wavelength being reproduced, the room response becomes dominated by standing waves, reflections, and boundary effects. With rooms the size of the average automobile compartment, the acoustics can be described as a "lossy" pressure field.[1]

A perfect pressure field would have rigid walls which would not transmit sound vibration (such as ones made of 12-inch-thick concrete). But cars have thin steel walls that flex and vibrate, hence the term "lossy." In a perfect pressure field, low-frequency SPL level would be constant; however, because some sections of the car body move more than others, an uncontrollable variation (3–6dB) in low-frequency SPL will exist. This is an acoustic affectation of automobiles of which to be aware.

The acoustic consequence of a speaker placed in a pressure field is the addition of enormous reinforcement or lift to the low-frequency part of the spectrum. To put this in perspective, rooms the size of the average home listening room (1,200–1,500ft³) will get 3–5dB of lift at 20Hz. In a car, however, this modest bass lift is considerably exaggerated. This aspect of loudspeaker design is one many people attempting to build a woofer enclosure for an automobile find confusing. Thiele/Small predictions, whether made using this book and a hand calculator or with a computer program, are really very accurate, but are based on free-field performance. They do not take into account the effects of the acoustic situation in which the speaker will be placed.

11.3 EFFECTS ON LOW-FREQUENCY PERFORMANCE OF WOOFERS IN CLOSED FIELDS.

The way in which a closed field changes loudspeaker performance can be determined by making a few simple measurements. The operating parameters of interest in a loudspeaker are impedance, excursion, and frequency response. If we know how these change going from a free field into a small, "lossy" pressure field, we can understand how to deal rationally with T/S parameters and box design.

Figure 11.1 shows the impedance of a vented speaker measured in a free-field environment. When the same speaker is placed inside the compartment of a small hatchback (a Nissan 240SX) with about 110ft³ of volume, and the impedance curve repeated and overlaid with the free-field impedance curve, the result can be seen in *Fig. 11.2*. Obviously, there is no change, and the closed field for all practical purposes has no effect on the speaker's impedance. Since driver Q is often calculated from impedance, it is safe to assume that Q_{ts} and box Qs also remain unchanged in a closed field.

Cone excursion, which is directly related to

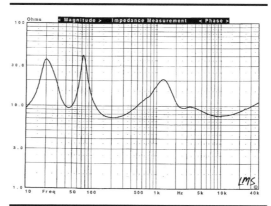

FIGURE 11.1: Free-field impedance of a vented loudspeaker.

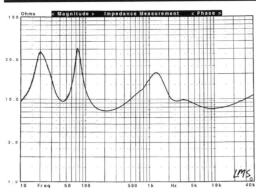

FIGURE 11.2: Comparison of free-field impedance (solid) with closed-field impedance (dot).

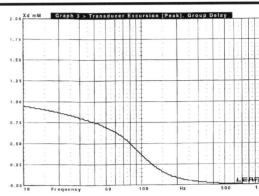

FIGURE 11.3: Simulated free-field cone excursion curve.

FIGURE 11.4: Measured free-field cone excursion curve.

FIGURE 11.5: Comparison of measured free-field excursion curve to closed-field excursion curve.

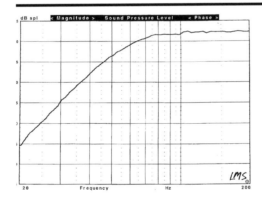

FIGURE 11.6: Free-field groundplane frequency response measurement of speaker in *Fig. 11.1*.

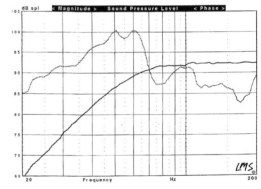

FIGURE 11.7: Comparison of free-field response in *Fig. 11.6* with closed-field frequency response.

power handling and distortion, is a very important aspect of loudspeaker performance. *Figure 11.3* is a LEAP simulation of a cone excursion curve for a Kenwood HQW300 12″ car subwoofer mounted in a small sealed enclosure. I attached a piezoelectric accelerometer similar to the one described in *Chapter 8* to the woofer cone, and used a LinearX LMS analyzer to measure the acceleration curve of the speaker in a free-field acoustic space. Dividing by radian frequency twice, I converted this to a cone excursion curve (uncalibrated), as depicted in *Fig. 11.4*. Since this looked reasonably like a valid excursion measurement, I proceeded to replace the woofer in the 110ft³ car compartment, remeasured the cone acceleration, and converted

this to cone excursion. *Figure 11.5* shows the comparison of the original free-field excursion curve with the closed-field curve, again demonstrating that the closed field has no effect on this aspect of woofer performance. So far, the T/S predictions of box Q and excursion appear valid for large or small acoustic spaces.

The last aspect of operational performance, and the easiest one for most people to relate to, is frequency response. *Figure 11.6* shows the free-field groundplane frequency response of the same 6.5″ vented box woofer used to measure impedance. (I used the 12″ woofer for the acceleration test, because the mass of the woofer cone was great compared to that of the piezo accelerometer; thus, the accelerometer mass

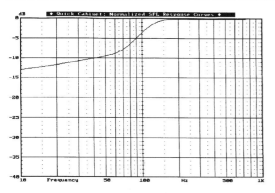

FIGURE 11.8: Computer simulation of closed-field response for a sealed box woofer.

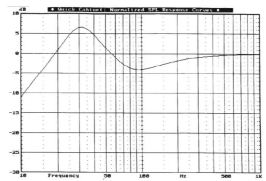

FIGURE 11.9: Computer simulation of closed-field response for a vented box woofer.

FIGURE 11.10: Comparison of free-field and closed-field frequency response measurements of speaker used for closed-field simulation in *Fig. 11.8.*

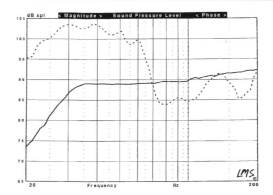

FIGURE 11.11: Comparison of free-field and closed-field frequency response measurements of speaker used for closed-field simulation in *Fig. 11.9.*

would not affect the woofer's performance.) You would get this kind of picture-perfect performance graph from a T/S computer program, and, indeed, they would be almost indistinguishable. Once this woofer and vented box combination is placed in the 110ft³ car compartment, the result is radically different from the free-field response. The frequency response graph depicted in *Fig. 11.7* (a computed average of several mike locations in and around the driver seat) shows the response in the car, sometimes referred to as the car's transfer function (minus the driver response), and has perhaps 7 or 8dB additional lift in the 40–50Hz region and an enormous 20dB boost at 20Hz. When you consider the amplifier headroom and dynamic range required to provide 20dB of boost at 20Hz electronically, this is nothing short of spectacular.

From these three experimental measurements, we can conclude that the car volume changes only the driver SPL. While it means the predicted f_3 and response shape obtained from a T/S design workup is totally out the window, you can still rely on the other aspects, such as damping and cone excursion, as accurate.

11.4 COMPUTER SIMULATION OF CLOSED-FIELD PERFORMANCE.

To fully understand the implications of the SPL changes made by the car's transfer function on

the free-field driver/box transfer function, it would help to look at the effect different-sized car compartments have on various types of enclosures, each having distinct rolloff frequencies and slopes. Doing this empirically by building a battery of different woofer boxes and measuring them in car compartments ranging in size from a Toyota MR-2 to a Dodge Caravan would be a fairly involved task. Fortunately, the LinearX LEAP software has a closed-field prediction option built into its Quick Cabinet box design subroutine. It is capable of giving a fairly good representation of the SPL changes resulting from a closed field. (At the time of publication, LinearX's LEAP was the only software available on the market which could perform this closed-field simulation function; however, another program, TermPro, allows the user to produce a measured car transfer function for a particular automobile and overlay this on different box calculations.)

LEAP's Quick Cabinet is not the main analysis mode, but a separate T/S-like calculator program. (The main routine is more complex and has a number of frequency-dependent variables not found in the T/S fourth-order model.) The program offers the choice of designing a box for a selected set of woofer parameters in either a free-field or closed-field acoustic space. The closed-field selection allows you to specify the volume of the enclosed space and any leakage

Sealed Box (f₃=40Hz)

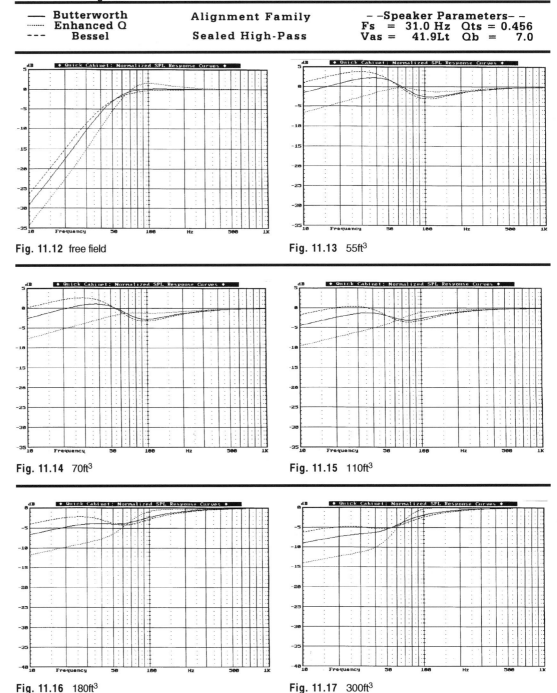

—— Butterworth	**Alignment Family**	– –Speaker Parameters– –	
······· Enhanced Q		Fs = 31.0 Hz	Qts = 0.456
- - - Bessel	**Sealed High-Pass**	Vas = 41.9Lt	Qb = 7.0

Fig. 11.12 free field

Fig. 11.13 55ft³

Fig. 11.14 70ft³

Fig. 11.15 110ft³

Fig. 11.16 180ft³

Fig. 11.17 300ft³

factor, such as that caused by rolling down a window or opening a door.

To verify the accuracy of this prediction, I used two different speakers: a 4″ woofer in a small sealed enclosure, having an anechoic response of −3dB at 100Hz, and an 8″ woofer in a vented box, with a −3dB frequency of 29Hz. This would provide two response profiles, a sealed box with a high rolloff and shallow slope and a vented box with a low rolloff frequency and steep slope, to test the closed-field effects on the bottom two octaves.

The Quick Cabinet closed-field predictions for the 110ft³ Nissan 240SX compartment (zero leakage) using the 4″ sealed box speaker are provided

in *Fig. 11.8*; *Fig. 11.9* shows them for the 8″ vented box. The actual measurements of the two enclosures' anechoic responses compared to the in-car measurements are given in *Figs. 11.10* and *11.11,* respectively. For the purposes of this demonstration, I positioned the mike facing forward in the driver seat at head height, with the woofer at the extreme rear position (a typical location for a car subwoofer).

As you can see, the correlation between the LEAP prediction and the actual measurement is reasonably good. In the case of the 4″ driver, the actual measurement had two major standing wave modes at 80Hz and 190Hz, which caused dips in the response. The prediction of the

Vented Box (f₃=40Hz)

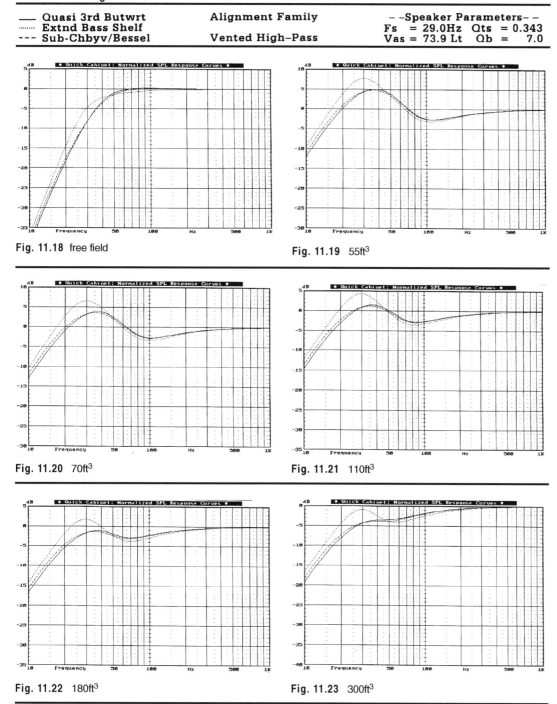

	Alignment Family	– –Speaker Parameters– –
—— Quasi 3rd Butwrt		Fs = 29.0Hz Qts = 0.343
······ Extnd Bass Shelf	Vented High–Pass	Vas = 73.9 Lt Qb = 7.0
--- Sub-Chbyv/Bessel		

Fig. 11.18 free field

Fig. 11.19 55ft³

Fig. 11.20 70ft³

Fig. 11.21 110ft³

Fig. 11.22 180ft³

Fig. 11.23 300ft³

response as down about −12dB, though, compares favorably with the actual measurement's indication of −11dB down at 20Hz. For the 8″ driver, the computer simulation in *Fig. 11.9* shows a peak in response centered on 40Hz, with a depression centered on about 90Hz. This also compares well with the shape measured in *Fig. 11.11*. While the predictions do not take into account the standing wave modes in the compartment, the Quick Cabinet simulations provide a sufficiently good picture of the approximate closed-field response shape, and are close enough to measured reality to draw some general conclusions.

Given that the computer depictions are at least in the ballpark of what happens acoustically in a closed field, the following simulations will give you a good idea of the effects of different car volumes on various types of enclosures. I derived this short study using four types of woofer enclosure: sealed, vented, sealed rear chamber bandpass, and vented rear chamber bandpass. The car compartments ranged from 55ft³ to 300ft³. Each enclosure type has over-damped, critically damped, and under-damped alignments. For a sealed box, this would be equivalent to a set of curves for a box Q_{tc} of 0.5 (over-damped), 0.7 (critically damped), and 1.1+ (under-damped). For a vented box, this would represent a Sub-Chebychev/Bessel alignment (over-damped), a

Sealed Bandpass (f₃=30Hz)

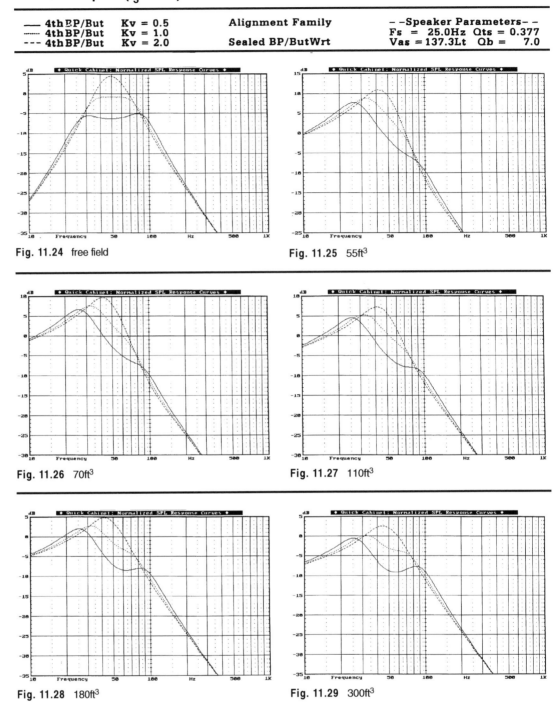

—— 4th BP/But Kv = 0.5	**Alignment Family**	– –Speaker Parameters– –
...... 4th BP/But Kv = 1.0		Fs = 25.0Hz Qts = 0.377
- - - 4th BP/But Kv = 2.0	**Sealed BP/ButWrt**	Vas = 137.3Lt Qb = 7.0

Fig. 11.24 free field

Fig. 11.25 55ft³

Fig. 11.26 70ft³

Fig. 11.27 110ft³

Fig. 11.28 180ft³

Fig. 11.29 300ft³

QB₃ alignment (critically damped), and an extended bass shelf alignment (under-damped). (Extended bass shelf is a class of alignment used in the LEAP software to describe a large, low-tuning frequency vented box whose corner frequency SPL level is lower than the driver's nominal SPL.) Bandpass examples use three efficiency bandwidth tradeoffs giving a narrow high-efficiency, medium bandwidth, medium efficiency, and wide bandwidth (lower f₃) low-efficiency.

The car volumes for each alignment set start at 55ft³ (a small two-seater such as a Toyota MR-2), 70ft³ (an import pickup), 110ft³ (a Toyota Camry or Honda Accord), 180ft³ (a Cadillac or Lincoln TownCar), and 300ft³ (the average minivan). (Acoustic volume for an automobile is rather difficult to determine, not only because of the unusual shapes and contours, but because the actual volume is probably somewhat frequency dependent. For instance, at low frequencies the barrier between a trunk area and main compartment in a sedan or coupe, which is separated by the back seat upholstery, is mostly transparent.) The four different curve sets for the series of woofer enclosures are numbered as shown in *Figs. 11.12–11.35* (f₃s are for the nominally damped example in the three-alignment set).

Vented Bandpass (f_3=25Hz)

—— 6thBP/But	Wide	**Alignment Family**	– –Speaker Parameters– –		
······· 6thBP/But	Medium		Fs = 25.0Hz	Qts = 0.377	
– – – 6thBP/But	Narrow	**Vented BP/ButWrt**	Vas = 137.3Lt	Qb = 7.0	

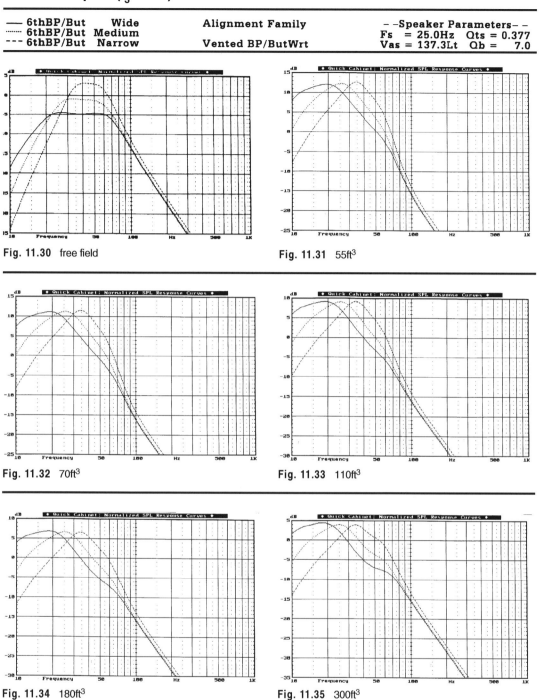

Fig. 11.30 free field

Fig. 11.31 55ft³

Fig. 11.32 70ft³

Fig. 11.33 110ft³

Fig. 11.34 180ft³

Fig. 11.35 300ft³

A number of general conclusions can be made from this series of simulations. For the sealed box set, it is obvious that a fairly flat and even response down to 10Hz can be obtained from a slightly over-damped woofer with a fairly high rolloff. This observation is somewhat substantiated by a current industry trend to use fairly large-diameter (10″ and 12″), low-Q woofers which normally would have served as larger vented boxes in small sealed boxes. Low-Q_{ts} woofers (Q_{ts} = 0.15–0.25) placed in small sealed enclosures produce not only an accurate and even frequency response but also relatively high power handling (if the suspension system can handle the pneu-

matic forces at high volumes), and, even more important, a small, easy to locate box volume.

The vented box examples show how the lower the enclosure's f_3 and the smaller the car volume, the more radical the low-frequency boost. In *Fig. 11.17*, the bass shelf alignment, which has an anechoic f_3=30Hz, yields a nearly 8dB peak centered on 40Hz. The result is exaggerated emphasis on bass guitar and kick drum program material, and probably accounts for the typical one-note bass phenomenon so common in automobiles with low-f_3 subwoofers. If you want exaggerated bottom end, this of course is the answer. Another point of view, however, is to equalize this 40Hz peak region to the nominal SPL level of the remaining system response, thereby giving your amplifier increased headroom and less clipping distortion at high volume levels.

A similar observation can be made of both bandpass-type alignments. Here it is obviously difficult, if not impossible, to rely on the bandpass shape as part of the crossover, as is often achieved in home speaker design. About all you can do in this regard is measure the response in the car and work out the network design accordingly; however, the low-frequency performance conclusion for the bandpass examples is about the same as for a vented speaker. The lower the f_3, the more exaggerated the bass lift. In *Fig. 11.27*, the 18Hz f_3 of the vented bandpass enclosure in the 55ft^3 car compartment (which is probably physically impossible unless the enclosure is mounted on the roof) has almost 20dB of lift at 20Hz and nearly 12dB at 20Hz in a 300ft^3 space.

If you know ahead of time what the acoustic effect of the car volume will be on the speaker's response (in other words, you have some idea of the transfer function of the car's acoustic response), then it is much easier to take advantage of the T/S box design technique. A closed-field computer simulation, such as the LEAP software, is certainly a valuable tool. Even taking a response measurement of a known speaker in an unknown volume before starting the design process can give you a good indication of what to expect.

11.5 DESIGNING FOR CENTER-CHANNEL IMAGE PERFORMANCE.

Phantom-image center-channel performance has almost become a fetish in high-end home loudspeakers, where a particular speaker's merits are largely judged by the height, width, depth, and focus of its image[2] and by overall soundstage quality. The ability to locate musicians playing on a stage is, of course, the intriguing aspect of stereo. I don't think this sonic ability is important to anyone whose lifestyle does not readily facilitate sitting in one position long enough to fully enjoy such a phenomenon (unless you happen to belong to the group of enthusiasts traditionally referred to as "audiophiles").

When doing time on a grid-locked freeway, however, few options exist other than sitting in a fixed position and listening to music. So a good center-channel image as part of the sound in a car is truly an outstanding experience. Center-channel imaging is also an important factor in judging car audio sound systems for events such as the IASCA competitions (in accordance with the 1993 IASCA rulebook). Unfortunately, the average automobile's acoustic environment makes performing this sonic trick more than a little difficult.

Understanding how to enhance image capability in the car acoustic environment starts with an awareness of the design criteria employed to enhance a home loudspeaker's spatial image quality. Some aspects of design which help create good image quality in a home loudspeaker are:

1. Locate the listener equidistant from the two speakers, with the speakers about 6–8' apart. From the listener to the speakers, the distance should be such that the area between the speakers creates an angle of about 60° with respect to the listener.

2. For any given frequency range, use no more than one driver, especially with tweeters. Subwoofers below 80–100Hz are an exception, as are multiple woofer/midranges in symmetrical arrays (mid/tweeter/mid).

3. Carefully match response curves for each speaker in the stereo pair. They should be as close to identical as possible.

If we adapt this criteria to the car compartment, we have a set of generic rules which can be used to enhance image performance:

1. The main speakers, which produce the above-100Hz information, need to be in front of the listener in order to produce a good center stage image.[3,4] This means rear deck full-range speakers are out, as they can produce a good diffuse overall sound quality but not good imaging.

2. The best imaging is generally produced using a two-way satellite for each channel, one midbass driver and one tweeter, mounted in close proximity. (Three-way configurations, while possible, usually require too much "real estate" to be practical.) Multiple mids and tweeters placed in different locations only confuse and destroy any possibility of creating a phantom center-channel image.[5,6,7]

There is one trick in car audio to control image that you almost never see in the home market. Home speakers always have tweeters located very close to mid range producing drivers. However,

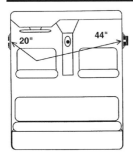

FIGURE 11.36: Relative distances from listening position to door mount location.

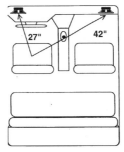

FIGURE 11.37: Relative distances from listening position to in-dash mount location.

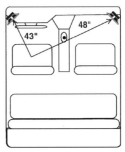

FIGURE 11.38: Relative distances from listening position to kick panel mount location.

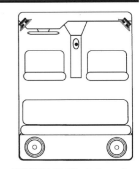

FIGURE 11.39: Subwoofer mounting location on rear deck.

the truth is that the ear locates sound primarily by high frequencies. This is why the dipole type speaker is used in home theater surrounds. With two tweeters operating out of phase, it produces a response null in the listening area and helps provide more diffuse and less localizing sound from the speaker. This acoustic phenomenon can also be taken advantage of in car audio installations. Instead of locating tweeters and mid woofers close to each other in a door or kick panel location, tweeters can be located by themselves in a dash corner near the junction of the windshield and the dash (the best windshield angle for imaging is, according to a study done by engineers at Harman-Motive [OEM car guys from the company who gives you JBL and Infinity] is something greater than 55°).[8] Mid/bass drivers can be located in a door panel (typical factory door panel installations are near the bottom front of the door) or in a kick panel location. With a setup like this, the center image is excellent and placed high, usually above the dashboard as opposed to below the dash for kick panel locations. This easy technique is frequently used in factory installations. In fact my last two cars, a Mitsubishi Eclipse and an Acura CL, both had tweeters installed in this fashion with 5.25″ woofers in the lower door panel location, and imaging was excellent in both cars.

3. The midrange and tweeter drivers operating above 100Hz should be positioned as equidistant as feasible from the listeners. Three possible locations for mounting drivers in the front section usually include the doors, the dash, or the kick panels. *Figures 11.36–11.38* show the approximate relative distances for each of these situations.[9] As you can see, the closest to equidistant mounting is in the corner kick panel, below the dash. Although this location may seem too low to provide the correct image height, the result is more than acceptable. The major drawback is that it usually requires custom plastic-molded kick panels which allow the drivers to be mounted at angles, so the sound is aimed toward the listening area.

4. Locating subwoofers in the front of the car so the center channel images properly at low frequencies is not necessary. Below 80–100Hz, the human ear has difficulty locating the direction from which sound is emanating. The brain tends to assign low frequencies to the same location it senses the high frequency is coming from, which in this case is the front of the car compartment.[10,11] Subwoofer low-pass networks, active or passive, at 80–100Hz should have at least a fourth-order acoustic slope. This will provide the maximum attenuation of upper frequencies, which give spatial clues as to the woofer's physical location. The best place for subwoofers is also the most practical: the rear deck near the trunk area *(Fig. 11.39)*. Although front-seat subs are available, they are generally not required for imaging purposes.

The overall system design resulting from these criteria would look a lot like the system pictured

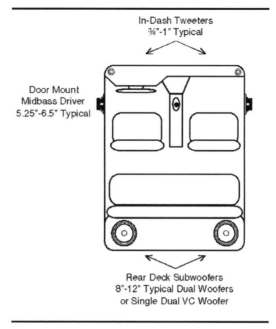

FIGURE 11.40: Typical transducer locations for a three-way biamped car audio speaker system.

in *Fig. 11.40*: Typically biamped electronics (usually two stereo amps and an electronic crossover) with a single DVC (dual voice coil) subwoofer or two subwoofers as depicted with a low-pass 18dB/ octave 80–100Hz crossover for the bottom end. The satellites would have 5.25″ or 6.5″ woofers located in the door panels with ¾″–1″ dome tweeters located in the dash corners and using a passive crossover. While this setup is only one of several good formats that will provide a good front-center image, when done correctly it is probably as easy an after-market layout to apply as there is.

At the time of publication, multichannel music DVD has not impacted the car audio market in any significant way. However, I would expect as this format begins to take hold for home loudspeakers (basically, your home theater type system), you will begin to see multichannel car systems with left, right, and center speakers with rear surround speakers in high performance car audio systems.

For more information on transducer locations and overall evaluation of speakers in cars, check out the following:

• E. Granier, "Comparing and Optimizing Audio Systems in Cars," 100th AES Convention, May 1996, preprint no. 4283.
• D. Mikat, "Subjective Evaluations of Automotive Audio Systems," 101st AES Convention, November 1996, preprint no. 4340.
• A. Farina and E. Ugolotti, "Automatic Measurement System for Car Audio Applications," 104th AES Convention, May 1998, preprint no. 4692.

11.6 NOISE CONTROL.
One of the most often overlooked methods for improving automobile sound quality is noise reduction (except for contest participants, who go to absolute extremes to deaden the noise levels of their cars). Ambient background noise levels in cars are quite high compared to home

listening situations, and greatly detract from sound system fidelity. Recent developments in noise cancellation technology suggest the use of microphones to sample ambient noise levels, followed by DSP circuits which produce an opposite-phase signal to be broadcast into the car compartment to lower noise levels.[12] A more practical method is to employ the same extensional damping techniques used to reduce vibration in wooden speaker enclosures. Removing cosmetic panels on doors and other locations and applying damping material, such as "Q" panels (self-adhesive bituminous felt panels used by auto body shops[13]), will reduce extraneous noise vibration and quiet the car compartment considerably. A sound level meter is helpful in tracking down other disturbing noises, such as air leaks through the firewall, which can be plugged with caulk or silicone. Since every car is different, tracking down squeaks and rattles requires a bit of creative sound sleuthing, but is well worth the effort.

REFERENCES

1. L. Klapproth, "Acoustic Characteristics of the Vehicle Environment," 77th AES Convention, March 1985, preprint no. 2185.

2. J. Atkinson, "As We See It," *Stereophile*, March 1990.

3. Clark and Navone, "Center Stage," Autosound 2000 Tech Briefs, June/July 1993.

4. W. Burton, "Spatial Effects," *Car Audio and Electronics*, February 1993.

5. Clark and Navone, "The Energy Time Curve," Autosound 2000 Tech Briefs, December/January 1992.

6. Clark and Navone, "An Ideal RTA Response," Autosound 2000 Tech Briefs, August/September 1991.

7. D. Staats, "Natural Sound," Autosound 2000 Tech Briefs, January 1993.

8. R. Shively and W. House, Harman-Motive, "Perceived Boundary Effects in an Automotive Vehicle Interior," 100th AES Convention, May 1996, AES preprint no. 4245.

9. Clark and Navone, "Speaker Placement and Center Channel," Autosound 2000 Tech Briefs, October/November 1991.

10. Clark and Navone, "Bass Up Front," Autosound 2000 Tech Briefs, November 1994.

11. W. Burton, "Lost in Space: How Do We Hear Where Sounds Are Coming From?," Autosound 2000 Tech Briefs, August/September 1992.

12. R. Bisping, "Psychoacoustic Shaping of Car Indoor Noises," 91st AES Convention, October 1991, preprint no. 3210.

13. M. Florian, "From Sad to Sparkle: A SAAB Story," *Speaker Builder* 2/95, p. 10.

TWO SYSTEM DESIGNS: HOME THEATER AND A STUDIO MONITOR

At the request of my publisher and good friend, Ed Dell, CEO of Audio Amateur Corp., this edition of the *LDC* includes a comprehensive system-design tutorial. The idea is that the tutorial will give the reader a good idea of how all the elements discussed in this book are brought together to produce truly competent completed designs. While the discussion that follows will provide considerable insight into the process, it does not have the depth of tutorial analysis presented in the companion volume to the *LDC, Loudspeaker Recipes: 1*, mostly because doing so would take more space than is practical for this volume. If you are interested in increasing your understanding of the loudspeaker system-design process beyond this latest edition of the *LDC, Loudspeaker Recipes: 1* is a good place to start.

I decided to produce two designs for this chapter: a complete home theater system, including the left-center-right (LCR), the surrounds, and a powered subwoofer; plus a studio monitor. The studio monitor is really just a high-end two-way that would work as well in a studio as it would as a high-end monitor in a two-channel music system. Actually, I needed a new monitor for my own recording efforts (I am also an amateur song writer and musician), and this speaker was already being designed when I started to update this edition of the *LDC*, so I included it as an example of a well-designed two-channel project.

These designs are not simply DIY projects for the amateur, but are pre-production prototypes for a system that could easily be implemented for manufacturing. This work is not different than any of the work I do for my manufacturing customers in my consulting business, and over the last several years, my designs have received a substantial number of excellent reviews from the audio press. And, like all my other work, it was designed exclusively using the LinearX LEAP simulation software and LMS analyzer. A note of caution, however: this material is both copyrighted and trademarked, so it's probably not a good idea to try to put these products into production without negotiating a license from the author.

12.10 THE LDC6 HOME THEATER SYSTEM.
Both systems in this design tutorial were done in conjunction with my friends at Parts Express (www.partsexpress.com), so assembling parts, drivers, and cabinets for this project—if you care to duplicate it for your own use—can be done from a single source if you like. Because of this, driver choices were limited by the selection available from Parts Express, but since this company

offers a very broad range of products, it was in no way a handicap.

A home theater system's principal loudspeakers are the LCR, left/center/right channel speakers, which ideally should have the exact same drivers to maintain identical timbre content for the front side of a home theater system (see *Chapter 10* for more complete criteria). The rest of the system consists of the left and right surround speakers (similar timbre is also a good idea here), and a subwoofer if the rest of the speakers are not full-range, which describes most home theater systems. Further, the subwoofer is generally powered, so the subwoofer in this project included a new 250W amp from Parts Express.

12.20 THE LEFT/RIGHT FRONT-CHANNEL SPEAKERS.
The LRs and the CTR for this system use the same drivers. My choices were the Audax AP130Z0 shielded 5.25″ woofers and the Morel MDT 40 neodymium 1″ cloth dome tweeters.

12.21 THE WOOFERS.
Audax's AP130Z0 woofer is a fairly recent offering and incorporates a lot of features that reflect the recent modernization of the company's plant. I visited Audax for the grand opening of their new plant in Tour, France (described in the June 1999 issue of *Voice Coil*) and saw this driver being produced. The AP130Z0 is built on a very good-looking injection-molded polymer frame that, when recessed, looks a lot like a powder-coated cast aluminum frame.

Plastic frames are ideal for home theater products, because you get rid of the extra flux leakage that is caused when the typical steel frame wicks the field away from the motor system. Using the polymer frame potentially provides somewhat better shielding when you place the speaker near direct-view screens, but this, of course, still requires a bucking magnet and shielding cup over the motor assembly. Other features for this 5.25″ woofer are an H.D.A. cone, rubber surround, 1″ diameter aluminum voice-coil former, flat spider, and soft PVC dustcap (for a review of this Audax woofer, see the *Voice Coil* Test Bench column, May 1999).

Since both speaker projects in this chapter are designed using the LEAP simulation software, each driver must have both impedance and fre-quency response data measured. I began this process using the LinearX LMS analyzer to measure the Audax AP130Z0 free-air and delta com-

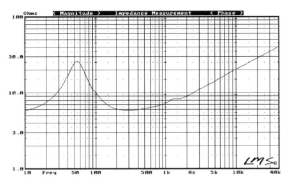

FIGURE 12.1: LR woofer free-air impedance plot.

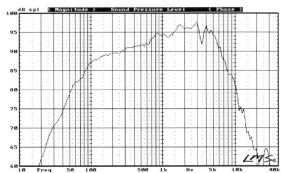

FIGURE 12.5: LR woofers on- and off-axis frequency response (solid = 0°; dot = 30°).

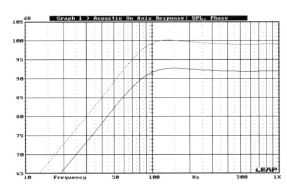

FIGURE 12.2: LR woofer box simulation (solid = 2.83V; dot = 7V).

FIGURE 12.6: LR woofers full-range anechoic frequency response.

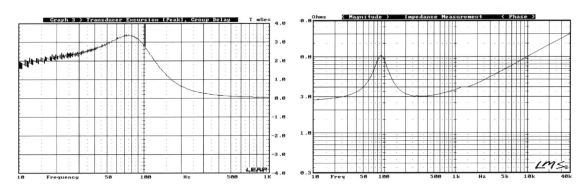

FIGURE 12.3: Group-delay curve for the 2.83V curve in *Fig. 12.2.*

FIGURE 12.7: LR woofers in-box impedance plot.

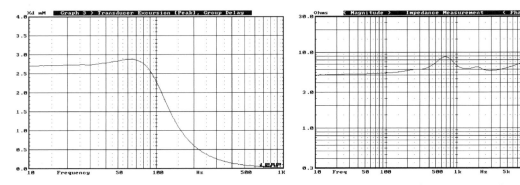

FIGURE 12.4: Cone-excursion curve for the 7V curve in *Fig. 12.2.*

FIGURE 12.8: LR tweeter impedance plot.

TABLE 12.1
AUDAX AP130Z0 PARAMETERS

f_S	57.8Hz
R_{EVC}	5.25
Q_{MS}	2.11
Q_{ES}	0.54
Q_{TS}	0.43
V_{AS}	75 ltr
Bl	4.9 TM
Sens.	87.7dB @ 2.83V
X_{MAX}	2.5mm

pliance (test box) impedance, with the free-air impedance shown in *Fig. 12.1*. I transported this data into the LinearX LEAP software and calculated the parameters as shown in *Table 12.1* (this data is for a typical driver out of the batch used for this home theater system).

Using the analysis mode in LEAP, I determined that a $0.31ft^3$ box would provide an f_3 of 81.5Hz with a box Q_{TC} of 0.83, which is ideal for a home theater satellite. *Figure 12.2* shows the LEAP box simulation in a $0.31ft^3$ box with 100% fiberglass fill material at both 2.83V and 7V (see *Fig. 12.3* for the 2.83V group-delay curve and *Fig. 12.4* for the 7V excursion curve), and the voltage required to increase the excursion to X_{MAX} + 15%. X_{MAX} + 15% represents the driver's maximum linear excursion limit that produces threshold distortion.

The output at the 7V level is 100dB, which is adequate for a home theater speaker in the aver-age living room. This is also a full-range simulation, and virtually all home theater receivers and preamps include a second-order high-pass for LCRs and surrounds usually at 80–100Hz, further increasing the excursion capability of this driver (these satellites are not designed for the "wide" setting on home theater electronics).

Next, I mounted the parallel-connected woofers in the cabinet designed for the LRs (see the cabinet section below for details) and measured the 2.83V/1m frequency response at 0 and 30° from 300–40kHz using the gated sine-wave feature of the LMS analyzer (*Fig. 12.5*). The response is smooth and even up to about 3kHz, more than sufficient for a 3kHz crossover frequency. I also measured the one-meter ground-plane response from 20–500Hz (not shown) and spliced that to the gated anechoic measurement in *Fig. 12.5*, resulting in the full-range anechoic response displayed in *Fig. 12.6*.

The next step is to tail-correct the high-pass and low-pass slopes to their nominal asymptotic value, which is 12dB/octave for both. The LMS calculated phase generated for a tail-corrected magnitude response is extremely accurate (it would require a two-channel sine-wave analyzer in an anechoic chamber to produce phase data as accurate as this). Validity of this method will be apparent later in this tutorial by the accuracy of the LEAP crossover simulations that are dependent upon quality phase data to produce accurate driver summations in a multi-way system design. The final data required for the LEAP simulation is the

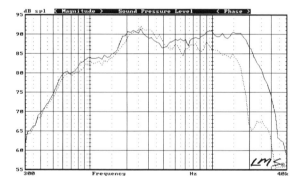

FIGURE 12.9: LR tweeter on- and off-axis frequency response (solid = 0°; dot = 30°).

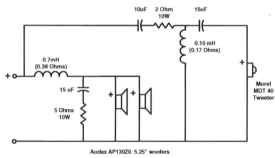

FIGURE 12.11: LR/CTR crossover schematic.

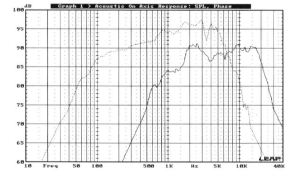

FIGURE 12.10: LR on-axis computer simulation without crossover (solid = tweeter; dot = woofer).

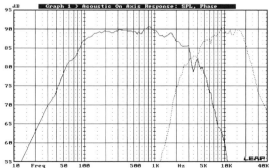

FIGURE 12.12: LR on-axis computer simulation with crossover (solid = tweeter; dot = woofer).

impedance of both woofers (connected in parallel) mounted in the enclosure (*Fig. 12.7*).

12.22 THE TWEETER.

The Morel MDT 40 is a very well-configured neodymium cloth dome chosen because of its small footprint (space it takes up on the front baffle), efficiency that would match up to the projected 87–88dB system efficiency from the Audax AP130Z0 woofers, and overall quality. Morel makes outstanding products, and the MDT 40 is an excellent neo tweeter that has the added advantage of having a vented magnet and cavity, which is unusual for a small neo tweeter. This tweeter also uses the Morel hexatech aluminum voice-coil technology that results in excellent

power handling for this type of device. The main reason, however, to use the MDT 40 is the small faceplate size that is necessary to place the center-channel woofers as close together as possible, which will be explained later in the CTR section.

The first step is to measure the tweeter's impedance (*Fig. 12.8*). As you can see, the 700Hz primary resonance of this device is really low for a small neo tweeter, which will mean virtually no interaction with the crossover that is planned to occur at 3kHz.

Next, I mounted the tweeter in the prototype LR cabinet along with two AP130Z0s. The 2.83V/1m gated sine-wave response curves taken at 0 and 30° off-axis are given in *Fig. 12.9*. While the response has a rise centered on 2.3kHz, the

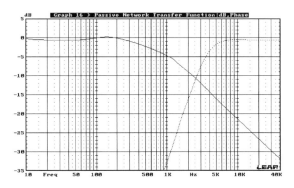

FIGURE 12.13: Transfer functions for networks simulated in *Fig. 12.12* (solid = woofer; dot = tweeter).

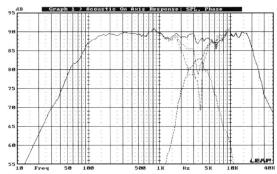

FIGURE 12.16: LR simulated on-axis summation, woofer, tweeter, and reverse phase summation (solid = on-axis summation; dot = woofer; dash = tweeter; dash/dot = reverse phase summation).

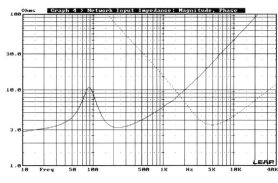

FIGURE 12.14: Impedance magnitude for networks simulated in *Fig. 12.12* (solid = woofer; dot = tweeter).

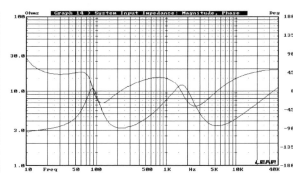

FIGURE 12.17: Computer-simulated system impedance plot.

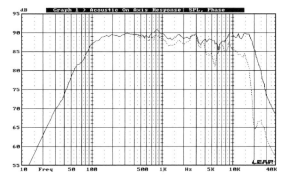

FIGURE 12.15: LR on- and off-axis computer simulation with crossover (solid = 0°; dot = 30°).

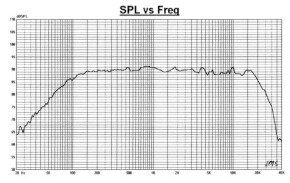

FIGURE 12.18: Measured LR on-axis frequency response.

response above about 2.8kHz is very even with no major anomalies. Off-axis is pretty typical of any 1″ cloth dome tweeter. As with all LMS-generated curves, these must be "tail corrected" and the phase response calculated (the phase calculation step actually occurs during the import process in the LEAP software).

12.23 THE CROSSOVER SIMULATION.

I then imported the data just produced into the LinearX LEAP software. Each driver response has three curves that are loaded into a driver entry in LEAP: an on-axis magnitude and phase, a 30° off-axis magnitude and phase, and an impedance magnitude and phase. All three curves are in the 10Hz–40kHz range. *Figure 12.10* displays the on-axis response for the two parallel Audax woofers and the Morel neo tweeter.

From this you can see that making the Audax woofers measure flat will require a crossover knee at about 150Hz and at the crossover frequency at 3kHz, which I used for a primary crossover frequency for two reasons. First, –3dB from 0–30° for the Audax AP130Z0 woofers occurs at about 4kHz, which means you could probably maintain a smooth off-axis at 30° and beyond at this high a crossover. However, the driver also exhibits a depression between 3–4kHz, which would likely become partially part of the on-axis response variations, so to avoid this, I chose 3kHz (for more in-depth explanation about how crossover frequencies are determined, see

Chapter 1, *Loudspeaker Recipes*). The final SPL for the woofer section should then be about 88–89dB, which although almost too high for the tweeter, simply means that the tweeter will not require much attenuation.

Since I have had much experience with crossover design, the process for me is pretty straightforward, and it is beyond the scope of this chapter to explain all the reasons for choosing a particular network topography (this is why I wrote *Loudspeaker Recipes*). The network topography that works for about 95% of all two-way designs you will ever encounter is shown in *Fig. 12.11*, a second-order topography on the woofer and a third-order topography on the tweeter. This allows the computer optimization of fourth-order Linkwitz-Riley high-pass and low-pass targets for both drivers. However, depending on the interdriver time delay, which is a differential 87µs from the Audax woofer voice coil to the Morel tweeter voice coil, these slopes never end up being symmetrical.

This was part of the story I put into the *Loudspeaker Recipes* book, using four discrete design examples to illustrate the different methodologies for achieving a flat summation using non-coincident drivers. As such, *Loudspeaker Recipes* is not just four construction projects, but a coherent story that flows from one design to another, demonstrating the various methodologies available to the designer.

However, generally speaking, you end up with

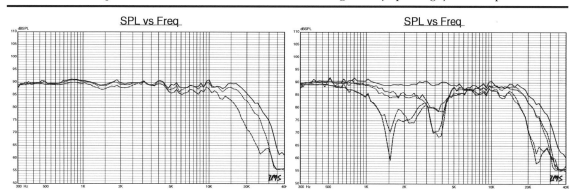

FIGURE 12.19: Measured LR horizontal on- and off-axis frequency response (solid = 0°; dot = 15°; dash = 30°).

FIGURE 12.21: Measured LR vertical on- and off-axis frequency response (solid = 0°; dot = 15° up; dash = 30° up; dash/dot = 15° down; dash/dotdot = 30° down).

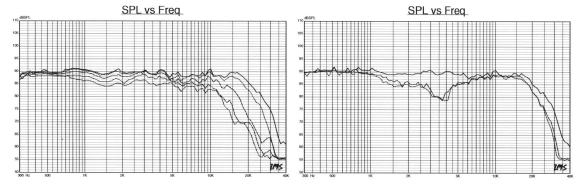

FIGURE 12.20: Measured LR horizontal on- and off-axis frequency response to 60° (solid = 0°; dot = 15°; dash = 30°; dash/dot = 45°; dash/dotdot = 60°).

FIGURE 12.22: Comparison of 15° curves in *Fig. 12.21* (solid = 0°; dot = 15° up; dash/dot = 15° down).

a shallower slope on the woofer and a fourth-order target slope on the tweeter (to maintain minimum excursion and increase power handling). This asymmetrical slope actually compensates for the interdriver time delay (again, this is explained in excruciating detail with pages of examples in *Loudspeaker Recipes*) and results in a flat summation. Note that what appears to be a CR conjugate on the woofer circuit is actually what is commonly referred to as a "Zobel" type circuit (but not by me; I refer to this type of CR and LCR circuit as a conjugate, which more accurately describes its function).

In other words, I didn't use LEAP and a "Zobel" to flatten the impedance before I added the second-order filter. That is the wrong way of

thinking about computer-optimized crossover design. I used that particular topography to manipulate the network-voltage transfer function so it would combine with the woofer-acoustic transfer function to produce the desired target transfer function. It's all a matter of mindset.

The results of the LEAP optimization (*Fig. 12.12*) show both woofer and tweeter connected to the simulated low-pass and high-pass filters described in the network diagram in *Fig. 12.11*. The final crossover frequency turned out to be about 3.2kHz. The transfer functions for these filters (*Fig. 12.13*) show a fairly shallow one for the tweeter and a dual break-point transfer function for the woofer. While I earlier suggested the final optimization would be around 150Hz and

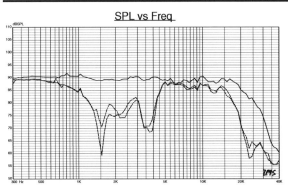

FIGURE 12.23: Comparison of 30° curves in *Fig. 12.21* (solid = 0°; dash = 30° up; dash/dotdot = 30° down).

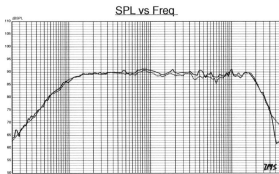

FIGURE 12.26: Comparison of measured LR on-axis frequency response and computer-simulated on-axis response (solid = measured LR; dot = simulated LR).

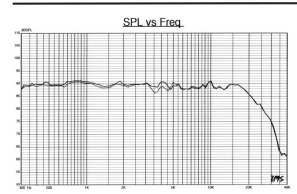

FIGURE 12.24: Comparison of both LR prototypes (LR1 = solid; LR2 = dot).

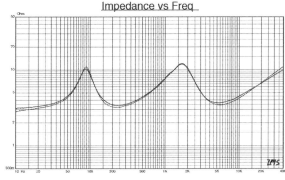

FIGURE 12.27: Comparison of measured LR system impedance and computer-simulated system impedance (solid = measured system impedance; dot = simulated system impedance).

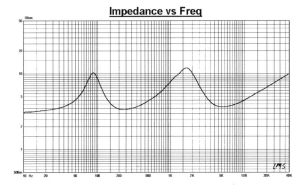

FIGURE 12.25: LR measured system impedance plot.

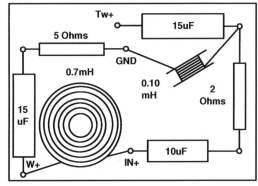

FIGURE 12.28: LR/CTR crossover mechanical layout.

Individual component losses that vary much from the idealized ones I programmed in LEAP could easily make the final result differ substantially from the simulation (discussed in *Loudspeaker Recipes*). However, years of experience have taught me that two-way designs are not very sensitive to this, at least not nearly as much as three-way designs (especially fourth-order band-pass filters), so it really isn't all that critical. However, this is the procedure you should follow if you have an LC bridge that can measure component reactance at different frequencies such as the one built into the LMS analyzer.

Figure 12.18 displays the on-axis frequency response of the LR at 2.83V/1m, measured using the new (at the time of publication) LMS 4 Windows software (see *Voice Coil* August 2000). Although you don't see the phase anomaly predicted in LEAP at 6.5kHz, the response is ±1.84dB from 112Hz–18.5kHz, which is quite good. The off-axis response is also quite good, and close to what LEAP predicted (see *Fig. 12.19* for the 0, 15, and 30° off-axis 2.83V/1m frequency responses). In fact, if you keep measuring off-axis out to 45 and 60° (*Fig. 12.20*), the overall power response of this speaker is obviously very smooth.

Figure 12.21 depicts the vertical responses of this driver on-axis (obviously the same as the horizontal on-axis) and 15° up/down and 30° up/down. *Figure 12.22* compares the 15° up/down measurements and shows them to be pretty symmetrical with no excessive lobing, which would be indicated with great differences in the two responses. Likewise, *Fig. 12.23* compares the 30° up/down vertical response measurements, again fairly symmetrical.

Notice that there are two nulls with the drivers placed in this physical location on the baffle

3kHz, it ended up about 160Hz and 2.2kHz to achieve the desired acoustic transfer function for the woofer in *Fig. 12.12*. *Figure 12.14* gives the impedance magnitude for each filter section.

In the final system summation both on-axis and at 30° off-axis (*Fig. 12.15*), except for the small anomaly at 6.5kHz, the response is close to ±2dB. The off-axis response looks fairly typical with some unavoidable loss in the 1kHz region and a phase anomaly at 6.5kHz. While LEAP is very good, models are almost never perfect presentations of reality, and the quality of the simulations can vary from example to example depending upon several factors. However, the final result should be sufficiently close to this simulation to proceed with any required final "tweaking." In truth, I have designed many loudspeaker products in the home, car, and pro sound markets using this system, and have never altered a design significantly from the LEAP simulation in the final production prototype.

One of my criteria for network summations is the degree to which they are in-phase at the crossover frequency. *Figure 12.16* depicts the on-axis response, plus both drivers' responses and on-axis summations with the tweeter polarity reversed. The result of the reverse polarity summation is a fairly deep null close to the crossover frequency, indicating that the crossover region is in-phase. Last, *Fig. 12.17* shows the LEAP-simulated system impedance magnitude and phase.

12.24 THE FINISHED PROTOTYPE.
Once you have optimized the design in LEAP, the next step is to prototype the network and test the speaker to make certain that the performance closely matches the simulation. Prior to measuring the response of the LR, I measured both inductors and capacitors used in this circuit at the frequency at which they were being used and also at an octave below this for high-pass components and an octave above that frequency for low-pass components.

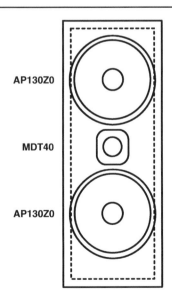

**Home theater
left/right speaker.**

AP130Z0

MDT40

AP130Z0

Int. Dim.= 15.75"x5.5"x6.5"
Ext. Dim.= 17.25"x7"x8"
Material= 0.75" MDF
Driver spacing= top woofer edge 1" from top of box;
1/4" spacing or less between tweeter and woofer perimeters.

board. The 3.5kHz null is near the crossover frequency and is caused by the path length difference from the microphone to the woofers and to the tweeter as you measure off-axis. The null located at 1.5kHz is caused by the different microphone path lengths to each of the two woofers. This will be important to note when looking at the center-channel horizontal off-axis performance.

One of the best indicators of good stereo imaging, or, in the case of home theater, symmetrical L/R response, is the matching of the two left/right speakers, which, as you can see in *Fig. 12.24*, is within less than 1dB, except for a small area centered on 4kHz. And last, the final measured system impedance magnitude response is illustrated in *Fig. 12.25*.

While there were some variations in the final measured prototype frequency response and impedance as indicated in the comparisons seen in *Figs. 12.26* and *12.27*, respectively, it is obvious that the simulations were within less than 1dB and 1Ω. This is really quite good, although I have often seen better examples than this.

12.25 CONSTRUCTION DETAILS.
The cabinets and all drivers and crossover parts for this home theater project are available from Parts Express (www.partsexpress.com). The net internal volume of the LR cabinet was approximately 0.31ft^3, with internal dimensions (HWD) of $15.75'' \times 5.5'' \times 6.5''$.

Built out of 0.75″ MDF with all drivers inset, the cabinet is 100% filled with R19 fiberglass (the pink stuff). You can observe driver layout in

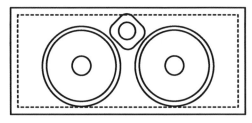

AP130Z0   MDT40   AP130Z0

Int. Dim.= 6.5"x15.75"x5.5"
Ext. Dim.= 8"x17.25"x7"
Material= 0.75" MDF
Driver spacing= woofers are separated 1" from each other, 0.5" from bottom of box. Tweeter is as close to woofers as practical, but must be minimum of 1/4" from top of box.

Home theater center channel.

the cabinet mechanical drawing; all drivers are basically centered and placed as close together as possible. The layout of the crossover board is pretty conventional (*Fig. 12.28*). Capacitors are all polypropylene types, inductors are air core, and the resistors are the non-inductive variety available from Parts Express.

12.30 THE CENTER-CHANNEL SPEAKER.
The CTR in this system is nearly identical to the LRs in every way: same drivers, same cabinet volume, and same crossover. The only way the two speakers are really different is in cabinet shape and driver layout and the fact that the LRs are acoustically oriented vertical WTWs (woofer/tweeter/woofer) and the CTR is an acoustically oriented horizontal WTW. This is mostly for cosmetic and physical reasons.

Since most center-channel speakers are placed on the top of a direct-view or rear-projection TV, having a speaker that is vertically oriented—that is, its height is substantially greater than its width—is visually distracting. Having the cabinet aspect ratio of the center channel more on the order of the aspect ratio of your TV cabinet is much more aesthetically pleasing. The original THX center channel was supposed to be placed with a vertical acoustic orientation the same as the LRs, and, from a sound presentation standpoint, this is absolutely correct. However, it looked like a haystack growing out of your TV and the American public just didn't like it.

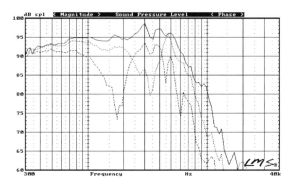

FIGURE 12.29: CTR horizontal on- and off-axis frequency response (solid = 0°; dot = 15°; dash = 30°).

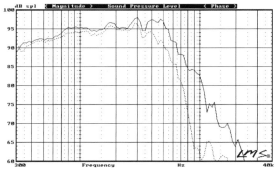

FIGURE 12.30: CTR vertical on- and off-axis frequency response (solid = 0°; dot = 30°).

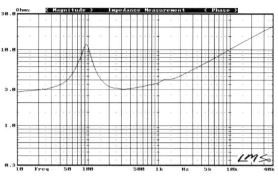

FIGURE 12.31: CTR woofers in-box impedance plot.

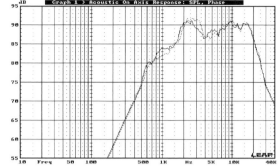

FIGURE 12.35: Comparison of the on-axis frequency response of the LR and CTR tweeters (solid = LR; dot = CTR).

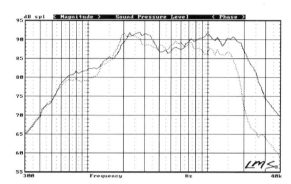

FIGURE 12.32: CTR tweeter on-axis frequency response (solid = 0°; dot = 30°).

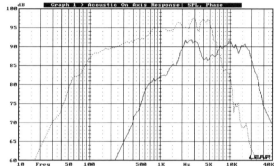

FIGURE 12.36: CTR on-axis computer simulation without crossover (solid = tweeter; dot = woofer).

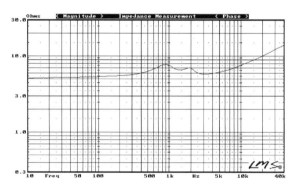

FIGURE 12.33: CTR tweeter impedance plot.

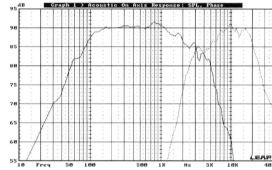

FIGURE 12.37: CTR on-axis computer simulation with crossover (solid = woofers; dot = tweeter).

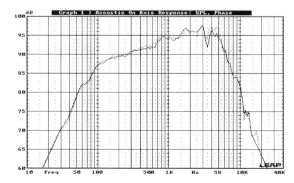

FIGURE 12.34: Comparison of the on-axis frequency response of the LR and CTR woofers (solid = LR; dot = CTR).

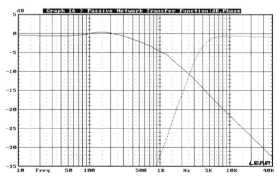

FIGURE 12.38: Transfer functions for networks simulated in *Fig. 12.37* (solid = woofers; dot = tweeter).

This was the key to the success of the Atlantic Technology 350/370 and 450 series I designed, in which the speakers were three-ways instead of two-ways. The reason for the greater degree of acceptance of these products was that the mid-tweeter arrays were vertically oriented on all three front speakers (LCR), but the center channel had the horizontal aspect ratio that matched the TV (*Chapter 10*). This became the preferred format for THX products prior to the new Select standard.

As a result of this focus on aspect ratio, nearly all center-channel speakers have an eye-pleasing aspect ratio and the wrong acoustic orientation. Fortunately, this is not quite as serious a problem as response curves seem to indicate, since these systems perform very adequately and the acoustic

disturbance is acceptable. You can, however, minimize this off-axis lobing problem caused by the separation of the two woofers typically used in these systems by locating the woofers as close together as possible.

Looking at the mechanical drawing of the center channel, you can see the tweeter mounted up high on the baffle and rotated 45° so that the woofers could be positioned as close together as possible. How this affects the performance will be obvious when we examine the final center-channel prototype.

12.31 THE WOOFER.
Since the woofers in the center channel are identical to the LRs, I will not repeat the T/S data

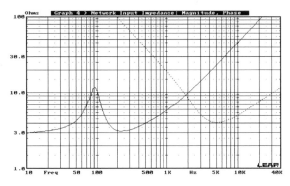

FIGURE 12.39: Impedance magnitude for networks simulated in *Fig. 12.37* (solid = woofers; dot = tweeter).

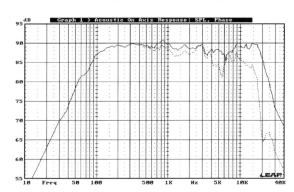

FIGURE 12.40: CTR on- and off-axis computer simulation with crossover (solid = 0°; dot = 30°).

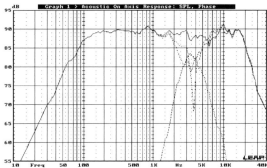

FIGURE 12.41: CTR simulated on-axis summation, woofer, tweeter, and reverse phase summation (solid = on-axis summation; dot = woofer; dash = tweeter; dash/dot = reverse phase summation).

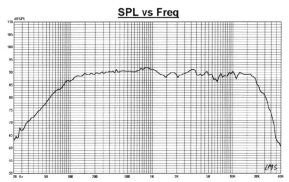

FIGURE 12.42: Computer-simulated system impedance plot.

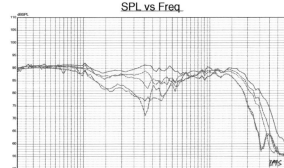

FIGURE 12.43: Measured CTR on-axis frequency response.

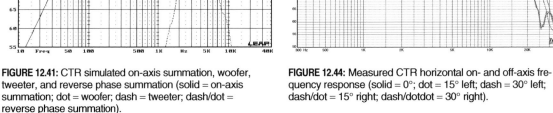

FIGURE 12.44: Measured CTR horizontal on- and off-axis frequency response (solid = 0°; dot = 15° left; dash = 30° left; dash/dot = 15° right; dash/dotdot = 30° right).

and LEAP box simulations, but rather move directly into the data that was collected to proceed with the network design. *Figure 12.29* depicts the on-axis and 15 and 30° off-axis response of the two Audax AP130Z0s mounted in the center-channel cabinet. I measured this at the usual 2.83V/1m but with the box mounted horizontally in the same position as it will be in actual operation sitting on the top of a television. As the measurement proceeds off-axis and the microphone distance to one driver becomes shorter than the other driver, phase cancellation occurs and you get the response nulls at 3.5kHz at 15° off-axis and 1.7kHz at 30° off-axis.

Figure 12.30 gives the same on- and off-axis data for 0 and 30°, but with the speaker placed in the vertical orientation, like the LR speakers. This is how I design center channels—in the vertical, not horizontal, orientation. You cannot compensate for the off-axis horizontal SPL anomalies with crossover parts.

My attempt to improve the off-axis horizontal response was done with the physical layout of the drivers. Using the vertical response curves to design the network results in a speaker that subjectively sounds more like the LRs, and this might not occur if I attempted to compensate for the horizontal lobing with the network design. *Figure 12.31* gives the woofer in-box impedance that will be transferred into LEAP for the optimization of the crossover.

12.32 THE TWEETER.

The data for the Morel MDT 40 is pretty much the same as with the LRs. Some differences are caused by the different baffle shape and driver locations for both woofer and tweeter, but these are relatively minor. The MDT 40's on- and off-axis response is given in *Fig. 12.32*, with the impedance plot shown in *Fig. 12.33*.

12.33 THE CROSSOVER SIMULATION.

As I just mentioned, the only difference between the response curves of the LR and CTR (besides simple QC differences between the samples) was the baffle shape. *Figure 12.34* gives the compari-

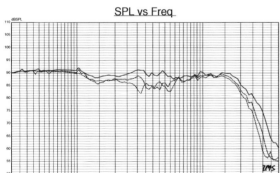

FIGURE 12.45: Comparison of 15° curves in *Fig. 12.44* (solid = 0°; dot = 15° left; dash/dot = 15° right).

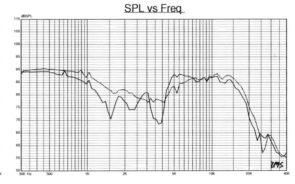

FIGURE 12.47: Comparison of CTR horizontal 30° and LR vertical 30° off-axis curves (solid = CTR; dash/dot = LR).

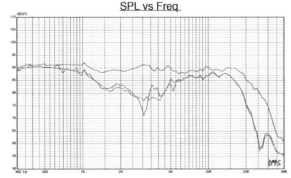

FIGURE 12.46: Comparison of 30° curves in *Fig. 12.21* (solid = 0°; dash = 30° left; dash/dotdot = 30° right).

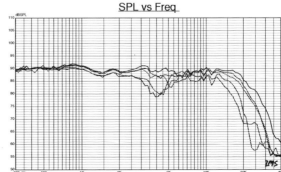

FIGURE 12.48: CTR vertical off-axis frequency response (solid = 0°; dot = 15° up; dash = 30° up; dash/dot = 15° down; dash/dotdot = 30° down).

son of the AP130Z0 response in the LR and the CTR cabinets, and *Fig. 12.35* depicts the comparison of the Morel MDT 40 in the two different baffles. As you can see, these differences were relatively minor.

Proceeding with the crossover optimization, I loaded all data into a LEAP design library, as displayed for the on-axis response in *Fig. 12.36*. Again, as with the LR, the woofer attenuation will start at about 120Hz with the actual crossover around 3kHz. As I expected, the crossover turned out to have the same values and topography as shown in *Fig. 12.11*. The individual driver response with the network simulation applied is illustrated in *Fig. 12.37*, with the driver transfer functions in *Fig. 12.38* and the driver impedance magnitudes in *Fig. 12.39*.

Summing woofer and tweeter responses together at 0° and 30° off-axis (including the 87μs time delay between the woofer and the tweeter) yielded the curves given in *Fig. 12.40*. The complete picture is depicted in *Fig. 12.41* with the on-axis response, the individual woofer and tweeter responses, and the on-axis summation with the polarity of the tweeter reversed. As with the LRs, the crossover exhibits a steep null in the crossover region indicating the transition is strongly in-phase. The last simulation (*Fig. 12.42*) is the system impedance magnitude and phase.

12.34 THE FINISHED PROTOTYPE.

The procedures for finishing a prototype are the same for all models, so I won't repeat them here. *Figure 12.43* shows the measured 2.83V/1m full-range anechoic on-axis frequency response for the CTR, which is the result of combining a gated sine-wave measurement from 300Hz–40kHz and a 20Hz–500Hz groundplane measurement. As you can see, the on-axis response of the CTR is very close to that of the LRs. The bigger concern is how the speaker will perform left and right off-axis in the horizontal orientation sitting on top of a television set.

Figure 12.44 gives the response on-axis plus the left/right 15° off-axis and the left/right 30° off-axis response curves. The separate 15° curves and 30° curves are compared in *Figs. 12.45* and *12.46*, respectively. If you compare the 30° CTR horizontal off-axis curves in *Fig. 12.44* with the 30° LR vertical off-axis curves (actually measured

in the same horizontal orientation as the CTR) in *Fig. 12.21* (depicted in *Fig. 12.47*), the CTR baffle arrangement actually produces definite improvement in the off-axis response.

If you look at the vertical off-axis of the CTR in *Fig. 12.48*, you can see that this baffle arrangement would not work well for a normal vertical-orientation speaker and is not nearly as "nice" as the horizontal (vertical orientation) off-axis curves of the LR shown in *Fig. 12.19*. And last, *Fig. 12.49* gives the system impedance for the CTR, which is identical to the LR, given that you are looking at the same drivers with the same crossover in the same box volume.

12.35 CONSTRUCTION DETAILS.

As with the LR cabinets, all cabinets, drivers, and crossover parts for this home theater project are available from Parts Express (www.partsexpress.com). The net internal volume of the CTR cabinet was approximately 0.31ft^3, with internal dimensions (HWD) of 6.5″ × 15.75″ × 5.5″. Built out of 0.75″ MDF with all drivers inset, the cabinet is 100% filled with R19 fiberglass (the pink stuff).

You can observe driver layout from the cabinet mechanical drawing. The layout of the crossover board is the same as the LR depicted in *Fig. 12.28*. Note that you should glue the crossover for the CTR (using silicone adhesive and hot glue) to the top of the inside of the CTR enclosure to position the inductors as far from a direct-view television as possible. The magnetic field emanating from inductors can conceivably cause screen interference with a large direct-view TV.

12.40 THE REAR-CHANNEL SURROUND SPEAKERS.

Surround speakers present a wholly different design problem from the rather standard LCR. Surrounds are intended to provide a more diffuse soundfield than standard direct radiators. This is supposed to mimic the effect of banks of speakers used on the left/right and rear walls of a full-sized theater (*Chapter 10*).

The typical dipole format generally uses two woofers at opposing angles connected out of phase (reverse polarity) and two tweeters at

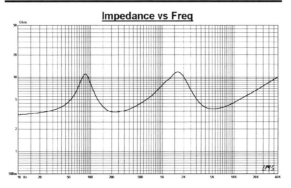

FIGURE 12.49: CTR system impedance plot.

opposing angles connected out of phase. In order for these full-dipole-type surround-sound speakers to have any low-end, you must separate the woofer cavities with an internal baffle and also use a high-pass filter on one of the woofers to prevent complete acoustic bottom-end cancellation. I have designed numerous dipole surrounds for manufacturers, including several THX certifications. They work well, but are much more difficult to design and are more expensive to manufacture, since they are essentially two complete two-way speakers in one box.

For this home theater system, I have opted for a sort of compromise surround speaker using a direct-radiating monopole woofer, but dipole tweeters. This compromise gives a null area for the high frequencies, but still allows for a fair

amount of localization for Dolby Digital movies that sometimes use discrete effects in the surround mix. The frequency null will extend to the crossover point for this speaker, at 3kHz, compared to between 400–1000Hz for a full-dipole-type of surround. So, choosing between using direct-radiator surrounds and the more diffuse full-dipole surrounds, the speaker I describe here is both a good performer and is less expensive to build.

12.41 THE WOOFER.

Instead of using two AP130Z0s in parallel, the surround in this system uses only one. As such, the output is 6dB less and the maximum (X_{MAX} + 15%) output is also less. However, given the fact that surrounds generally don't receive nearly

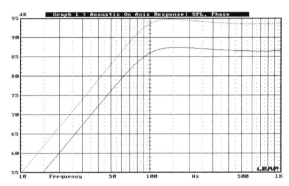

FIGURE 12.50: Surround woofer box simulation (solid = 2.83V; dot = 7V).

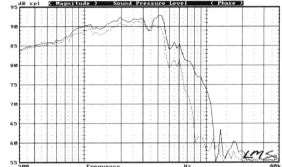

FIGURE 12.53: Surround woofer on- and off-axis frequency response (solid = 0°; dot = 30°).

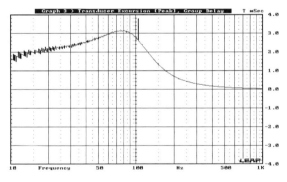

FIGURE 12.51: Group-delay curve for the 2.83V curve in *Fig. 12.50.*

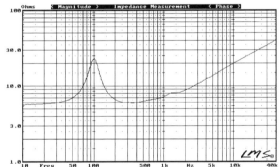

FIGURE 12.54: Surround woofer in-box impedance plot.

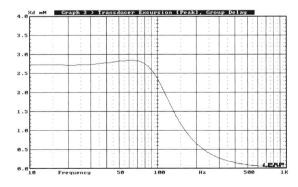

FIGURE 12.52: Cone-excursion curve for the 7V curve in *Fig. 12.50.*

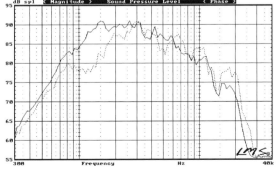

FIGURE 12.55: Surround tweeters measured on- and off-axis to woofer and in-phase (solid = 0°; dot = 30°).

as much information as the CTR, this will be more than adequate, especially considering the usually second-order high-pass filter at 80–100Hz on the surround "narrow" channel setting for most home theater electronics.

The LEAP box simulation for a single woofer is given in *Fig. 12.50*, with the associated 2.83V group delay in *Fig. 12.51* and the 7V cone excur-

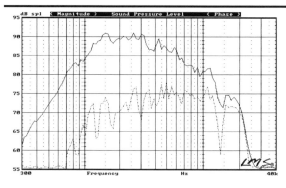

FIGURE 12.56: Comparison of surround tweeters measured on-axis in-phase and out-of-phase (solid = in-phase; dot = out-of-phase).

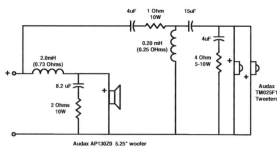

FIGURE 12.60: Surround crossover schematic.

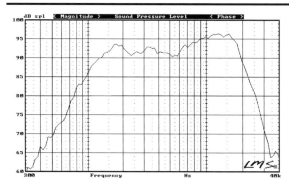

FIGURE 12.57: Surround single tweeter on-axis frequency response.

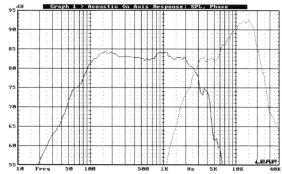

FIGURE 12.61: Surround on-axis computer simulation with crossover (solid = woofer; dot = tweeter, no conjugate).

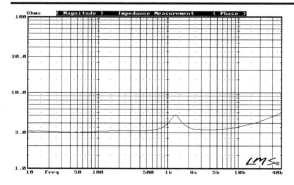

FIGURE 12.58: Surround two tweeters in parallel impedance plot.

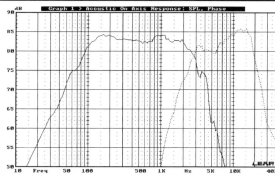

FIGURE 12.62: Same as *Fig. 12.61*, but with tweeter conjugate circuit (solid = woofer; dot = tweeter, with conjugate).

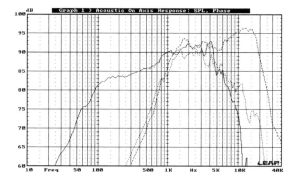

FIGURE 12.59: Curves used for the surround crossover optimization (solid = woofer; dot = two tweeters in-phase; dash = single tweeter SPL, two-tweeter impedance).

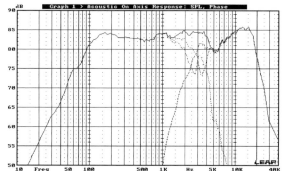

FIGURE 12.63: Surround simulated on-axis summation, woofer, tweeter, and reverse phase summation (solid = on-axis summation; dot = woofer; dash = tweeter; dash/dot = reverse phase summation).

sion curve in *Fig. 12.52*. f_3 is close to the same as the LR/CTR at 86Hz, and the maximum output about 6dB lower at 95dB.

Next, I mounted the drivers in the prototype enclosure and measured the frequency response at 0 and 30° (*Fig. 12.53*). The response is, as expected, similar to that seen for the LRs and CTR. The single driver impedance is given in *Fig. 12.54*.

12.42 THE TWEETERS.

The tweeters for the surrounds are different from those used in the LRs and CTR. Since there are two of them, I chose not to use a pair of the fairly expensive Morel tweeters and instead opted for a pair of the less expensive Audax TM025F1 neo 1″ cloth dome tweeters (two of these cost about one-third less than one Morel MDT 40). Since both domes—the Morel and Audax—are doped cloth types, the timbre will not be different enough to cause any kind of serious problem with the presentation.

I began collecting data for the network design by measuring the 2.83V/1m response of the TM025F1 tweeters mounted in the surround enclosure. Surround tweeters configured as a dipole require special design considerations and a host of different measurements to get an idea of how this speaker will function. *Figure 12.55* is the first iteration of measurements made on-axis to the woofer and in the same location that will be the null area of the tweeters. However, this curve is taken with both tweeters connected with the same polarity to get an idea of what the power response would be with the drivers in-phase.

Next, I compared the on-axis to the woofer microphone location with the tweeters connected in-phase and with the polarity reversed, out of phase (*Fig. 12.56*). This shows the null area effect you are looking for in a dipole surround tweeter. The last tweeter response measurement was on-axis to a single tweeter; the null area measurements are at the summation point off-axis 90° to each of two tweeters (*Fig. 12.57*). Finally, the impedance magnitude for two tweeters connected in parallel is depicted in *Fig. 12.58*.

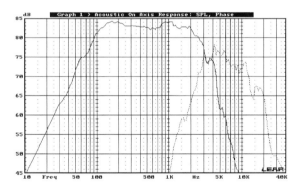

FIGURE 12.64: Same as *Fig. 12.62*, but using two-tweeter SPL (solid = woofer; dot = tweeter).

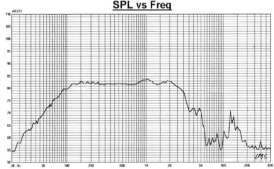

FIGURE 12.66: Surround measured on-axis null frequency response.

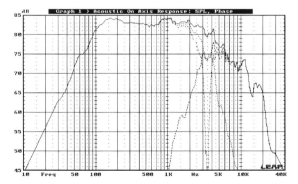

FIGURE 12.65: Same as *Fig. 12.63*, but using two-tweeter SPL (solid = on-axis summation; dot = woofer; dash = tweeter; dash/dot = reverse phase summation).

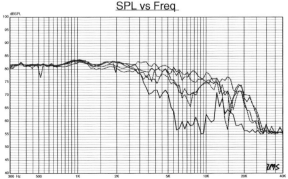

FIGURE 12.67: Surround measured on-axis null and left/right off-axis (solid = on-axis null; dot = 15° right; dash = 30° right; dash/dot = 15° left; dash/dotdot = 30° left).

12.43 THE CROSSOVER SIMULATION.

I worked with three curves for the design of this surround—the woofer, and both the single tweeter on-axis (combined with the two-paralleled tweeter impedance), and the on-axis of the two tweeters connected in-phase (*Fig. 12.59*). The primary crossover values (*Fig. 12.60*) were optimized using the single tweeter on-axis response data combined with the two-tweeter impedance data. This resulted in the individual driver response curves (*Fig. 12.61*) minus the conjugate circuit you see on the tweeter. This tweeter has a fair amount of horn loading, and when a network is placed on it, without compensating for the rise above 6kHz, the rise becomes exaggerated.

This was corrected with the 4Ω/4μF conjugate circuit (*Fig. 12.60*). *Figure 12.62* shows the corrected result, with *Fig. 12.63* giving the on-axis summation (bearing in mind that these drivers were individually measured on a different axis), plus the individual woofer and tweeter responses and the on-axis summation with the tweeter polarity reversed. The out-of-phase null is moderate, but still acceptably indicates the network is mostly in-phase at the crossover region.

I next looked at the same data with the on-axis two-tweeter in-phase/two-tweeter imped-

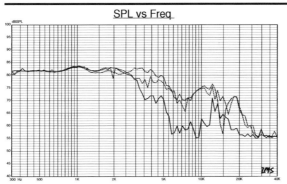

ance measurements. The individual driver responses are shown in *Fig. 12.64*. *Figure 12.65* gives the on-axis in-phase and out-of-phase on-axis summation plus the individual driver responses. Again, this results in a response that is reasonably in-phase in the crossover region.

12.44 THE FINISHED PROTOTYPE.

Figure 12.66 shows the measured on-axis null area response for the surround speaker placed in the "stand" position. As predicted, the null begins at 3kHz and is fully developed by 5kHz. This represents the placement of the speaker with the woofer facing forward and placed on a tall speaker stand. Note that this box design is also configured to be mounted on the wall 15–20″ from a ceiling with the woofer firing upward toward the ceiling at an angle.

Figure 12.67 gives the null axis measurement and the left/right 15 and 30° off-axis measurements, plus the two off-axis 15° (*Fig. 12.68*) and the two 30° off-axis curves (*Fig. 12.69*). From this you can see how narrow the null area is for a surround speaker, but this is how the effect is supposed to measure and what makes it work.

Figure 12.70 compares the two surround samples, which are closely matched. Please note that the location of the normal polarity

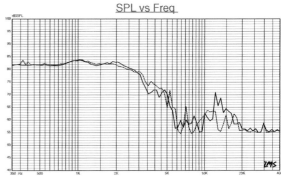

FIGURE 12.68: Surround measured on-axis null and left/right off-axis (solid = on-axis null; dot = 15° right; dash/dot = 15° left).

FIGURE 12.70: Comparison of both surrounds on-axis (null) (solid = surround 1; dot = surround 2).

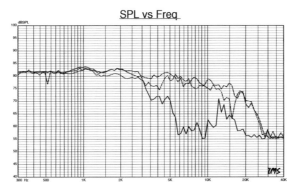

FIGURE 12.69: Surround measured on-axis null and left/right off-axis (solid = on-axis null; dash = 30° right; dash/dotdot = 30° left).

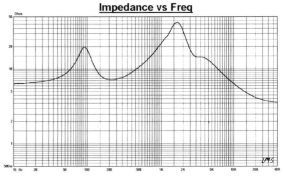

FIGURE 12.71: Surround system impedance.

tweeter (dipole tweeters are connected with reverse polarity) is on opposite sides of the enclosure so that each surround can have the positive-going tweeter diaphragm facing toward the LCR speakers. This technique is discussed in *Chapter 10* and illustrated in *Fig. 10.27*. Last, *Fig. 12.71* depicts the system impedance magnitude for the surround speaker.

12.45 CONSTRUCTION DETAILS.

As with the rest of the speakers in this chapter, all cabinets, drivers, and crossover parts are available from Parts Express (www.partsexpress.com). The net internal volume of the surround cabinet was approximately 0.12ft^3. For internal dimensions, refer to the mechanical drawing for this cabinet. Built of 0.75″ MDF with all drivers inset, the cabinet is 100% filled with R19 fiberglass (the pink stuff).

You can observe driver layout in the cabinet mechanical drawing. The layout of the crossover board is pretty conventional (*Fig. 12.72*). Inductors are the air-core types, capacitors are polypropylene solid dialectic types, and the resistors are the non-inductive variety (available from Parts Express). Note that the low-pass and high-pass sections were built on separate boards to facilitate inside installation. I placed these on opposite walls so the inductors would have the least amount of coupling.

12.50 THE POWERED SUBWOOFER.

This part of the project was fairly simple and straightforward. Take a great woofer—in this case the Parts Express Titanic 10″—mount it in the right small box, add an amp that has a little boost on the bottom end, and you have a good-performing home theater sub.

Regardless of the relative ease of designing subwoofers (at least in comparison to surrounds and center channels), the process involves measuring the woofer and simulating its performance. The Titanic woofer is built for Parts

Express by NCA, which is a small but extremely knowledgeable OEM driver manufacturer.

Their drivers not only have excellent engineering, but first-rate quality control, and are built to be about as bullet-proof as you will find in a hi-fi woofer. Built on a cast frame, the Titanic 10″ features a stiff talc-filled poly cone,

Front View   **Bottom View**

AP130Z0

TM025F1

Side View

Terminal Cup

Int. Dim.= 5.5"x6.0"x7.625" (5.625")
Ext. Dim.= 7"x7.5"x9.125" (7.125")
Material= 0.75" MDF
Driver spacing= woofers centered on front of box.
Tweeters located on both sides 1" from top and front of box.

Home theater surround speaker.

Titanic 10"

Int. **Dim**.= **12.75"x12.75"x11.75"**
Ext. **Dim**.= **14**.25"x**14.25"x13.25"**
Material= 0.75" MDF
Driver spacing= center woofer on front baffle.
Use **0.75" "H"** brace in center of enclosure.

Home theater subwoofer.

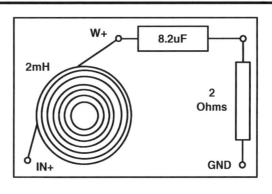

FIGURE 12.72: Surround crossover mechanical layout.

poly dustcap, Apical high-temperature voice-coil former, heat-formed Santoprene surround, motor parts coated with a black high-emissivity coating (for enhanced heat dissipation), bumpout for excursion, and pole vent for cooling, plus five-way binding-post terminals.

12.51 THE WOOFER SIMULATION.

I began analyzing this woofer using the LinearX LMS analyzer to perform both free-air and delta-compliance impedance measurements. This was done, however, using the LinearX VIBox with an amp connected directly to the woofer. This allows you to measure the voltage and current (admittance) separately, then later divide the two (Ohm's Law V/I = Z) to get the impedance.

TABLE 12.2
TITANTIC 10″ PARAMETERS

f_S	28.6Hz
R_{EVC}	3.1
Q_{MS}	8.29
Q_{ES}	0.50
Q_{TS}	0.47
V_{AS}	57 ltr
Bl	9.2 TM
Sens.	89.9dB @ 2.83V
X_{MAX}	12mm

Figure 12.73 shows the VIBox curves prior to being divided in the post-processing section of the LMS software. *Figure 12.74* depicts the final

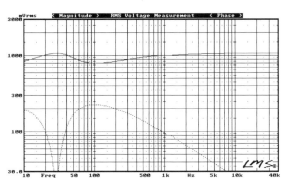

FIGURE 12.73: Subwoofer voltage and admittance impedance curves (solid = voltage curve for impedance; dot = current curve for impedance).

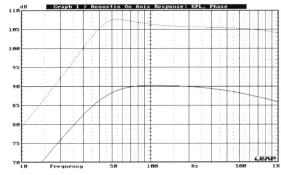

FIGURE 12.75: Subwoofer box simulation (solid = 2.83V; dot = 27.5V).

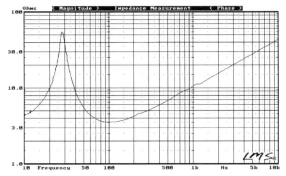

FIGURE 12.74: Subwoofer free-air impedance plot.

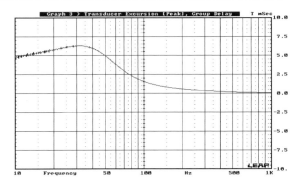

FIGURE 12.76: Group-delay curve for the 2.83V curve in *Fig. 12.75.*

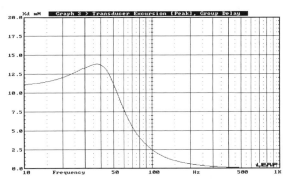

FIGURE 12.77: Cone-excursion curve for the 27.5V curve in *Fig. 12.75*.

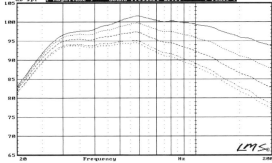

FIGURE 12.78: Subwoofer groundplane measurement with different low-pass settings (solid = 160Hz; dot = 120Hz; dash = 80Hz; dash/dot = 60Hz; short dash = 40Hz).

free-air impedance curve.

I imported this data into the LEAP software and calculated the T/S parameters as shown in *Table 12.2*.

This woofer has excursion for days, which is very easy to see in the simulation. With this data in LEAP, I performed a sealed box simulation for this woofer placed in a "virtual" 1.1ft³ box with 100% fiberglass fill (again, R19, the pink stuff). The results are shown in *Fig. 12.75* (see *Fig. 12.76* for the 2.83V group-delay curve for this graph, and *Fig. 12.77* for the 27.5V cone-excursion curves). This yielded a simulated f_3 of 44.2Hz with a Q_{TC} of 0.71.

This represents a fairly low frequency high-pass rolloff and is appropriately damped. Increasing the simulation input voltage to 27.5V, which pushes the cone excursion to X_{MAX} +15% (13.8mm in this case), resulted in an SPL of 107.8dB. This means that you can play this woofer to over 105dB before it just barely begins to audibly distort. With program material, you can exceed this limit by several dB before noticing any problems.

12.52 THE FINISHED PROTOTYPE.
The measured f_3 without the amp connected was 46Hz with a Q_{TC} of 0.73, very close to the LEAP simulation. Next, I connected the woofer amp and made groundplane measurements with successive iterations of different settings on the subwoofer amp low-pass control (*Fig. 12.78*). The Parts Express 250W amp has a 6dB boost at about 30Hz, which results in a combined f_3, amp boost, and woofer/box of 27Hz, a little less than an additional 2/3 of an octave. *Figure 12.79* shows the comparison of the near-field response of the woofer with and without the amp.

12.53 CONSTRUCTION DETAILS.
There's not much to be said beyond the drawing presented. The box has a single cross

brace and the woofer is surface mounted, with 100% R19 fiberglass fill in the enclosure.

12.60 THE LDC6 HOME THEATER SYSTEM PERFORMANCE.

I had my dear friend and business associate Nancy Weiner assist me in evaluating this product. Nancy has worked with me in voicing the entire Atlantic Technology line of products (she's the Director of Sales and Marketing at Atlantic Technology), all of which have received spectacular reviews in the home theater press. Basically, if Nancy thinks it sounds good, it is. After we both spent considerable time listening to this system, we thought you would have to spend at least $2000 on the open market to get something any better.

12.70 THE LDC6 STUDIO MONITOR.

This project is close to my heart, since I still write music and play a lot (I played professionally in a successful Colorado rock band about 1965–1970). My current studio uses a keyboard performance speaker I designed. However, it was the only speaker in my studio, so I also used it for CD playback and as a mix monitor for the 12-track Akai digital recorder that I use. While this is a great speaker through which to play my Kawai MP9000 stage piano (it's a three-way with a 3″ voice coil EV 12″ woofer, an Audax pro sound 6.5″ mid, and a 3″ × 9″ horn), it's not quite what the doctor ordered for mix-down with a multi-track recorder.

What I really needed was a compact high-end two-way design that could play fairly loudly and supply sufficient neutral detail for mix and other monitoring applications. I have designed studio monitors for manufacturers in the past and received some rather positive reviews in magazines such as *Electronic Musician*, so what I needed was something on that order. The format I chose was a 6.5″ woofer and a 1″ cloth dome tweeter. Drivers I decided would work well are the Scan Speak 18W/8545K00 and the Scan Speak D2905 Revelator.

12.71 THE WOOFER.

The 6.5″ Scan Speak 18W/8545K00 woofer is a very strong unit. Features include a cast frame, rubber surround, paper cone with inverted paper dustcap, 42mm diameter voice coil, and pole vent. I began analysis by measuring the free-air and delta-compliance (test box) impedance magnitude (see *Fig. 12.80* for the free-air impedance plot). I then imported these files into the

TABLE 12.3
SCAN SPEAK 18W/8545K00

f_S	31.3Hz
R_{EVC}	5.5
Q_{MS}	9.26
Q_{ES}	0.46
Q_{TS}	0.44
V_{AS}	36 ltr
Bl	6.8 TM
Sens.	86.9dB @ 2.83V
X_{MAX}	6.5mm

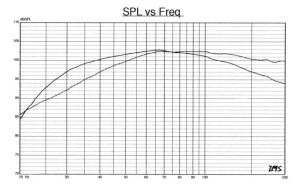

FIGURE 12.79: Comparison of nearfield subwoofer response with and without amplifier (solid = with amplifier; dot = without amplifier).

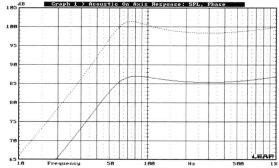

FIGURE 12.81: Studio monitor woofer box Q_{TC} = 0.92 simulation (solid = 2.83V; dot = 15.5V).

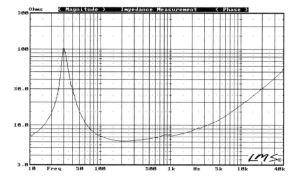

FIGURE 12.80: Studio monitor woofer free-air impedance plot.

FIGURE 12.82: Group-delay curve for the 2.83V curve in *Fig. 12.81*.

LEAP software and calculated and displayed the T/S parameters in *Table 12.3*.

Applying the LEAP analysis mode, I determined that a 0.28ft³ box with 100% fiberglass fill material (R19) would provide an f₃ of 54Hz with a box Q_{TC} of 0.92, which would add a bit of warmth for the 50Hz rolloff and also decrease excursion somewhat at high listening levels. *Figure 12.81* shows the 2.83V and 15.5V curves for this box simulation, with the 2.83V group-delay curves shown in *Fig.*

12.82 and the 15.5V cone excursion curve in *Fig. 12.83*.

However, I decided to also investigate a lower Q box stuffing arrangement and see if the difference was worthwhile. To do this in the simulation, I increased the LEAP FGEF (Fiberglass Equivalency Factor) to 2.5, which closely simulates the effects of Owens-Corning 703 fiberglass (R19 = 1 in the LEAP FGEF scenario) and re-ran the analysis mode of the program.

Figure 12.84 compares SPL

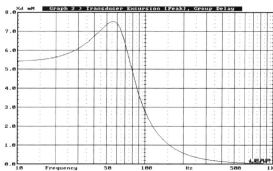

FIGURE 12.83: Cone-excursion curve for the 15.5V curve in *Fig. 12.81*.

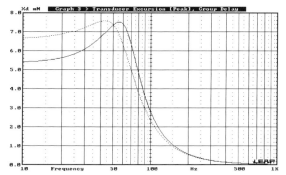

FIGURE 12.86: Comparison of cone-excursion curves for the 15.5V curve in *Fig. 12.84*.

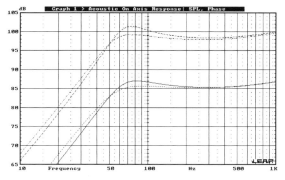

FIGURE 12.84: Studio monitor woofer comparison box simulation with different Q_{TC}s (solid = 2.83V with Q_{TC} 0.92; dot = 2.83V with Q_{TC} 0.75; dash = 15.5V with Q_{TC} 0.92; dash/dot = 15.5V with Q_{TC} 0.75).

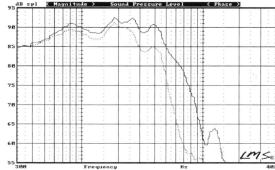

FIGURE 12.87: Studio monitor woofer on- and off-axis frequency response (solid = 0°; dot = 30°).

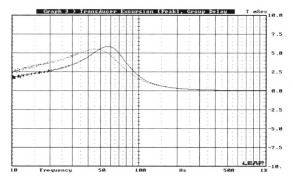

FIGURE 12.85: Comparison of group-delay curves for the 2.83V curve in *Fig. 12.84*.

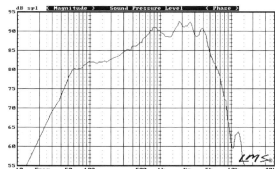

FIGURE 12.88: Studio monitor woofer full-range anechoic frequency response.

with the box stuffed with R19 and with 703. The 703-filled box had a slightly lower f_3 of 49.6Hz and a box Q_{TC} of 0.75. You can see the Q differences in the group-delay curves in *Fig. 12.85*.

However, the excursion curve with a voltage increase to achieve X_{MAX} + 15% (7.5mm for this woofer) for box woofers (*Fig. 12.86*) didn't result in a very significant change in peak SPL, so there really isn't that much difference. After carefully listening to the two iterations, Nancy and I both decided that the R19 stuffed version was all-around better-sounding, although we also noticed a bit more clarity in the 703-stuffed version.

Next I mounted the woofer in the cabinet (along with the D2905 tweeter) and measured the 2.83V/1m frequency response at 0 and 30° from 300–40kHz using the gated sine-wave fea-

ture of the LMS analyzer (*Fig. 12.87*). The response is not as smooth as the Audax AP130Z0's but is reasonably even up to about 2.8Hz. However, above that frequency it varies by up to 5dB, which will be difficutlt to make a really flat transition to the tweeter.

If you have ever read the introduction to the *Loudspeaker Recipes* book, you understand that absolute flatness of a speaker system response is not as critical as many would like to think. Many outstanding speakers have had considerable variation in system frequency response, while many ruler-flat designs have not sounded so great. In and of itself, an absolutely flat magnitude response does not determine the sound quality of the speaker.

I also measured the 1m groundplane response

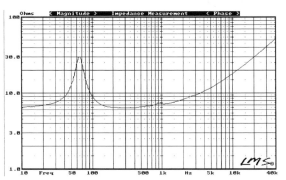

FIGURE 12.89: Studio monitor woofer in-box impedance plot.

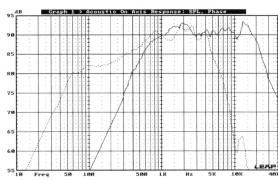

FIGURE 12.92: Studio monitor on-axis computer simulation without crossover (solid = tweeter; dot = woofer).

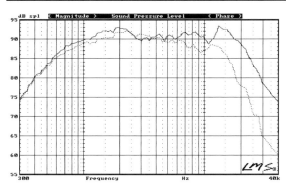

FIGURE 12.90: Studio monitor tweeter on- and off-axis frequency response (solid = 0°; dot = 30°).

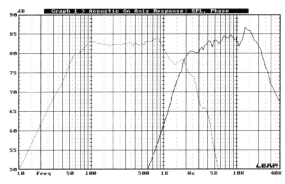

FIGURE 12.93: Studio monitor on-axis computer simulation with crossover (solid = tweeter; dot = woofer).

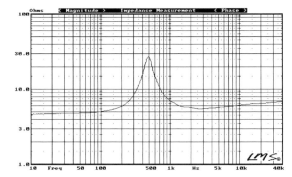

FIGURE 12.91: Studio monitor tweeter impedance plot.

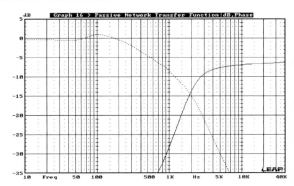

FIGURE 12.94: Transfer functions for networks simulated in *Fig. 12.93* (solid = tweeter; dot = woofer).

The final data required for the LEAP simulation is the impedance of the Scan Speak 18/8545K00 mounted in the enclosure (*Fig. 12.89*).

12.72 THE TWEETER.

The Scan Speak D2905 Revelator is a really exceptional device and ideal for use as a monitor. Unfortunately, the woofer cannot take advantage of the tweeter's sensitivity, which is 91dB, but has a very low resonance of 500Hz, allowing for a low-crossover frequency without having to deal with the resonance. Other features include a doped cloth 1″ dome, 28mm voice coil, and an RMS power-handling rating of 225W using a 12dB/octave crossover at 2.8kHz. The other interesting feature is a large shallow waveguide that should reinforce the low-frequencies somewhat and make it more appropriate for a low crossover frequency, which is what I had planned.

I started analyzing this device by measuring the 2.83V/1m at 0° and 30° off-axis response curves (*Fig. 12.90*). The overall response of this tweeter is good, with some unevenness above 10kHz, but this is really not of much consequence. The off-axis of this tweeter is quite good for a 1″ soft dome. The last data needed for the crossover simulation was the impedance (*Fig. 12.91*).

12.73 THE CROSSOVER SIMULATION.

I loaded the on- and off-axis frequency response and impedance phase and magnitude data (*Fig.*

from 20–500Hz (not shown) and spliced that to the gated anechoic measurement in *Fig. 12.87*, resulting in the full-range anechoic response displayed in *Fig. 12.88*. The next step was to tail-correct the high-pass and low-pass slopes to their nominal asymptotic value, which like the other speakers in this chapter is 12dB/octave for both.

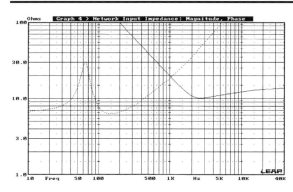

FIGURE 12.95: Impedance magnitude for networks simulated in *Fig. 12.93* (solid = tweeter; dot = woofer).

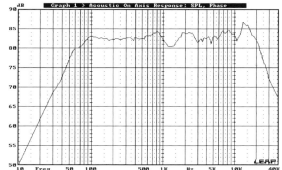

FIGURE 12.97: Studio monitor on-axis full-range anechoic frequency response.

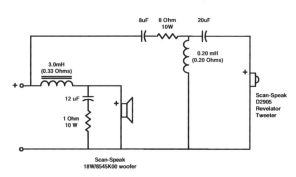

FIGURE 12.96: Studio monitor crossover schematic.

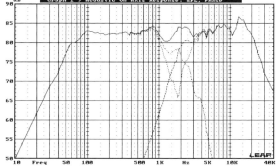

FIGURE 12.98: Studio monitor simulated on-axis summation, woofer, tweeter, and reverse phase summation (solid = on-axis summation; dot = reverse phase summation; dash = tweeter; dash/dot = woofer).

12.92) into the program. The last detail prior to commencing the network design was to enter the 95μs interdriver time delay into the LEAP Active Filter Library. The format that almost always works well for two-way designs like this is a second-order topography on the woofer and a third-order topography on the tweeter, with fourth-order Linkwitz-Riley target filter functions. As with the home theater LR and CTR, the interdriver time delay meant that the low-pass slope would actually be somewhat shallower to achieve a flat summation.

Most high-end speaker designers take advantage of the fact that tweeters tend to sound more open and detailed crossed over at 2kHz or lower. The only problem is that it takes a robust tweeter to keep from self-destructing at high SPL levels at this low a cross point, even with a steep filter high-pass slope. But this is why I chose the D2905, since it can handle reasonable SPL crossed over this low with a high-order filter.

The 2kHz target was optimized for both woofer and tweeter to achieve a flat summation and resulted in the individual woofer and tweeter response curves in *Fig. 12.93*, with the associated network transfer functions seen in *Fig. 12.94* and the driver/network impedance given in *Fig. 12.95*. These responses were for the network values depicted in the crossover schematic in *Fig. 12.96*. Note that the woofer inductor is 3mH, which caused a choice in terms of inductor types. My first choice for a high-end product such as

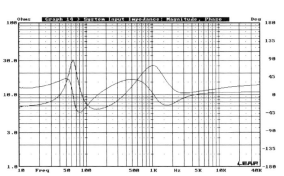

FIGURE 12.99: Studio monitor simulated system impedance plot.

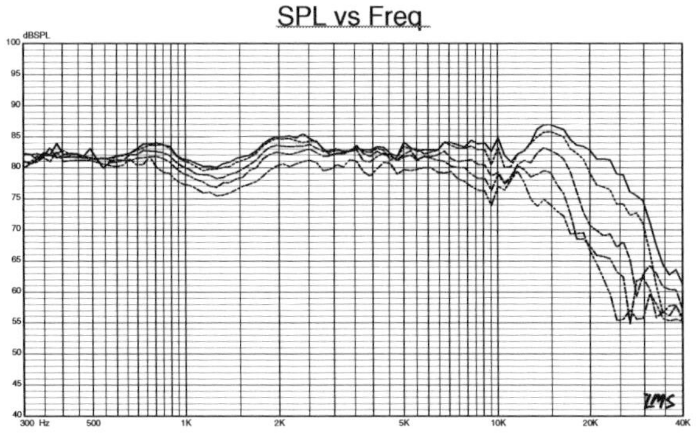

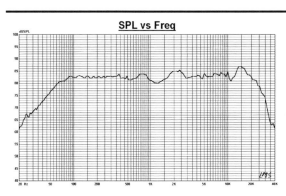

FIGURE 12.100: Measured studio monitor on-axis frequency response.

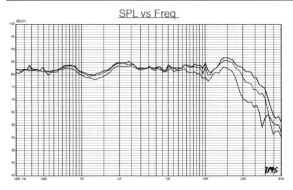

FIGURE 12.101: Measured studio monitor horizontal on- and off-axis frequency response (solid = 0°; dot = 15°; dash = 30°).

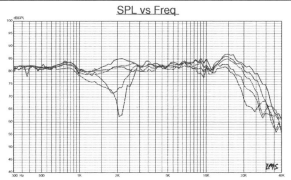

FIGURE 12.102: Measured studio monitor horizontal on- and off-axis frequency response to 60° (solid = 0°; dot = 15°; dash = 30°; dash/dot = 45°; dash/dotdot = 60°).

FIGURE 12.103: Measured LR vertical on- and off-axis frequency response (solid = 0°; dot = 15° up; dash = 30° up; dash/dot = 15° down; dash/dotdot = 30° down).

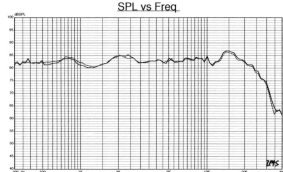

FIGURE 12.104: Comparison of both studio monitor prototypes (SM1 = solid; SM2 = dot).

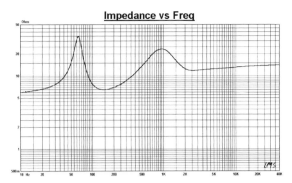

FIGURE 12.105: LR measured system impedance plot.

FIGURE 12.106: Comparison of system impedance with different box Q_{TC}s (solid = impedance with R19 fill material; dot = impedance with 703 fill material).

this would have been a 16-gauge air-core inductor, but there really is not room in the enclosure for this size inductor.

If I were manufacturing this speaker for the high-end two-channel or studio monitor market, I would probably have built a false back and mounted the network outside of the box volume to allow plenty of room for a full-size air core (actually, for the pro-sound monitor market, this speaker would have ended up biamped with an electronic crossover). So, the compromise in this situation was to use a laminate core inductor because its size would allow the network to fit in the cabinet. Other than that, all capacitors used were Solen and the resistors were the non-inductive variety, available from Parts Express.

Figure 12.97 shows the on-axis full-range anechoic summation for the Scan Speak woofer and tweeter combination. Overall, the response is not wickedly flat, but does manage to come out to about ±2.15 from 100Hz–10kHz, which is not all that bad. *Figure 12.98* illustrates the on-axis response with the individual driver responses, plus the on-axis summation with the tweeter polarity reversed (out of phase connection). This produces a fairly deep null at the crossover frequency, indicating the response is in-phase in the crossover region. Last, *Fig. 12.99* gives the simulated impedance magnitude and phase.

12.74 THE FINISHED PROTOTYPE.
Figure 12.100 shows the prototype's 2.83V/1m measured response, which was about ±2.5dB, about 0.35dB out from the LEAP simulation, which is typical of the results achieved with the LEAP/LMS workstation. (Although I have on a few occasions been accused of being financially involved with LinearX, the reality is that they are just good friends and the only entanglement is one of respect and friendship. I use these products in my work because of their functionality and for no other reason. I think that should be obvious from this tutorial.)

Examining the prototype further, *Fig. 12.101* gives the horizontal response out to 30°, and it is obvious that the critical off-axis response is nearly as flat as on-axis. *Figure 12.102* shows the off-axis response out to 60°, which implies a rather flat

power response overall, an important criteria for a high-quality loudspeaker. Looking at the other global axis, the vertical off-axis response for 15° up/down and 30° up/down is depicted in *Fig. 12.103*. These curves are moderately symmetrical, indicating only minor lobing in the vertical response.

In terms of imaging, matching the left and right speakers in a stereo pair is very important. *Figure 12.104* compares the on-axis response of both prototypes and as is obvious, they are very closely matched to within 0.5dB. Last, *Fig. 12.105* gives the measured impedance magnitude of the LDC6 Studio Monitor with the R19 stuffing, while *Fig. 12.106* compares both the R19 and the 703 stuffing effects on the impedance.

12.75 CONSTRUCTION DETAILS.
The cabinets and all drivers and crossover parts for the studio monitor (or two-channel hi-fi moni-

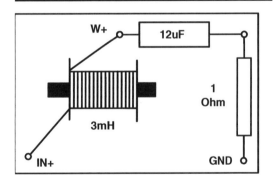

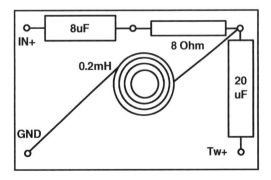

FIGURE 12.107: Studio monitor crossover mechanical layout.

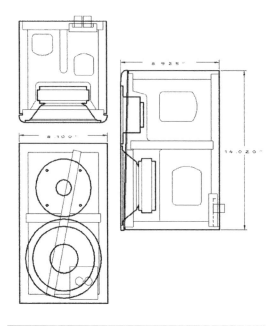

tor) project are available from Portland Audio Lab, e-mail dbnelson@teleport.com. If you are building your own cabinets, the net internal volume of the studio monitor cabinet is approximately 0.28ft^3, with internal dimensions (HWD) of 12.5″ × 6.5″ × 7.125″ (which in my prototype includes substantial bracing). Built out of 0.75″ MDF sides and rear baffle and 1.125″ front baffle, with all drivers inset, the cabinet is 100% filled with R19 fiberglass.

You can observe driver layout in the cabinet mechanical drawing, but all drivers are basically centered and placed as close together as possible. The layout of the crossover board is pretty conventional (*Fig. 12.107*). Since this cabinet has extensive bracing, it was not possible to place the entire crossover on one board. As a result, I built the network with one board for the tweeter high-pass, mounted in the top of the box near the tweeter, and a separate board for the woofer low-pass mounted in the bottom of the enclosure near the woofer. This method also ensures that no cross-coupling occurs between high-pass and low-pass inductors.

12.80 THE LDC6 STUDIO MONITOR'S SUBJECTIVE PERFORMANCE.

I asked Nancy to evaluate the two stuffing materials used in the monitors, which meant listening not only to the low-end sound quality, but also to midrange clarity. After extensive listening, we both agreed that the bass with the $Q_{TC} = 0.92$ R19 version was better-suited for standalone listening, although we both noticed the somewhat enhanced clarity of the 703-stuffed version. This version is probably best-suited for use with a subwoofer. After an hour or so of listening, we concluded that this was an outstanding two-way in terms of presentation, clarity, detail, and overall musicality. Not a bad place to be for a studio monitor.

Afterword

At the close of this 7th edition of *LDC*, I have several important details to address. First, I would like to extend my sincere appreciation to those of you who have been reading the monthly issues of *Voice Coil*, which has been an important part of my work over the last 28 years.

Not long after I finished assembling the 5th edition of the *LDC* in June 1995, the Audio Engineering Society presented me with an award citing my long-term "contributions to audio education and the industry, through the dissemination of information related to loudspeaker measurements, systems, and components." This award would not have been possible if all of you did not have the amazing thirst and interest in learning more about the art of recreating music and sound. For this I would like to take this opportunity to thank you all for giving me the opportunity to involve myself in such a rewarding profession.

Finally, I would at long last wish to dispel a possible myth about the *Loudspeaker Design Cookbook*—that it is written by an engineer who thinks only about analyzers, mathematics, and science and believes that all the answers are objective and probably never listens to music. Well, I not only listen to music, in my own humble way, I even try to write it!

Vance Dickason
October 2005

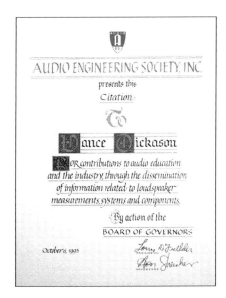

Printed in Great Britain
by Amazon